Instant Notes

Psychology

The INSTANT NOTES series

Series editor

B.D. Hames

School of Biochemistry and Molecular Biology, University of Leeds, Leeds, UK

Animal Biology
Ecology
Microbiology
Genetics
Chemistry for Biologists
Immunology
Biochemistry 2nd edition
Molecular Biology 2nd edition
Neuroscience
Psychology

Forthcoming titles
Developmental Biology
Plant Biology

The INSTANT NOTES Chemistry series
Consulting editor: Howard Stanbury

Organic Chemistry
Inorganic Chemistry
Physical Chemistry

Forthcoming titles
Analytical Chemistry
Medicinal Chemistry

Psychology

I.P. Christensen

Department of Psychology, University of Manchester, Manchester, UK

H.L. Wagner

Department of Psychology, The University of Central Lancashire, Preston, UK

and

M.S. Halliday

St Ives, Cornwall, UK

First published 2001

A CIP catalogue record for this book is available from the British Library.

ISBN 1 85996 097 9

BIOS Scientific Publishers Ltd
9 Newtec Place, Magdalen Road, Oxford OX4 1RE, UK
Tel. +44 (0)1865 726286. Fax +44 (0)1865 246823
World Wide Web home page: http://www.bios.co.uk/

Published in the United States of America, its dependent territories and Canada by Springer-Verlag New York Inc., 175 Fifth Avenue, New York, NY 10010-7858, in association with BIOS Scientific Publishers Ltd.

Published in Hong Kong, Taiwan, Cambodia, Korea, The Philippines, Brunei, Laos and Macau only, by Springer-Verlag Hong Kong Ltd, Unit 1702, Tower 1, Enterprise Square, 9 Sheung Yuet Road, Kowloon Bay, Kowloon, Hong Kong, in association with BIOS Scientific Publishers Ltd.

Production Editor: Fran Kingston
Typeset and illustrated by Phoenix Photosetting, Chatham, Kent, UK
Printed by Biddles Ltd, Guildford, UK, www.biddles.co.uk

CONTENTS

ABBREVIATIONS

ACTH	adrenocorticotropic hormone
ADH	antidiuretic hormone
ANS	autonomic nervous system
AMH	anti-Mullerian hormone
β	decision criterion
CAT	computerized axial tomography
CCK	cholecystokinin
CHD	coronary heart disease
CNS	central nervous system
CR	conditioned response
CRF	continuous reinforcement schedule
CS	conditioned stimulus
CSF	cerebrospinal fluid
d′	(d-prime) discriminability
DNA	deoxyribonucleic acid
DSM-IV	*Diagnostic and Statistical Manual of Mental Disorders*
EEG	electroencephalogram
FI	fixed interval (reinforcement schedule)
fMRI	functional magnetic resonance imaging
FR	fixed ratio (reinforcement schedule)
FSH	follicle-stimulating hormone
GABA	γ-aminobutyric acid
5-HT	5-hydroxytryptamine (serotonin)
IQ	intelligence quotient
LGN	lateral geniculate nucleus
LTM	long-term memory
LH	lateral hypothalamus
MLU	mean length of utterance
MMPI	Minnesota Multiphasic Personality Inventory
MPA	medial preoptic area
NPY	neuropeptide Y
PET	positron emission tomography
PNS	parasympathetic nervous system
PRE	partial reinforcement effect
PTSD	post-traumatic stress disorder
RAS	reticular activating system
REM	rapid eye-movement sleep
SCN	suprachiasmatic nucleus
SDN	sexually dimorphic nucleus
SNS	sympathetic nervous system
S-R	stimulus-response
STM	short-term memory
SW	slow-wave (sleep)
UCR	unconditioned response
UCS	unconditioned stimulus
VI	variable interval (reinforcement schedule)
VMH	ventromedial hypothalamus
VR	variable ratio (reinforcement schedule)
WAIS	Wechsler Adult Intelligence Scale
ZPD	zone of proximal development

PREFACE

The popularity of psychology as a subject of study continues to increase. Psychology attracts those who want to go on to become professional psychologists, those who need (or want) to understand psychology as part of their training in other fields (medicine, nursing, the law, teaching, business, etc.), and those who just want to study an interesting subject.

As the popularity of psychology grows, so too does the number of introductory textbooks on the subject. Many of these are profusely illustrated (mainly with cartoons and news photographs) and provide a wealth of detail and examples. *Instant Notes in Psychology* takes a different approach. It is essentially a revision aid, rather than an elaborate textbook. We have extracted from the material generally covered in introductory psychology courses the facts and theories that are essential to the student facing examinations and tests. The illustrations are limited to those that aid the understanding of the material and they are presented in a way that should make them easy for the student to reproduce.

The book is divided into 13 sections with a total of 75 topics. Each topic begins with a Key Notes panel containing concise summaries of the key points covered, which are expanded in the main text of the topic. To get the most from the book, first learn the material in the main text of a topic, then use the Key Notes as a rapid revision aid. Although each topic stands alone, it is the nature of psychology that topics are interrelated. To help the student see these interrelationships we have provided numerous cross-references between topics.

We have based our approach in this book on our many years of experience teaching psychology at introductory level, as well as our own specialist areas. Section A sets psychology in its evolutionary and biological contexts and outlines the essentials of physiology. Motivation is fundamental to behavior and thinking and Section B deals with basic motivational processes (with more complex motives covered in later sections). We cannot act effectively without obtaining information from the environment, and the sensory and perceptual processes through which we achieve this are discussed in Sections C and D. Adaptation to the environment is largely based on learning and this is the subject of Section E. Section F covers memory, which allows us to retain adaptively important information, and thinking, which is the mental process with which we manipulate perceived and remembered information.

In Section G, we consider the production, reception and learning of language as cognitive processes. Section H covers human development and considers how our abilities and performance change over our life span. We live in a social world and social psychology, which studies this, is dealt with in two sections: Section I concerns the cognitive processes underlying how we understand the social world, and Section J looks at how we interact with others. The wide variety of approaches to understanding the individual is examined in Section K, which covers personality and intelligence. While we are mostly concerned with normal psychology, dysfunction is of interest in its own right and for what it tells us about normal psychology. Psychopathology is the subject of Section L, and Section M deals with how our behavior and thinking influence our health and how we respond to illness.

While this book should help the student to pass exams, for some courses it may not be sufficiently detailed to completely replace the comprehensive course text. In the further reading section we suggest a number of general introductory texts, and others will no doubt be recommended by course tutors. We also suggest more specialized sources for each section for those whose interest is awakened, and who want to learn more about particular aspects of the subject.

Ian Christensen, Hugh Wagner and Sebastian Halliday

A1 EVOLUTION AND BIOLOGICAL PSYCHOLOGY

Key Notes

The biological context of psychology

Psychology is a biological science: we are products of evolution, and our mental processes and behavior depend on physiological processes. Some approaches to psychology emphasize this biological basis, examining human behavior in an evolutionary context, comparing human behavior with that of other species, or examining its physiological bases.

Evolution

Modern species, including humans, have evolved by processes of natural selection. The relationships amongst species are described by a phylogenetic tree.

Genetics and the 'nature–nurture' issue

We inherit genes that code for particular molecules, and these form our genetic makeup (genotype). Our genotype influences our structural and psychological phenotype (final individual characteristics). Extreme positions on the 'nature–nurture' issue have been largely replaced by a recognition of the importance of both genetic and environmental influences on psychological characteristics.

Related topics

The nature of development (H1)

Heredity, the environment and intelligence (K8)

The biological context of psychology

Whatever else it may be, psychology is a biological science. This is, first, because we are the products of **evolution**. Second, mental processes and behavior are dependent on the **nervous system** and the **endocrine system**. **Biological psychology** (or **psychobiology**) includes any approach to psychology that places it in its biological context; that is, examining the biological bases of behavior. In its broadest sense, this can be the examination of the evolution of human behavior, in a field known as **evolutionary psychology**. Principles from studies of the adaptive functions of structures and behaviors in other species are extended to try to explain human behavior. The approach to social behavior that this leads to is known as **sociobiology**. This applies Darwinian principles of evolution to the study of social, and particularly reproductive, behavior. A key concept is the **selfish gene**. In this view, all animal behavior is interpreted as having the goal of increasing the proportion of the individual's genes in the next generation. The direct comparison of human behavior with that of other species is known as **comparative psychology**. Usually, it is assumed that the underlying processes examined in other species are the same as, or similar to, those in humans, and this itself implies evolutionary continuity. **Physiological psychology** is the study of the physiological mechanisms, especially those of the nervous and endocrine systems, underlying behavior and mental processes. Since much of our information about physiological mechanisms comes from studies of laboratory animals, physiological psychology assumes similarity of mechanisms, and evolutionary continuity. **Neuropsychology** is the study of the neural mechanisms, especially in the cerebral cortex, that underlie psychological, particularly cognitive, processes (see Topic A4).

The aim of biological psychology is to explain human mental processes and behavior in biological terms, either evolutionary or physiological. The practice of explaining complex events in terms of simpler ones is known as **reductionism**. While many biological psychologists do not concern themselves with other approaches to psychology, a fuller picture will come from a broader view. Since human beings live in social groups, we must consider the social context of human mental processes and behavior, and this makes psychology a **social science**, as well as a biological one. But our social groups and our responses to them are themselves influenced by evolutionary processes (see Sections I and J). We could argue that all psychological phenomena have underlying physiological mechanisms. However, it is not always useful to attempt to explain them at the physiological level.

Evolution

In 1859, in his book *The Origin of Species*, Charles Darwin proposed a mechanism for evolution, namely that it proceeded by a process of **natural selection**: the 'survival of the fittest'. Naturalists had, for a long time before that, classified animals and plants into groups of like species, a process called **taxonomy**. Taxonomy placed into a 'tree' species related to one another to different degrees by similarities of their structural features. Some taxonomists speculated that species had evolved from other species, and the concept of evolution made sense of these relationships amongst species, now known as the **phylogenetic tree** (see *Fig. 1*). More similar animals are so because they have diverged from a

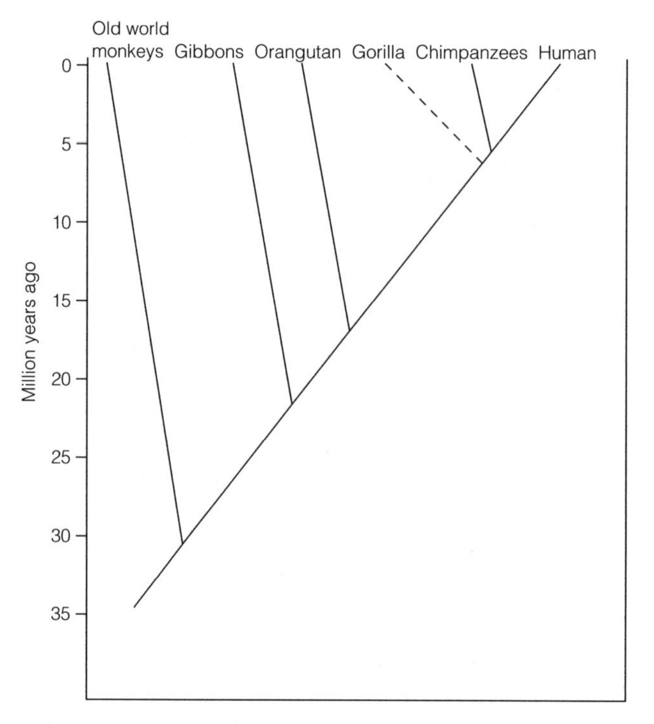

Fig. 1. Part of the phylogenetic tree of humans and apes, derived from DNA and morphological data. The dashed lines for the gorilla indicate uncertainty about its point of separation.

common ancestor more recently in evolutionary history. The techniques of molecular biology are now being applied to taxonomy, and have clarified some relationships.

Genetics and the 'nature–nurture' issue

All of the tissues in our bodies derive from **genes** which we inherit from our parents. Each gene carries the code for the production of a particular protein or other molecule. These molecules are directly or indirectly the basis of cell and tissue development. Each gene is part of a very large molecule called **deoxyribonucleic acid** (or **DNA**). Each DNA molecule is looped and bound to a central matrix to form a **chromosome**. We have 23 pairs of chromosomes, and inherit one of each pair from each parent. The genetic makeup of the individual organism is called the **genotype**. The extent to which a gene exerts its effect depends on environmental influences, and the resulting characteristics of the individual are known as the **phenotype**. For example, genes that specify physical characteristics like height will only ensure that you are tall if you are adequately nourished during growth. Genetic determinants are not only highly dependent on environmental factors, but genetic influences on behavior are **polygenic**, meaning that two or more genes help to produce variations in a particular behavior or characteristic.

For a large part of the 20th century, psychologists debated the roles of innate and environmental factors on psychological processes: the **nature–nurture** problem. Many took extreme positions, dictated more by ideology than by science. It has become clear that for most psychological processes, genetic and environmental factors interact to produce the behavioral phenotype. There are many ways of trying to establish the extent to which a particular behavior is determined by genetic factors. In humans, most studies look at how similar are people of different degrees of family relationship. **Monozygotic** (identical) **twins**, who have an identical genotype, should be most similar. **Dizygotic** (fraternal) **twins** have 50% of their genes in common, just as any other siblings. The proportion of shared genes decreases as relationships become more distant. Of course, environmental similarities usually vary in a parallel manner, making it difficult to disentangle the two influences. The best studies of this type have compared monozygotic twins reared together with those reared apart after adoption. In other species, selective breeding produces strains that differ in behavioral characteristics, demonstrating a genetic basis for those characteristics. Ultimately, the aim of **behavioral genetics** is to identify the genes that affect psychological characteristics.

Molecular biology can identify the genetic makeup of a particular organism, and the sites of the genes which code for particular molecules, thereby providing the means to control the development of particular structures. In this way, behavioral genetics should be able to locate genes significant for the production of particular behaviors or psychological traits. Biologists can now define phylogenetic relationships in terms of the similarity of the DNA of different species. We share more than 98% of the DNA of chimpanzees, and slightly less than 98% of that of gorillas. While this common inheritance establishes phylogenetic continuity, the 1.6% or so of our DNA which we *do not* share with other species is clearly extremely important, and defines our differences from the other primates. Primates, and in particular human beings, are characterized by the evolution of progressively larger brain size, especially with regard to the **cerebral hemispheres** (see Topic A4). This provides enhanced individual adaptability, increasing along the phylogenetic scale to the human brain.

Natural selection applies to the individual, not the group. That is, it is the fittest *individual* who survives, not the fittest *species*. Eventually, enhanced individual fitness may lead to the establishment of a new species. All present day species are equally 'evolved', all being the latest step in their line of descent. We are *not* descended from any other existing species. Our line of descent diverged from that of the chimpanzees at a common ancestor some 5–7 million years ago.

A2 ACTIONS OF THE NERVOUS SYSTEM

Key Notes

The nature of nervous tissue

The main functional cells in the nervous system are neurons. These typically consist of a cell body (soma), dendrites (through which they receive information), and an axon (through which they transmit information).

The action potential

A neuron is normally electrically polarized. Information is transmitted along the axon by depolarization: a change in polarity of charge across the cell membrane following transport of sodium and potassium ions. This action potential is always of the same size, regardless of the strength of the stimulus.

Synapses

Information is passed between neurons across synapses: clefts between the terminal buttons of the presynaptic neuron and the dendrites or soma of the postsynaptic neuron. Chemical neurotransmitters are released into the cleft from vesicles in the presynaptic axon, attach to receptors in the postsynaptic membrane, and change its polarization. The effect may be inhibitory or excitatory.

Neurotransmitters

Over 100 neurotransmitters are known. Neural circuits in the brain use particular neurotransmitters to control particular functions.

Altering neurotransmitter action

Many drugs interfere with neurotransmitter action, either enhancing it (agonists) or opposing it (antagonists).

Related topics

Divisions of the nervous system (A3)
The cerebral cortex (A4)
The endocrine system (A5)

The nature of nervous tissue

The nervous system is one of the two major systems through which the actions of the body are controlled and coordinated (the other being the endocrine system: see Topic A5). The main type of cell forming nervous tissues is the **neuron** (nerve cell). Other cells are different types of **glia**, which perform various functions, including physical support, assisting neural conduction, and carrying nutrients and waste products between the neurons and blood vessels. Neurons in different parts of the nervous system have some different characteristics, but the main common features are shown in *Fig. 1*. The **cell body** (or **soma**) contains the **nucleus** of the cell (in which are located the chromosomes). Neurons receive information from other neurons, mostly through **dendrites**, of which each neuron has a large number, mostly connected directly to the cell body. The neuron has an **axon** (which may have collateral, or side branches), an elongated fiber that branches towards the end into **terminal buttons**, through

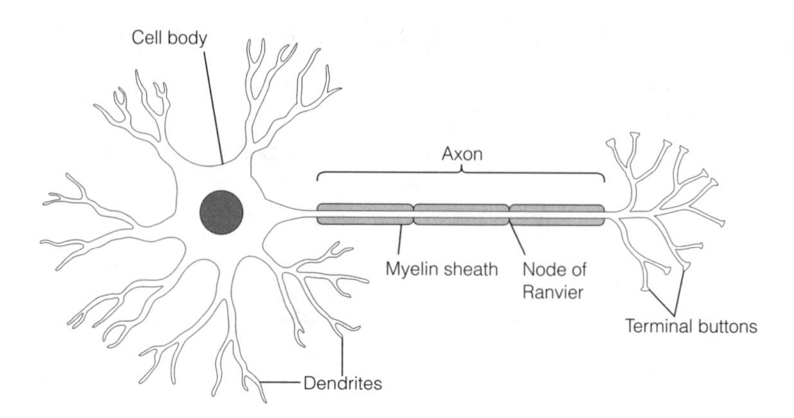

Fig. 1. The main features of a typical neuron.

which it communicates with other neurons, or with **effector cells**, such as muscles and glands. Many axons are surrounded by a fatty **myelin sheath**, which is interrupted at regular intervals by uncoated gaps called **nodes of Ranvier**. A neuron is small in diameter (5–100 microns, or millionths of a meter), but axons may be very long (some extend from the spine to the tips of the toes). Differences between neurons reflect adaptation to perform specialized tasks. Some are modified to act as **receptors** for sensory information (see Topic C1). Others that carry sensory (afferent) information (**sensory neurons**) are described as **unipolar**, since the dendrites and axon are continuous with one another, the cell body being attached at the side. **Motor neurons** carry **efferent** commands to effector cells.

The action potential

The neuron has a **resting potential** of about –70 mV (millivolts; thousandths of a volt). The inside of the cell is negative compared with the outside, and is described as being **polarized**. Signals do not travel along axons by simple electrical conduction, as in a wire, but are propagated by **depolarization**, the active transport of the ions of sodium and potassium across the **cell membrane**, the outer covering of the axon. This results in the inside of the neuron becoming positive compared with the outside. This lasts for only about 1 msec (milliseconds; thousandths of a second), but it initiates depolarization of the next part of the axon. This nerve impulse is known as the **action potential** of the neuron, and in this way is propagated along the axon. The action potential is always of the same size (the **all-or-none law**), regardless of the strength of the stimulus, provided the stimulus reaches a **threshold**, a minimum intensity required to produce sufficient depolarization to initiate the action potential. Typically, stronger stimuli increase the firing rate of a neuron (see Topics C3 and C4). The myelin sheath surrounding many axons permits faster conduction of the action potential, which 'jumps' from one node of Ranvier to the next.

Synapses

Most 'connections' between neurons are, again, not like electrical connections, but are made by way of chemical transmission across a gap called a **synapse**. In these chemical synapses, the terminal buttons of the axons of the **presynaptic neuron** are separated from the **postsynaptic neuron** by narrow **synaptic clefts** (see *Fig. 2*). There will be many terminal buttons on each postsynaptic neuron,

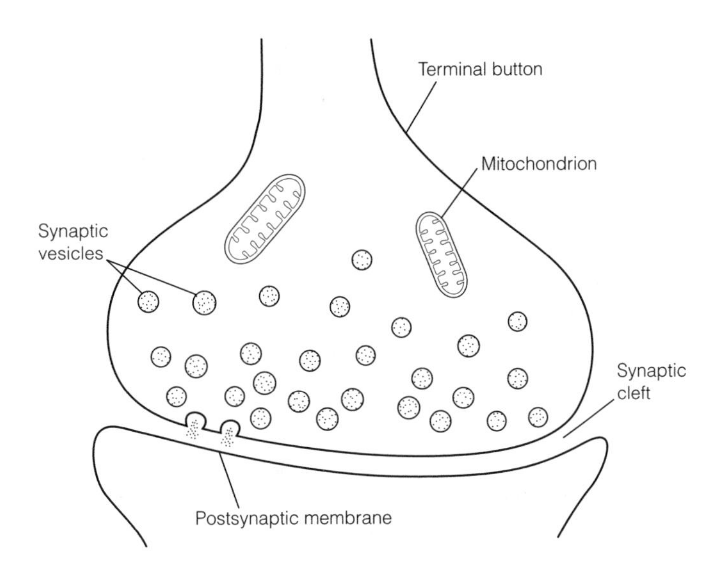

Fig. 2. A chemical synapse. Neurotransmitters are synthesized and stored in vesicles which fuse with the cell membrane to release the transmitter into the synaptic cleft. Mitochondria are the site of cellular metabolism.

and these can come from many different presynaptic neurons. The presynaptic neuron contains **synaptic vesicles** which contain a chemical **neurotransmitter**. When an action potential reaches the synapse, it causes the vesicles to attach to the cell membrane and release the neurotransmitter into the cleft. The neurotransmitter attaches to **receptor molecules** in the membrane of the postsynaptic neuron. For each type of neurotransmitter there is a specific receptor molecule, which has the right shape for the neurotransmitter molecule to fit (a 'lock-and-key' mechanism). Once a transmitter molecule has attached to a receptor, it opens **ion channels** which allow ions to pass across the postsynaptic membrane, changing the **polarization** of the postsynaptic neuron. Some ion channels allow positively charged sodium ions to pass into the neuron, which depolarizes it, increasing its probability of firing. This is known as an **excitatory** effect. Other ion channels allow positive potassium ions to leave the cell, and others permit negative chloride ions to enter the cell. These hyperpolarize the postsynaptic cell, making it less likely to fire; an **inhibitory** effect. The balance of excitatory and inhibitory influences determines whether or not the neuron is sufficiently depolarized to start an action potential travelling along the postsynaptic neuron.

The controlling functions of the nervous system are rapid and may be terminated quickly. The rapid onset of neural responses results from the fast transmission of action potentials. In order for neural action to stop quickly, neurotransmitters released into the synaptic cleft are removed immediately. This is accomplished by two mechanisms. First, **reuptake**: Neurotransmitter molecules are actively reabsorbed into the presynaptic neuron. Second, neurotransmitter molecules are deactivated by **enzymes** in the synaptic cleft.

Neurotransmitters

Over 100 neurotransmitters have been discovered since the 1950s. **Acetylcholine** is the transmitter between somatic nerves and muscles (which have acetylcholine receptors in their membranes), and in the **autonomic**

nervous system (see Topic A3). **Norepinephrine** is a neurotransmitter in the **sympathetic nervous system**. Each of these also acts as a neurotransmitter within the central nervous system (CNS). The other transmitters are all found in the CNS. These include **5-hydroxytryptamine (5-HT)** or **serotonin** (involved, for instance, in emotional responses), and **dopamine** (involved in reward circuits, and in some mental illnesses, see Topic M7).

The brain operates on the basis of **neural circuits**, with each circuit controlling, or helping to control, a particular function. The brain is a very dense medium, with different neural circuits close to one another. One way that activity in adjacent circuits is isolated from other circuits is by the use of different neurotransmitters. The existence of so many different neurotransmitters permits synaptic transmission to be very specific. In addition, different neurotransmitters have different effects. Some (e.g. γ-aminobutyric acid, **GABA**) are generally inhibitory, while others are usually excitatory (e.g. **glutamate**). More than one type of receptor has been identified for many neurotransmitters, so the same neurotransmitter can have different effects, depending on the receptor type. Acetylcholine has three types of receptor, two excitatory and one inhibitory; norepinephrine has four types, and serotonin at least six. Synapses and neural circuits are often referred to by a name derived from the transmitter they use. Thus, synapses using acetylcholine are **cholinergic**, those using norepinephrine are **noradrenergic** (from the older name, noradrenalin), circuits involved in schizophrenia (see Topic L2) are **dopaminergic**, and so on.

Altering neurotransmitter action

The chemical nature of neural communication means that chemical substances can alter synaptic activity, and consequently can modify physiological and psychological states. Many **drugs** act on neurotransmitter systems, and can act on them in different ways. In general, a drug that enhances the action of a neurotransmitter is called an **agonist**, while one that opposes its action is an **antagonist**. Agonistic and antagonistic effects can result from influences at different stages of neurotransmitter action. Some agonists block the reuptake of the transmitter, some interfere with the enzymes that break the transmitter down, while others increase the amount of **precursor** substances (molecules from which the transmitter is synthesized). The result of each of these is an increase in the amount of transmitter in the synaptic cleft. Some agonists mimic the effects of neurotransmitters by attaching to, and 'unlocking', the receptors. Conversely, some antagonistic substances increase the reuptake of transmitter molecules, some activate enzyme systems, some decrease the amount of precursor molecules, and others, **receptor blockers**, attach to receptors, but do not cause them to open their ion channels.

A3 DIVISIONS OF THE NERVOUS SYSTEM

Key Notes

The nervous system

Cutting across the division of the nervous system into central and peripheral parts is a functional division into autonomic and somatic parts.

The central nervous system

The central nervous system comprises the brain and spinal cord. The brain is a complex structure, conveniently divided into three main regions: the forebrain (cerebral hemispheres, thalamus, hypothalamus, limbic system); the midbrain (reticular formation, and various centers controlling movement); and the hindbrain (cerebellum, pons, medulla). Most of the brain is protected from the effects of blood-borne substances by the blood–brain barrier.

The peripheral nervous system

The nerves of the peripheral nervous system connect sensory systems and effector organs to the central nervous system.

The autonomic nervous system

The autonomic nervous system controls the internal environment and basic bodily functions. Its two branches, the parasympathetic and sympathetic nervous systems, generally act antagonistically. It has central components in the hypothalamus.

Related topics

The cerebral cortex (A4)
The endocrine system (A5)
Psychophysics (C1)

The nervous system

The nervous system may be functionally divided in two different ways. First, the **central nervous system (CNS)** is distinguished from the **peripheral nervous system**. In general, the CNS is concerned with coordinating and controlling the actions of all the organs of the body. The peripheral nervous system carries commands *from* the CNS to the various **effector organs**, and carries sensory information *into* the CNS. Cutting across this division is a distinction between the **autonomic nervous system (ANS)** and the **somatic nervous system**, each of which has both peripheral and central components. Broadly, the ANS is concerned with the control of the functions of internal organs, while the somatic nervous system controls the activities of the **voluntary muscles**, and handles information from the **special senses** (see Topics C2 and C3).

The central nervous system

The CNS comprises the **brain** and the **spinal cord**, each of which is composed of neurons and various other tissues. Most of the control and organizing functions of the CNS are carried out by neurons. The spinal cord mainly serves to channel sensory information to the brain, and motor commands from the brain. In addition, many **reflexes** use neural pathways that involve a sensory neuron and a motor neuron synapsing in the spinal cord. Although these **spinal reflexes** occur even if the spinal cord is disconnected from the brain, they are modified and integrated by the brain in intact animals.

The brain is usefully subdivided into three regions (see *Fig. 1*): the **forebrain**, the **midbrain**, and the **hindbrain**. The forebrain comprises the two **cerebral hemispheres**, the **thalamus**, the **hypothalamus**, and the **limbic system**. The outer layers of the cerebral hemispheres are the **cerebral cortex** (see Topic A4), which is the site of awareness of sensory input, of voluntary action, and of symbolic activity like language (see Section G). Corresponding parts of the left and right cortex are connected by a fiber bundle called the **corpus callosum**. The thalamus is a relay station in most of the sensory pathways into the brain. The hypothalamus is vitally important in the control of motivated behavior; it coordinates ANS activity, and is an important controller of the endocrine system (see Topic A5). The limbic system is involved in emotion-related sensory processing and in memory (see Topic F3). *Figure 2* shows the location of some of the areas mentioned.

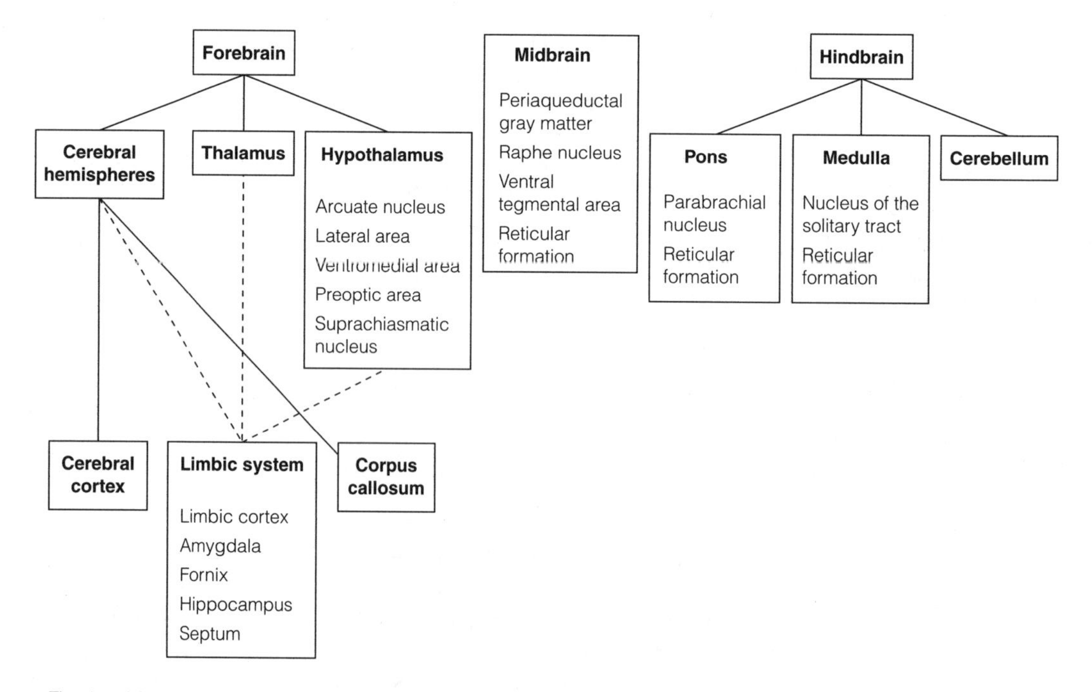

Fig. 1. Major divisions of the brain, with some important centers.

The midbrain contains neural centers that integrate reflexes into coherent actions, as well as part of the **reticular formation**: a network of neurons that is important in maintaining the arousal of the CNS (see Topic B6). The hindbrain consists of the **cerebellum** (which controls motor coordination and balance), the **pons** (which contains nuclei controlling movement, as well as sensory inputs from some **cranial nerves**), and the **medulla** (which connects the brain to the spinal cord). The reticular formation extends from the midbrain into the medulla. The term **brain stem** is often applied to all of the midbrain and hindbrain structures apart from the cerebellum.

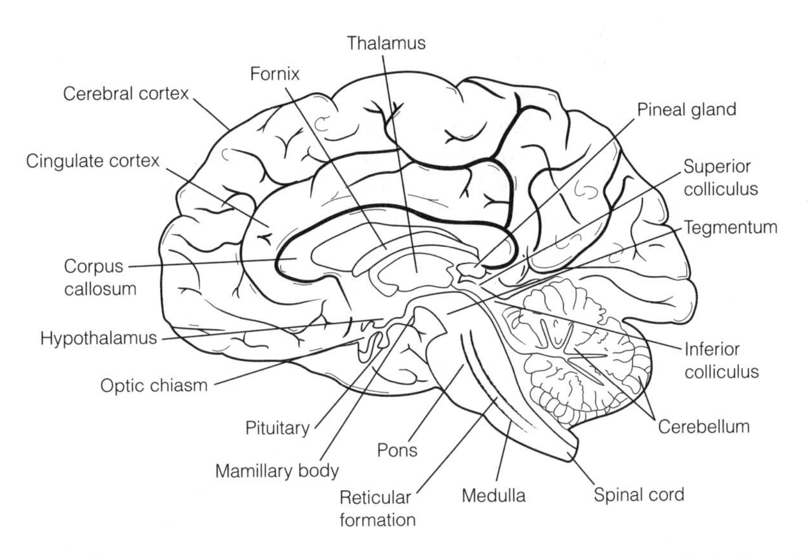

Fig. 2. Cross-section of the human brain, showing locations of some major brain regions.

Centrally located throughout the CNS is an arrangement of canals, and chambers known as **cerebral ventricles**, which are filled with **cerebrospinal fluid (CSF)**. This acts as a protective cushion against damage to the brain during movement, and provides a way in which some hormones act on brain centers. The blood capillaries that supply most of the brain are less permeable than those in other parts of the body, providing a **blood–brain barrier**. Oxygen, carbon dioxide, and water cross this readily, glucose quite easily, ions like sodium and potassium more slowly, and many hormones and proteins to a very limited degree. This protects the brain from circulating substances that have effects on peripheral organs, maintaining a constant chemical environment.

The peripheral nervous system

The peripheral nervous system consists of **nerves**, which are bundles of individual **axons** of **neurons** (see Topic A2). These connect the sensory systems and the effector organs (e.g. muscles) to the CNS. Nerves carry commands away from the CNS to the various organs (**efferent** signals), and information back to the CNS from sensory systems (**afferent** signals), both from specialized organs like the eyes and ears, and sensory receptors in the skin, muscles, joints, and so on. The motor functions that the somatic nervous system controls are mostly under voluntary control. The exceptions are spinal reflexes, like the knee jerk, and those which coordinate the opposed contraction and relaxation of muscles in an action such as walking, which operate at the spinal level.

The autonomic nervous system

The ANS coordinates the control of the internal environment, and the bodily functions that are basic for survival. It has afferent and efferent components, and operates involuntarily, and largely without our awareness. The ANS has components in the CNS: in the hypothalamus, the brain stem, and the spinal cord. The peripheral ANS is divided into the **sympathetic nervous system (SNS)** and the **parasympathetic nervous system (PNS)**. Most organs are innervated (supplied) by both branches, and the two generally act in an antagonistic manner, so that, for example, the state of constriction or dilation of blood vessels is controlled by an equilibrium of activity in the SNS and the PNS. This allows the resources of the body to be matched to its varying needs. In general, the PNS maintains the body in a steady state, while SNS activity prepares it to respond to emergency situations.

The afferent fibers of the ANS pass information about the state of the internal organs to the CNS, where they initiate corrective responses at either the spinal level or the hypothalamic level. Hypothalamic responses are generally more coordinated responses over a number of organ systems. Pain information from internal organs is not separately represented in the cerebral cortex. Visceral pain fibers synapse with somatic pain fibers that enter the spinal cord at the same level. For example, pain fibers from the heart enter the spinal cord with those from the upper left arm and chest. Damage to the heart is experienced as pain in the left upper arm and chest. This is called **referred pain**.

Almost all efferent ANS fibers have a synapse between the spinal cord and the effector organs. **Preganglionic fibers** of the SNS leave the spinal cord in the dorsal and lumbar regions, and pass to **ganglia** close to the spinal cord. After cholinergic synapses in the ganglia, the **postganglionic fibers** pass to the effector organs, with which they communicate by noradrenergic synapses (see Topic A2). Preganglionic fibers of the PNS leave the brain stem and the sacral region of the spinal cord and pass directly to synapses near the target organs, which they influence through cholinergic synapses.

A4 THE CEREBRAL CORTEX

Key Notes

Projection areas

The cerebral cortex consists of the outer regions of the cerebral hemispheres. Each hemisphere is divided by fissures into four lobes. Certain areas of the cortex have clearly defined motor or sensory functions, controlling actions of, or receiving information from, the opposite side of the body.

Localization of function

The functions of cortical areas have been mapped by stimulation studies, lesion studies, computerized axial tomography (CAT scan), positron emission tomography (PET scan), and functional magnetic resonance imaging (fMRI).

Association areas

Areas of cortex close to the projection areas are involved with single sensory or motor functions, while those further from the projection areas coordinate different senses, and sensory–motor functions. Lesions in the motor association areas result in apraxias (disorders of movement). Agnosias, disorders of perceptual organization without sensory loss, result from lesions in the sensory association areas.

Hemispheric lateralization

Hemispheric lateralization is shown by handedness. The left hemisphere is relatively specialized for analytic/temporal functions; the right for synthetic/spatial functions. The two hemispheres communicate by a number of tracts, the main one of which is the corpus callosum. Lesions of these bundles produce split-brain patients, who have been used to study lateralization and localization.

Recovery after brain damage

Recovery from brain lesions results from four mechanisms: removal of cause (when neurons have not been destroyed); growth of collateral fibers from undamaged neurons, which may be promoted by brain tissue transplanted from another animal; assumption of a function by another brain region; and a brain-injured person may learn a different way of performing a function. Recovery is best in younger organisms.

Related topics

Divisions of the nervous system (A3)
Perception (section D)
Speech recognition and production (G2)
Reading (G3)

Projection areas

The cerebral cortex consists of the outer regions of the cerebral hemispheres. It is about 3 mm thick but, being deeply convoluted, forms a large part of the human brain. It consists mainly of cell bodies, dendrites, short axons, and glia (see Topic A2). It has a gray appearance, and is known as **gray matter**. Beneath the cortex is a mass of myelinated axons, connecting cortical regions with each other and with other brain centers. Because of the pale appearance of the myelin it is called **white matter**. Each hemisphere is divided by fissures into

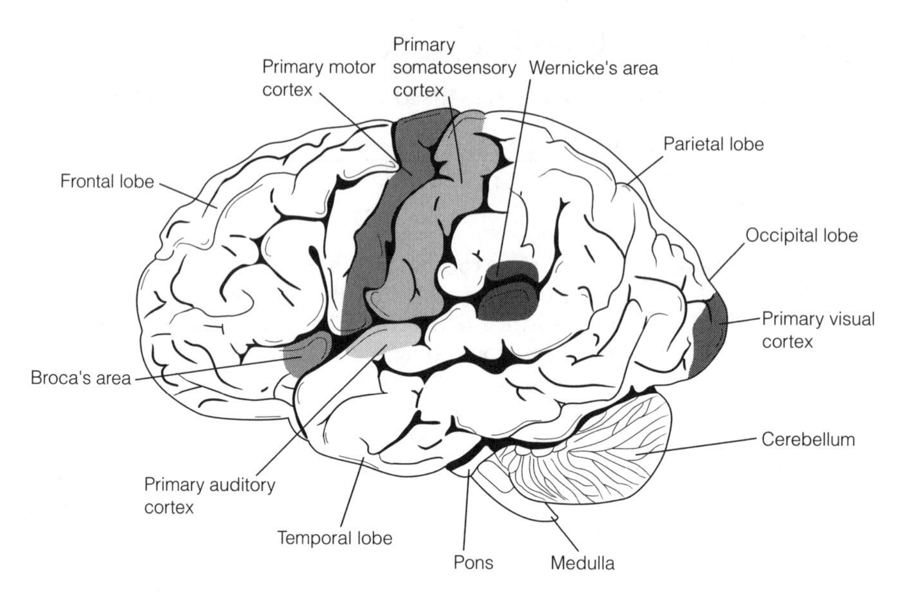

Fig. 1. Primary sensory and motor areas of the human brain with the association areas involved in language.

four lobes (see *Fig. 1*). Certain areas of the cortex have clearly defined motor or sensory functions. The rear of the frontal lobe is the **motor projection area** (**motor cortex**). Each point within this region controls a particular voluntary action on the opposite side of the body. Parts of the body with more finely controlled movements (e.g. the fingers) have larger areas of the motor projection area devoted to them. Three cortical regions are primary **sensory projection areas** (see *Fig. 1*). The **somatosensory cortex**, at the front of the parietal lobe, receives touch, temperature, and movement information from receptors in the skin and elsewhere, again on the opposite side of the body. Body areas are mapped onto the somatosensory cortex in a way parallel to that of the motor cortex. The **visual projection cortex** is located in the occipital lobe, and the **auditory projection cortex** at the top of the temporal lobe, much of it in the fissure between the temporal and parietal lobes. These receive direct inputs from the eyes and ears, respectively (see Topics C2 and C3).

Localization of function

The functions of the projection areas were originally discovered by **stimulation** studies; applying electrical stimuli to the exposed cortex of animals and humans permitted the mapping of motor and sensory projection areas. Other information about the functions of parts of the brain has come from **lesion** studies, in which the effects of experimentally induced damage, accidental damage, or damage caused by disease to localized parts of the brain are observed. While, in animal experimental studies, the extent and location of lesions can be precisely defined, injuries and disease processes in the human brain tend to be more diffuse, and until recently could only be defined by post-mortem examination. The modern technique of **computerized axial tomography** (**CAT**) scans the living brain with X-rays, and allows a detailed picture of brain lesions to be produced. Two newer techniques reveal changes in the brain during its normal functioning. **Positron emission tomography** (**PET**) relies on more active parts of the brain taking up more of an injected radioactive substance **2-deoxyglucose**, the distribution of which in the brain can be imaged. **Functional magnetic resonance imaging** (**fMRI**) assesses the state of excitement of atomic nuclei which, when the image is scanned and formed quickly enough, gives detailed information about the activity of particular brain regions.

Association areas

Areas of cortex close to the projection areas are involved with single sensory or motor functions, while those further from the projection areas coordinate different senses, and sensory–motor functions. The area of the frontal lobe adjacent to the motor cortex is the **motor association area**. Lesions there result in **apraxias** (disorders of movement): patients are unable to perform coordinated, previously well-practiced movements such as those involved in dressing. However, they show normal sensory function, and can perform the isolated actions that make up the coordinated movement. Neurons in this area organize patterns and sequences of motor movements, and are themselves controlled by prefrontal areas that plan the movements. Other apraxias result from lesions in parts of the cortex that coordinate sensory and motor activities. For example, lesions in the left posterior parietal lobe prevent the coordination of somatosensory and visual (e.g. hand position), auditory (e.g. spoken commands) and motor actions (e.g. drawing), preventing a person from being able to follow instructions (e.g. to draw an object). Lesions in the **auditory association area**, in the left parietal lobe, prevent the translation of a command into planned actions. Lesions in the right parietal lobe lead to **constructional apraxia**, in which a person has difficulties with spatial tasks, such as drawing pictures.

Agnosias, disorders of perceptual organization without sensory loss, result from lesions in the sensory association areas that are adjacent to the sensory projection areas. In visual agnosia, patients may be able to identify simple components of a complex object, such as the parts of a clock, but be unable to identify it as a clock. If asked to copy a drawing, they may reproduce parts, but not put them together properly. **Prosopagnosia** is a specific inability to recognize faces, or even to tell that a face *is* a face. In **unilateral neglect syndrome**, people with a lesion in the lower part of the right parietal lobe will ignore the left side of the world. This is shown when they are asked to copy a drawing, and extends also to their own bodies, so that they will fail to dress their left limbs or even to identify them as their own. This seems to reflect damage to a system involved in maintaining spatial attention.

Expressive (or **Broca's**) **aphasia** is an apraxia of language, resulting from damage in and around an area of the (usually left) frontal lobe called **Broca's area**, adjacent to the auditory projection area (see *Fig. 1*). Damage to a region of the (usually left) temporal cortex, **Wernicke's area**, immediately behind the auditory projection area leads to **receptive** (or **Wernicke's**) **aphasia** (see Topic G2).

Hemispheric lateralization

While the gross structures of the cerebral hemispheres are duplicated in the left and right hemispheres, there is marked differentiation of certain functions between the two, known as **hemispheric lateralization**. One obvious illustration of this differentiation is **handedness**, which is the tendency to favor the use of one side of the body over the other. The majority of people are right-handed, meaning that functions such as manual dexterity are more finely controlled by the left hemisphere. In right-handed and many left-handed people, language functions are clearly lateralized in the left hemisphere (see above). In general, the left hemisphere is relatively specialized for *analytic* (or perhaps *temporal*) functions. The right hemisphere, in most people, is specialized for *synthetic* (or *spatial*) functions, such as complex drawing, interpreting maps, and spatial localization.

Lateralization of function implies communication of sensory input and associative output between the two hemispheres. This is achieved by a number of tracts that connect the hemispheres, the main one of which is the **corpus callosum** (see Topic A3). Lesions of these bundles, sometimes made surgically to relieve severe epilepsy, produce **split-brain** patients. Sending information to just one hemisphere in such patients means it is unavailable to the other, except by indirect means. A patient is unable to name an object shown briefly to the right hemisphere (showing it in the left part of the visual field), because the right hemisphere cannot communicate with the language areas in the left hemisphere. However, the patient can respond appropriately with a nonlinguistic response, such as pressing a button (with the left hand) to indicate the object seen.

Recovery after brain damage

The effects of brain lesions are not necessarily permanent. Recovery results from one or more of four mechanisms. First, if the damage does not involve destruction of neurons, for example resulting from pressure from a tumor, function may often be at least partly recovered when the cause of the damage is removed. Second, while neurons that are destroyed are not replaced, some function may return if **collateral fibers** from undamaged neurons grow to replace the missing neurons, especially in infants. Brain tissue transplanted from one animal to another may restore lost function, either by directly replacing lost tissue or encouraging the growth of collaterals. Third, the brain shows *plasticity* of function, also greater in younger brains. Damage to Broca's area in a child younger than 5 years has little effect on language; however, in an adult it results in aphasia. In younger adults, slow recovery from aphasia may take place, and this is accompanied by increased circulation in the equivalent area in the right cortex. Fourth, a brain-injured person may learn a different way of performing a function.

A5 THE ENDOCRINE SYSTEM

Key Notes

Hormones

The endocrine system acts as a longer-term control system. Hormones are usually released into the blood and circulate to target cells, where they act by attaching to specific receptors. The endocrine system works on the basis of negative feedback.

Pituitary gland

The posterior pituitary is an outgrowth of the hypothalamus. Its main hormones are vasopressin, involved in maintaining water and electrolyte (dissolved substances) balance, and oxytocin (lactation and labor). The anterior pituitary produces a variety of hormones, including the gonadotropic hormones, which cause the gonads to produce hormones; adrenocorticotropic hormone, which stimulates the adrenal cortex; and prolactin, which is involved in reproductive behavior.

Adrenal medulla

The adrenal medulla produces mainly epinephrine. Its secretory cells are analogous to the postganglionic fibers of the sympathetic nervous system, and are themselves innervated directly by preganglionic fibers. It serves longer-term emergency responses.

Adrenal cortex

The adrenal cortex produces steroid hormones. Glucocorticoids (cortisol) are involved in glucose metabolism and in stress reactions. Mineralocorticoids (aldosterone) primarily affect electrolyte balances by actions on the kidney. Androgens (testosterone) have effects to do with sexual characteristics and behavior.

Pancreas

The hormones of the pancreas include the peptide hormones insulin and glucagon, which are key factors in the control of metabolic processes.

Ovaries

The ovaries exert endocrine control over reproduction in females. The hormones estrogen and progesterone are produced by developing follicles in the menstrual cycle, which is controlled by a feedback loop involving the hypothalamus, the pituitary gland, the ovaries, and the uterus.

Testes

The testes produce the androgen hormones, mainly testosterone which is responsible for the embryonic differentiation of the male reproductive organs, and the development of the secondary sexual characteristics of males. The testes also produce anti-Müllerian hormone, which prevents the development of female internal genitalia in male fetuses.

Pineal gland

The pineal gland secretes melatonin, which matches bodily rhythms with seasonal variations.

Related topics

Divisions of the nervous system (A3)
Sex (B4)
Aggression (B5)

Hormones

The **endocrine system** acts as a longer-term control system. Hormones are produced in many places in the body, including specialized **endocrine glands**. Hormones are usually released into the blood and circulate to target cells, where they act by attaching to specific receptors, in the same way as neurotransmitters, producing a change in the activity of the target cells. Hormones can have **organizing effects** (preparing the body for particular behaviors) and **activating effects** (triggering behaviors). The endocrine system works on the basis of **negative feedback**. The magnitude of the change in the target cells is fed back to the endocrine gland. If the change is too small, more hormone is produced; if it is too great, less hormone is produced. The main hormones that are known to affect psychologically relevant processes, and their functions are shown in *Table 1*.

Table 1. The origins and effects of some hormones

Gland	Hormone	Effects on
Posterior pituitary	Vasopressin	Water and electrolyte balance
	Oxytocin	Lactation and labor; male and female sexual behavior
Anterior pituitary	Luteinizing hormone	Ovulation; sperm and testosterone production
	Follicle stimulating hormone	Development of ovarian follicles; testes
	ACTH	Control of adrenal cortex
	Prolactin	Lactation; inhibits male sexual behavior
Adrenal medulla	Epinephrine	Stimulates cardiovascular function; regulates metabolism
Adrenal cortex	Glucocorticoids (cortisol)	Regulates metabolism in liver, muscles, and adipose tissues
	Mineralocorticoids (aldosterone)	Water and electrolyte balance
	Androgens (testosterone)	Growth and development; sex and aggression
Pancreas	Insulin	Energy storage; glucose uptake by cells
	Glucagon	Energy release
Ovaries	Estrogens (estradiol)	Female reproduction
	Progesterone	Female reproduction
Testes	Androgens (testosterone)	Sex differences, sex and aggression
	AMH	Male sexual differentiation
Pineal	Melatonin	Coordinating body rhythms

Pituitary gland

The pituitary gland releases many hormones that act on other endocrine glands, controlling their hormone release. It is situated immediately below, and is controlled by, the hypothalamus. The pituitary gland is really three glands, two of which concern us. The **posterior pituitary gland** is an outgrowth of the hypothalamus. Its main hormones are produced by neurons in the hypothalamus, and are transported along their axons to the pituitary, where they are secreted into capillaries. These hormones are **vasopressin** (also called **antidiuretic hormone, ADH**), which is involved in maintaining water and **electrolyte** balance (see Topic B2); and **oxytocin**, which is involved in lactation and labor. The **anterior pituitary gland** has a circulatory link with the hypothalamus and produces a variety of hormones, most of which control other glands. Among these are the **gonadotropic hormones**, which cause the gonads to produce hormones (see *Ovaries* and *Testes* below); **adrenocorticotropic hormone (ACTH)**, which causes the secretion of **glucocorticoids** by the adrenal cortex (see below); **prolactin**, which is involved in reproductive behavior (see Topic B4); and others, including growth hormone. The secretion of the anterior pituitary hormones is under the control of releasing and inhibiting hormones produced by the hypothalamus.

Adrenal medulla

The adrenal glands are located immediately above the kidneys. The central part, the adrenal medulla, produces mainly **epinephrine**. Its secretory cells are analogous to the postganglionic fibers of the sympathetic nervous system (SNS), and are themselves innervated directly by preganglionic fibers (see Topic A3). Just as the SNS serves the rapid responses of the body to emergency situations, so the adrenal medulla serves longer-term emergency responses. Epinephrine and norepinephrine act through different receptor types so do not have identical physiological effects.

Adrenal cortex

Surrounding the adrenal medulla is the adrenal cortex, which produces a large number of different **steroid** hormones, falling into three classes. **Glucocorticoids**, primarily **cortisol**, act mainly on glucose metabolism, but are also involved in stress reactions. Their main physiological effects support the actions of epinephrine: the release of glucose and other energy sources, and the promotion of the uptake of glucose by muscles. They reduce some immune system functions, so have anti-inflammatory and immunosuppressive properties. **Mineralocorticoids**, the main one of which is **aldosterone**, primarily affect electrolyte balances by actions on the kidney (see Topic B2). **Androgens** have effects to do with sexual characteristics and behavior, after conversion to **testosterone** in target organs (see Topic B4). In men most testosterone derives from the testes, but in women most circulating androgens are secreted by the adrenal cortex. The secretion of these **corticosteroids** is controlled by ACTH produced by the anterior pituitary gland.

Pancreas

The pancreas has a variety of functions, all related to the digestion, absorption, and use of food and its products (see Topic B3). Its endocrine functions are the secretion of four hormones, of which we will focus on two: the peptide hormones **insulin** and **glucagon**, which are key factors in metabolic processes, controlling the storage and release of carbohydrates and fats.

Ovaries

The ovaries exert endocrine control over reproduction in females. The hormones are produced cyclically by developing follicles (see Topic B4), which also produce the ovum. This cycle is controlled by a feedback loop involving the hypothalamus, the pituitary gland, the ovaries, and the uterus. The hypothalamus stimulates the release of **luteinizing hormone** and **follicle-stimulating hormone (FSH)** by the posterior pituitary. The effects of these hormones depend on the stage of the cycle. Just after menstruation, the beginning of which is Day 1 of the cycle, FSH stimulates growth of a number of ovarian follicles (the **follicular phase**). These produce **estrogen**; then all but one of the follicles cease to grow. The increasing estrogen level of the blood, peaking about Day 12, stimulates secretion by the pituitary, which in turn causes the follicles to reduce their estrogen secretion and to secrete **progesterone**. It also stimulates the release by the follicle, on Day 14, of an ovum. The high level of luteinizing hormone at ovulation stimulates the follicle to develop into the **corpus luteum** (the **luteal phase**). The corpus luteum secretes a large amount of progesterone, and a smaller amount of estrogen. This combination suppresses the secretion of luteinizing hormone and FSH, preventing growth of further follicles (and is the basis of oral birth control). The corpus luteum grows for 7 or 8 days; then, if the ovum is not fertilized, it starts to degenerate so that, after about Day 23, the blood level of progesterone falls. Blood levels of luteinizing hormone and FSH rise again, starting a new cycle of follicle growth.

Another of the effects of estrogen secreted during the follicular phase is to cause the growth of the **endometrium**, the lining of the uterus. This continues during the luteal phase. If the ovum is not fertilized, the drop in circulating estrogen and progesterone causes the degeneration of the endometrium, leading to menstruation. If the ovum is fertilized, it embeds itself in the endometrium, and forms a placenta. This secretes gonadotropic hormones which cause the corpus luteum to enlarge, and continue to secrete progesterone and estrogen. This prevents the endometrium from degenerating, and allows the fetus to develop.

Testes

The testes produce primarily androgen hormones, mainly testosterone. Testosterone is an **anabolic steroid**, promoting tissue growth, and has some effect on almost every tissue in the body. It is responsible for the embryonic differentiation of the male reproductive organs, and the development of the secondary sexual characteristics of males at puberty (see Topics B4 and B5). Its secretion by the testes in the adult male is controlled by the hypothalamus; however, the controlling hormone is secreted steadily rather than cyclically. The testes also produce **anti-Müllerian hormone (AMH)**, which prevents the development of female internal genitalia in male fetuses (see Topic B4).

Pineal gland

The pineal gland, which is situated between the brain stem and the cerebral cortex, secretes **melatonin**, mainly during darkness. The effect of this hormone is to match bodily rhythms with seasonal variations. In birds and some reptiles, light reaches the pineal gland, and acts directly on it. In humans, the effects of light are mediated by neural connections. (See Topic B6.)

B1 APPROACHES TO MOTIVATION

Key Notes

What is motivation?

Motivation energizes and directs all behavior apart from the simplest reflexes. Homeostatic motives such as thirst and hunger are physiological motives that derive from bodily imbalances. Other types of motivation are nonhomeostatic physiological, social, self-presentation, and self-integrative motives.

Homeostasis

Homeostasis is the maintenance of the internal environment close to optimal conditions. It is maintained by physiological negative feedback systems assisted by homeostatic behaviors.

Drive-reduction theories

These approaches explain human motivation as being based on mechanisms analogous to those in homeostatic motivation. Motivation is nonspecific arousal which is directly related to performance. Nonphysiological motives are secondary drives, learned by association with the satisfaction of primary (physiological) drives.

Curiosity

Curiosity is a widespread motive throughout the animal world. Unlike homeostatic drives it does not show generalized satiation, and so serves the adaptive function of keeping the animal aware of its surroundings.

Optimal arousal

Some theorists argue that underlying all motivation is a need to maintain the level of arousal optimal for the task being performed. The optimal level of arousal is lower for more complex tasks. Others have proposed that we seek always to minimize arousal.

Related topics

Instrumental learning (E2)
Self perception (I4)
Groups (J2)

What is motivation?

All behavior except the simplest reflexes is considered to be motivated. Motivation controls behavior, and is usually regarded as having two aspects: it *energizes* behavior and *directs* it towards some goal. We distinguish two types of **physiological motive**: those that derive from imbalances in the body (**homeostatic motives**; see Topics B2 and B3); and those that do not have the function of maintaining bodily equilibrium (**nonhomeostatic** motives, see Topics B4, B5, and B6). Other motives are not clearly related to physiological needs. We will discuss **social motives**, based on group membership, in Topic J2; **self-presentation motives**, concerned with our desire to appear to others in the best light, in Topic I3; and **self-integration motives**, concerned with protecting and enhancing one's identity, in Topic I4.

Homeostasis

Animal cells and organs will only work optimally when their operating environment is maintained within a very narrow range. The internal environment has to be controlled to provide optimal conditions of, for example, temperature, electrolyte concentrations, pH (acidity) of body fluids, oxygen level, and carbohydrate concentrations of tissues. The physiological process that produces this stability in the face of fluctuations in the demands that the environment makes on the body is known as **homeostasis**. Homeostasis operates through negative feedback (see Topic B2). Animals also control their internal environment by engaging in particular activities: **homeostatic behavior**. For the homeostatic motives a need arises from a specific tissue deficit. This leads to a **drive** that energizes (and perhaps directs) the animal to **consummatory behavior** that satisfies the need. The main approach adopted to homeostatic drives is to search for the physiological mechanisms through which the needs are assessed and their satisfaction is achieved.

Drive-reduction theories

These are an attempt to explain other types of motivation in a way parallel to homeostatic motivation. People are assumed to pass into a state of deprivation, similar to the tissue deficit of a homeostatic need. This leads to a drive state that is either general, or specific to the particular state of deprivation. This drive leads the person into 'consummatory' behavior that relieves the state of deprivation. Because the reduction of a drive is pleasant, it rewards behavior that immediately precedes it; that is, it makes that behavior more likely to occur. Learning theorists such as Hull described drive as nonspecific **arousal** that is capable of energizing any behavior, and supporting any kind of learning. Drive-reduction theory predicts a linear relation between motivation (arousal) and performance (*Fig. 1a*).

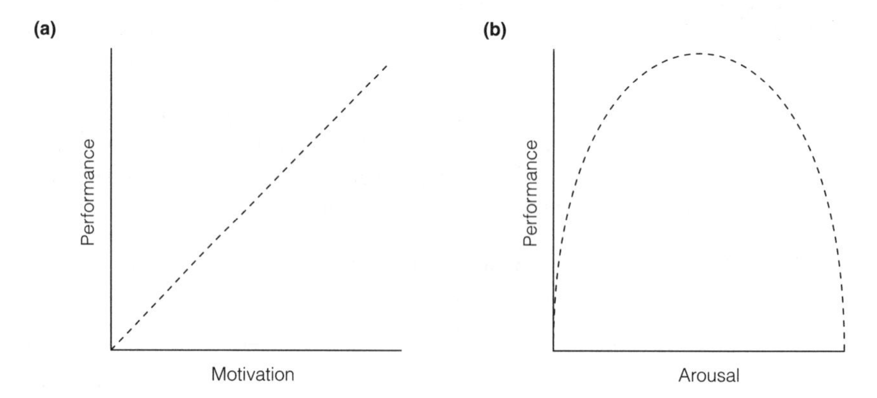

Fig. 1. (a) The relation between performance and motivation hypothesized by drive-reduction theorists. (b) The 'inverted U' relation between performance and arousal hypothesized by optimal arousal theorists.

Learning theorists described the physiological drives as **primary drives**, in the sense that the drive arises internally, but is directed by external stimuli. Learning can modify the external stimuli with which the drive is associated. They attempted to apply these principles to other types of motivation, which they described as **secondary drives**. These were learned in the sense that the source of the motivation is learned, but the drive state itself is the same as for a primary drive. When initially neutral stimuli are paired with the satisfaction of a primary drive, that stimulus itself comes to evoke a similar drive state. This was said to be demonstrated by **conditioned emotional responses**. Pairing of a frightening stimulus (an electric shock, or a loud noise) with a previously neutral stimulus leads to the neutral stimulus becoming aversive. However, it was impossible to demonstrate empirically this sort of process taking place for the 'higher' motives of human beings.

Modern approaches to human motivation borrow the language of drive-reduction theory, but apply it in a more general way. Thus, the term 'drive' is used to describe the state of feeling motivated to perform some action. A 'need' does not relate to a tissue deficit, but to a feeling that something is missing. Underlying these approaches is the view that the satisfaction of needs or drives is pleasant, either intrinsically, or because it reduces an unpleasant feeling of need.

Curiosity

Curiosity, or an exploratory drive, is common to humans and other animals. Animals will usually explore a novel environment. Unlike drive-reduction motives like hunger, curiosity seems not to be reduced when the 'consummatory behavior' of exploration takes place; that is, it does not **satiate**. This presents a problem for drive-reduction theories; if there is a drive for exploration, then it should be reduced when exploration takes place. However, the satiation of drives can be, at least partly, stimulus specific; an apparently sated animal or person will start to eat again when a different food is offered (**sensory-specific satiety**, see Topic B3). The same thing happens with curiosity. A rat will explore a maze with no external reward, but will soon stop doing so when it becomes familiar with it. In the same way, humans become bored when the novelty of a new situation wears off. However, put the rat in a new maze, or give the person something new to think about, and exploratory behavior, or interest, is rekindled. What *is* different is that there appears to be no longer-term (generalized) satiety. We cannot go on introducing new food indefinitely; eventually, general satiety occurs. This does not appear to be the case with curiosity.

This makes biological sense. The needs underlying most biological drives can be satisfied episodically, and may correct specific tissue deficits. Curiosity represents a need to be constantly aware of the environment so as to be able to respond efficiently, for example, to threat. So, whenever an aspect of the environment changes, it needs to be explored.

Optimal arousal

In the 1950s, Hebb and others put forward the view that there is an **optimal level of arousal** for the performance of any activity, and that we actively seek to maintain that level of arousal. If our arousal level falls below the optimum, we experience boredom, and seek novelty or other types of stimulation that will increase arousal. This might be the basis of the curiosity motive. If arousal rises above the optimum, then we experience anxiety and distress, and will engage in activities that reduce it. One consequence of the optimal arousal view of motivation is that high levels of motivation may disrupt behavior, if the associated arousal is above the optimum. Super-optimal arousal is supposed to be the common mechanism by which such states as anxiety, or external conditions such as heat or noise produce a deterioration in performance. This effect is shown in the 'inverted U' curve that relates performance to arousal (*Fig. 1b*). The more complex the task to be performed, the lower the optimal level of arousal (the **Yerkes–Dodson Law**). In performing a skilled task, we have to choose the correct response from all available responses. Higher levels of arousal can promote the performance of incorrect dominant responses, leading to poorer performance. More complex tasks have more competing responses, so their optimal arousal level is lower.

B2 NEGATIVE FEEDBACK AND DRINKING

Key Notes

Negative feedback systems

The elements of negative feedback control systems are a system variable, a set point, a sensor, a comparator, a control, and a correctional process. All of these may be identified in homeostatic systems, but physiological systems typically have multiple components of each type.

Volumetric and osmometric thirst

Both the volume of water in the body and electrolyte concentrations need to be maintained. Water is continuously lost from the body by processes such as urination, which also loses electrolytes, and evaporation, which does not. Blood-flow receptors in the kidneys and baroreceptors in the heart detect changes in blood volume and pressure, while osmoreceptors in the circumventricular organs detect changes in electrolyte balance. Signals from the specialized receptors are integrated in the circumventricular organs which, through connections with other brain areas, coordinate hormonal, autonomic, and behavioral corrective processes. Satiety is signaled by osmoreceptors in the gastrointestinal tract.

Related topics

Approaches to motivation (B1) Hunger and eating (B3)

Negative feedback systems

The essential features of control systems such as those involved in homeostasis are shown in *Fig. 1*. We will use the analogy of thermostatic control of room temperature to explain them. The **system variable**, the property that is to be controlled (e.g., room temperature) has a **set point**, which is the target value of the property (e.g., the desired temperature). The system needs a **sensor** (e.g., thermometer) to detect and report the current state of the system to a **comparator**, which tests if the system variable is different from the set point. A **control** (e.g., a switch) starts and stops a **correctional process** (e.g., a heater). A simple system like a room thermostat has a number of disadvantages which could be fatal in a homeostatic mechanism. First, thermostats cannot maintain an exact temperature: the temperature at which they switch the heater on is always lower than the temperature at which they switch it off. In a physiological system this could lead to an unacceptably wide variation in metabolic efficiency. This could be improved on by having a heater capable of continuously variable output, controlled by detected variations in temperature. Physiological control systems have this characteristic. Second, a simple heating system permits correction only in one direction. This could be overcome by adding a process to cool the room if it gets too hot. Bidirectional control is a feature of physiological systems. Third, if any of the components or the connections between them fails, the thermostat system fails. Physiological systems have built-in **redundancy,** having more than one of each component, with more than one connection

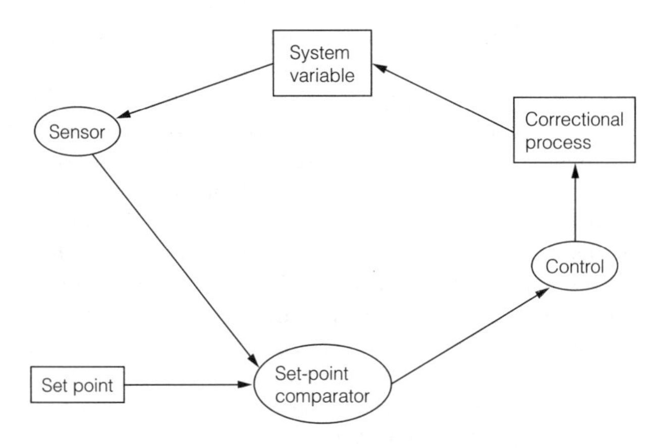

Fig. 1. The components of a simple negative feedback control system.

between components. Similarly, homeostatic systems have correctional processes of different types. These involve the *autonomic nervous system* (Topic A3) which, for example, controls changes in the diameter of blood vessels in the skin, allowing increased or decreased heat loss; the *endocrine system* (Topic A5) which, for example, causes changes in heat production by increased secretion of thyroid hormones; and *behavioral responses*, for example, seeking sunlight or shade to increase or decrease heat absorption.

Volumetric and osmometric thirst

The amount of water contained in our bodies has to be controlled for two reasons. First, some physical processes, in particular the circulation of the blood, require the maintenance of **volume** and pressure within a narrow range. Second, the biochemical reactions that are the basis of life, and which take place in the **intracellular fluid** within cells, require that the concentration of dissolved substances (**electrolyte concentrations**) are within a particular range. The intracellular fluid is separated from the **interstitial fluid** (the fluid immediately surrounding all cells) by complex **semipermeable** membranes. These resist the passage of certain inorganic **ions** (charged particles that make up chemical compounds) such as sodium, while permitting the passage of water and waste products. Normally, the various fluids maintain a balance of concentration of dissolved substances (**solutes**) inside and outside the cells (they are **isotonic**). If the concentration of a solute inside a cell rises (becoming **hypertonic**), water will pass into the cell to re-establish the equilibrium. Conversely, if the concentration within the cell falls (becoming **hypotonic**), water will pass out of the cell until the intracellular and extracellular fluids are again isotonic. This process is **osmosis**. The interstitial fluid is in contact, through semipermeable membranes, with the blood plasma in the capillaries. In order for ingested water to influence the intracellular fluid, it has to be absorbed from the gastrointestinal tract into the blood stream, and from there into the interstitial fluid.

The way we normally lose most water from the body is by urination, which also depletes the body of electrolytes, as do sweating and bleeding. We lose water also through processes that do not involve loss of electrolytes (e.g., breathing). None of these processes can be completely switched off, so water needs to be replaced for the two reasons mentioned earlier. There are two sensory mechanisms involved in the control of water: **volumetric thirst**, based on loss of water volume, and **osmometric thirst**, based on changes in electrolyte concentrations. Rapid decrease in the volume of the extracellular fluid (e.g. following hemorrhage) results in increased drinking. **Blood-flow receptors** in the kidneys detect a decrease in blood flow, resulting in the circulation of the hormone **angiotensin**. In response to this there is peripheral **vasoconstriction** (producing an immediate increase in blood pressure); the adrenal cortex secretes **aldosterone** (which causes salt retention by the kidneys, preventing further loss of electrolytes); and the posterior pituitary gland secretes **vasopressin** (which causes water retention by the kidneys). Blood-pressure receptors (**baroreceptors**) located in the walls of the atria of the heart detect reduced pressure in the blood returning to the heart, and this also results in drinking. Loss of water alone changes electrolyte concentrations in cells, particularly in **osmoreceptors** located in the **circumventricular organs** in the center of the brain, where they can be rapidly affected by electrolytes and hormones in the blood. The autonomic, hormonal, and behavioral pathways for the control of electrolyte and water balance are shown in *Fig. 2*.

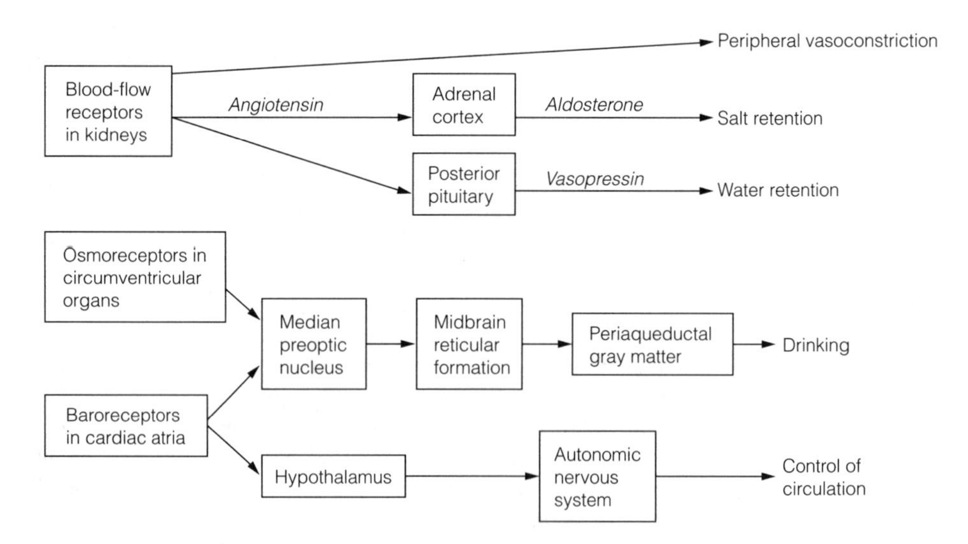

Fig. 2. Simplified diagram of the main mechanisms for the control of fluid balance.

The mechanisms we have looked at would have a considerable lag before they could change the system sufficiently to stop ingestion. For example, drinking stops several minutes before the ingested water passes from the gastrointestinal tract to rehydrate the extracellular fluid. Yet humans and animals drink just about the right amount of water to correct a deficit. What mechanisms signal satiety? Wetting of the mouth quenches thirst temporarily, but drinking is quickly resumed. This suggests that there are two satiety mechanisms: a short-term mechanism, which is strongest in the mouth and weakest further down the gastrointestinal tract; and a long-term mechanism, based on osmoreceptors in the throat, in later parts of the gastrointestinal tract, and in the liver. These inhibit vasopressin secretion when stimulated by water. Passing water into different levels of the gastrointestinal tract in animals has shown that the long-term inhibitory effect on drinking increases the further down the tract the water is introduced.

B3 HUNGER AND EATING

Key Notes

Central mechanisms

The long accepted dual-center set-point theory has been discredited. Lesions in the lateral hypothalamus and ventromedial hypothalamus do not prevent control of eating, but change the set-point for body weight.

The gastrointestinal tract

Short-term feeding and satiety are controlled by interactions between centers in the hypothalamus, influenced by peripheral and central actions of peptide hormones produced in the walls of the gastrointestinal tract, liver, and pancreas. Long-term control of weight is via leptin secreted by adipose tissues.

Psychological influences on eating

Eating and food selection are influenced by psychological factors such as palatability, innate taste preferences, and various types of learning including sensory-specific satiety, flavor–flavor learning, flavor–nutrient learning, taste-aversion learning, and family influences.

Obesity

Obesity results from ingesting more calories than are used. It is influenced by personal and environmental factors such as learning and the availability of high-calorie foods. Genetic predispositions, possibly through a deficit of leptin production, and changes in metabolic efficiency (the 'yo-yo effect') make obesity difficult to treat.

Related topics

Negative feedback and drinking (B2)

Eating disorders (L5)

Central mechanisms

The dominant theory of feeding for some 25 years was the **dual-center set-point theory**. According to this view, eating commences with the stimulation of a **feeding center** in the **lateral hypothalamus** (LH) by a decrease in blood glucose below a **set point**, and ceases with the stimulation of a **satiety center** in the **ventromedial hypothalamus** (VMH) when blood glucose rises above the set point. Lesions in the LH produce **aphagia** (failure to eat), and the animals starve to death unless tube fed. Electrical stimulation of the LH can produce eating. Conversely, lesions in the VMH, specifically destruction of neurons sensitive to glucose concentration, result in **hyperphagia** (overeating); the rats eat until they become extremely obese.

Closer examination of the behavior of animals following VMH lesions reveals that the result is not simple hyperphagia (*Fig. 1*). The initial hyperphagia decreases after 12 days or so, and weight is then maintained at a higher level, but with food intake only slightly higher than normal. Starving or force feeding these animals is followed by a return to this new body weight. These results only occur with highly palatable diets. VMH lesions do not prevent the control of food intake, they seem to increase the set point for body weight, probably through disruption of hormonal control of fat metabolism. Similarly, LH-lesioned rats start to eat and drink again, as long as they are kept alive long

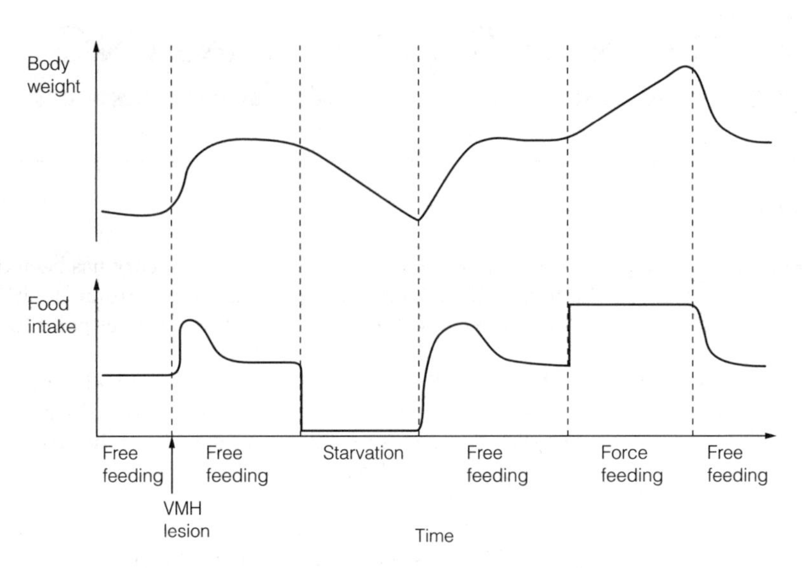

Fig. 1. Following VMH lesions, hyperphagic rats stabilize their body weight at a new, higher level, which is maintained even after starvation or force feeding.

enough by being tube fed, and maintain a lower weight. So, although centers in the VMH and the LH exert control over eating, there must be other control mechanisms that can take over their functions if necessary, and may normally act in parallel with them.

The gastrointestinal tract

Sham feeding, in which animals are given fistulas (openings between internal organs and the body surface) preventing food from reaching their stomachs, results in copious eating which is stopped by injections of food directly into the stomach. Conversely, removal of food from the stomach starts an animal eating again. At first, it was thought that this indicated that feeding stopped when the stomach is full, following stimulation of stretch receptors in the stomach wall. However, the pressure in the stomach does not increase following eating, and filling the stomach with inert material does not stop eating. It is now clear that a number of **peptides** (chains of amino acids smaller than proteins) are produced by endocrine cells in the lining of the stomach and intestines, and in the pancreas and liver, in response to food. These peptides include **cholecystokinin (CCK)**, insulin, and bombesin. Peripheral injection of peptides into rats and humans causes them to stop eating, and they have satiety actions in the hypothalamus. Peptides are also involved in the *commencement* of feeding. **Neuropeptide Y (NPY)** is found in many parts of the brain, including the hypothalamus, and injection of NPY into the hypothalamus causes sated rats to start feeding. The physiological control of weight is a complex interaction of neurotransmitters, hormones, peripheral chemical and sensory factors, and brain nuclei.

The longer-term control of body weight (which is disturbed, for example, in hyperphagic rats) is achieved by the secretion of a hormone **leptin** by fatty tissues. Leptin probably acts on the satiety mechanisms in the hypothalamus. The **ob mouse**, a strain bred to be obese, does not possess the gene that produces leptin.

Psychological influences on eating

Eating is also controlled by external factors, both sensory and psychological. The first, obvious, factor is the **palatability** of food. Humans, and other animals, will eat more when the food tastes nice than when it is made less palatable, for example by the addition of quinine. The administration of opioid antagonists reduces pleasantness ratings of foods, suggesting that the rewarding properties of eating are due to the release of **endogenous opioids** in the hypothalamus. Various learning processes are also important for eating. During the consumption of a particular food, the rate of eating slows, and this is followed for up to an hour by a decrease in palatability. These effects are specific to the particular food eaten, and are known as **sensory-specific satiety**.

We have innate taste preferences, and newborn infants show distinctive facial responses to sweet, sour, bitter, and possibly salt stimuli. The responses to sour and bitter tastes involve expulsive mouth movements. These aversive responses to sour and bitter tastes may have evolved to protect against eating substances that may be injurious: many poisons have a bitter taste, and spoiled foods may be sour. The learning of taste preferences starts soon after birth. Flavors of foods eaten by a mother are transferred to her milk, and in animal and human research have been shown to influence the offspring's food preferences. This may be a way in which young mammals learn which foods are good to eat. **Flavor–flavor learning** results from the association of a new flavor with one that is already preferred. The most obvious example of this is the association of flavors with sweetness, which is innately preferred. Since 'healthy' foods are less palatable, the adoption of a healthy diet must be based on learning, much of which takes place within the family. The eating of less palatable foods more often has a negative association (threats or punishment), while inherently palatable foods, high in carbohydrates often combined with fats, are presented in pleasant contexts (parties, treats, etc.). This makes it difficult for the child to learn healthy eating habits.

We also learn what *not* to eat, through **taste-aversion learning**. Many people develop an aversion to foods that have made them ill, or have been associated with nausea. Aversion can occur with just one pairing of a particular food with nausea or illness. The significance of learned aversion is that a food that has once made an animal ill is likely to do so again, and so must be avoided. While learned aversion has some similarities to standard classical (Pavlovian) conditioning (Topic E1), it is different in that it occurs in a single trial, and can take place when the food and the sickness are separated by a relatively long interval, of up to 12 hours. These properties are crucial if conditioned aversion is to have its protective effect. An animal could not learn to avoid poisonous food over a number of pairings of food and sickness, and the ill-effects of poisonous food are frequently delayed by hours.

Obesity

In general, the term **obesity** may be used simply to refer to having more than average body fat. Obesity results only, in the last analysis, from taking in more calories than we expend. However, there is a large number of potential reasons for the existence of such an imbalance. One factor is the easy availability of palatable foods in Western societies. But not everybody exposed to such a potential diet becomes obese. Another factor is **learning**. Children in many families are encouraged or even forced to eat all the food placed in front of them. This can lead to eating more calories than required, and can become an habitual pattern. The division of meals into more than one course provides the opportunity for sensory-specific satiety to increase total calorie intake, particularly if the meal pattern reserves energy-dense desserts until the end of the meal. The preference for energy-dense foods may be based on endogenous **opioid** circuits (presumably **β-endorphins**). This would account for phenomena related to 'comfort eating', when food is used to elevate mood.

Genetic factors are also important. The weight of people who were adopted in infancy is more highly correlated with the weight of their birth parents than with that of their adoptive parents. Strains of mice and of rats have been bred with a genetic predisposition to obesity. A genetic predisposition to obesity in humans interacts with exposure to particular family, social, and cultural environments. There is some evidence that at least some obese persons do not produce normal amounts of leptin in their adipose tissues. It is partly because of the hereditary contribution that obesity, with its implications for health, is difficult to treat. Whatever method is used, weight loss is rarely maintained in the long term. Another factor maintaining obesity has been called the **'yo-yo' effect**. This is an increased metabolic efficiency that occurs after severe calorie restriction, and makes it successively harder to maintain a lower body weight. Disorders such as **bulimia** and **anorexia nervosa** are best not considered primarily as disorders of eating or weight control (Topic M5).

B4 SEX

Key Notes

Sexual differentiation

Testosterone secretion before birth causes the differentiation of male sexual organs and characteristics. At puberty, increased testosterone production in males, and estrogen production in females produces further sexual dimorphism.

Activating effects of sex hormones

In most mammals, sexual behavior depends on the presence of sex-appropriate hormones. This is not always true in men, and is not true at all of women, whose sexual behavior is more related to androgens.

Organizing effects of testosterone

The presence of perinatal testosterone causes the development of masculine behavior, through changes in the central nervous system.

Brain mechanisms

The medial preoptic area is essential for sexual behavior. It has a sexually dimorphic nucleus that is larger in males. The amygdala integrates sensory information and internal hormonal factors, and influences centers in the hypothalamus which stimulate brain stem areas controlling sexual behavior.

Sexual orientation

Male homosexuality is not associated with low levels of testosterone, but may follow lower levels of prenatal testosterone following stress to the mother, affecting brain development. There are genetic predisposing factors, the survival of which might result from bisexuality conferring some reproductive advantage.

Related topics

The endocrine system (A5) Aggression (B5)

Sexual differentiation

Males and females are structurally and behaviorally differentiated. This **sexual dimorphism** starts at about 6 weeks of gestation, when testosterone produced by the **testes** causes the male embryo to develop the male genitalia. Further differentiation takes place at puberty, when increased testosterone secretion in males leads to the secondary sex characteristics of the adult male body (beard growth, increased muscle bulk, growth of external genitalia, enlargement of vocal cords and larynx), while estrogens from the ovaries produce female changes (breast development, subcutaneous fat deposits, onset of menstruation).

Activating effects of sex hormones

The species-typical sexual behavior of female mammals other than primates is triggered by the presence and behavior of a male of the same species, but only when she is at the receptive stage of her **estrus cycle**. At other times, she will defend herself as if she were being attacked. The receptive phase coincides with the time when her ova are ready to be fertilized. The receptivity and the physiological changes preparing the animal for pregnancy are controlled by cyclical secretion of hormones from the pituitary gland, which in turn cause ovarian follicles to produce **estrogens,** ova, and progesterone in sequence (Topic A5).

Removal of the ovaries causes a rapid decline into complete absence of sexual behavior, which is reinstated by administering hormones matching the natural sequence. Similarly, castration of male animals leads quickly to cessation of sexual activity, which can be reinstated by testosterone injection. The amount of testosterone is unimportant, so long as there is a certain, minimal blood level. The presence of a female animal in the receptive stage of the estrus cycle produces an increase in the male's testosterone level.

Castration of adult men leads in most to a reduction or cessation of sexual activity. But the effect is much more variable than in other species. A number of studies, using both surgical and chemical castration (administration of a substance that blocks androgen receptors), have shown that many men remain sexually active even years after the treatment. Testosterone levels in men also *respond* to sexual behavior: viewing erotic films increases blood testosterone levels. However, most research finds no relationship between the amount of testosterone and the intensity of sexual desire or amount of sexual activity. Women's sexual behavior is much less determined by hormones than is that of men or of female rats. The menstrual cycle is a cycle of hormonal changes, including the sequential secretion of estradiol and progesterone. Levels of estradiol peak at around the time of ovulation, and progesterone peaks a few days later (*Fig. 1*). However, research has failed to show a peak in sexual desire or activity at these times. While there are many inconsistent results, the consensus is that sexual activity is fairly evenly distributed, apart from a dip in the menstrual phase. Ovariectomy in adult women, and the menopause, sometimes decrease sexual interest and behavior, but the effects are reversed by adminis-

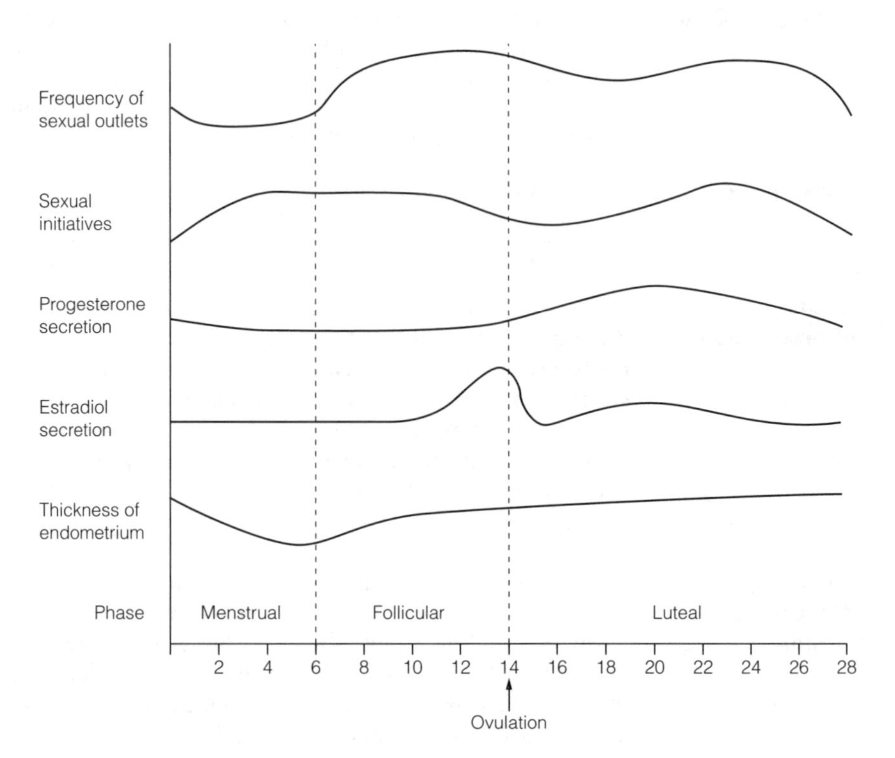

Fig. 1. Behavioral, endocrine, and physiological changes during the human menstrual cycle.

tration of androgens, not estrogens. So, women's sexual drive is more closely affected by androgens (secreted by the adrenal cortex and the ovaries) than by estrogens.

Organizing effects of testosterone

The development of male or female sexual behavior depends on the perinatal presence or absence of androgens. Exposing female guinea pigs in the uterus to testosterone has a **masculinizing** and **defeminizing** effect on them as adults. They show a masculine response to injections of androgens, attempting to mount other females, and they do not show the normal receptive responses to estrogen injections. The reverse pattern is shown in male animals: preventing testosterone exposure by castrating rats at birth produces a **feminizing** and **demasculinizing** effect. Thus, perinatal testosterone levels have sex-specific organizing effects on behavior and brain structures in rodents.

Brain mechanisms

An area of the forebrain just in front of the hypothalamus, the **medial preoptic area** (MPA), has long been known to be essential for sexual behavior in male laboratory animals. Destruction of this area in a wide range of species permanently stops them engaging in sexual activity. Electrical stimulation of this region produces mounting behaviour in male rats, and activity of the cells there increases during normal copulation. The MPA has a **sexually dimorphic nucleus (SDN)** that is three to five times larger in males than in females. Castration of animals immediately after birth prevents the SDN from retaining its greater size in males, so this sexual dimorphism is the result of an organizing effect of perinatal testosterone. The MPA controls the cyclical and constant rates of secretion of hormones that are typical of females and males, respectively. It also seems to coordinate sensory information and hormonal signals, triggering midbrain areas that organize the motor control of sexual activity.

Sexual orientation

Despite some early claims, homosexual men do not have lower levels of circulating testosterone than do heterosexual men, and neither does the injection of androgens affect their orientation. A series of studies in the 1980s showed that stress in pregnancy leads to increased likelihood of homosexuality in offspring. This appears to be caused by the testosterone-reducing effect of cortisol. This supports a **deandrogenization** theory of male homosexuality, in which the developing male brain is affected by decreased levels of androgens. As we saw above, perinatal androgen levels produce sex differences in the brain. In the 1990s a number of structural differences were reported between the brains of straight and gay men. For example, LeVay in 1991 showed that a particular nucleus in the hypothalamus, known to influence sexual behavior in laboratory animals, is on average the same size in gay men as in women, which is half the size it is in straight men. However, this result remains unreplicated.

A number of studies have suggested a genetic influence on homosexuality. The chances of the monozygotic twin of a homosexual person also being homosexual are at least twice as high as for a dizygotic twin. It might seem that a genetic basis for homosexuality is impossible, as it should directly reduce the rate of reproduction, leading to the gene dying out. While this is obviously true of exclusive homosexuality, bisexual persons do reproduce; bisexual women produce nearly as many children as exclusively heterosexual women. There are several alternative explanations for how genes for homosexuality might be maintained in successive generations. One explanation is **balanced superior heterozygotic fitness**. An individual having one copy of a gene (being heterozygous) might gain some advantage, even if having two copies of the gene is disadvantageous. Having two copies of a 'gay' gene might lead to exclusive homosexuality, while one copy might result in bisexuality, which is not reproductively disadvantageous. But what advantage might such a gene or genes confer on heterozygous persons? Female bisexuals tend to have children when they are younger, which might confer a survival advantage. Alternatively, heterozygocity might confer some advantage in sexual performance, such as greater sexual desire, fertility, or potency.

B5 AGGRESSION

Key Notes

Biology of aggression

Offensive and defensive aggression are deployed in the furtherance of reproductive goals, often serving to maintain territories or dominance hierarchies. Predation is aggressive behavior directed at obtaining food.

Hormones in aggression

Testosterone has perinatal organizing, and adult activating, effects on offensive aggression. These hormonal effects interact with social and environmental factors, especially in primates.

Human aggression

These factors also influence human aggression, although there is evidence that testosterone is related to dominance rather than aggression as such. Humans use aggression to achieve or maintain dominance, and to protect their personal space.

Sex and aggression

There are many links between sex and aggression, beyond the similar effects of testosterone. Both are associated with extreme sympathetic arousal, and have similar effects on cognition. Aggression is used directly in the achievement of sexual goals.

Related topics The endocrine system (A5) Sex (B4)

Biology of aggression

Aggression is probably universal in the animal kingdom. Aggressive behavior can be considered as being of three types:

- **offensive aggression** is aggressive behavior involving attacks of one animal on another;
- **defensive aggression** is aggressive behavior when attacked or threatened;
- **predation** is aggressive behavior directed at a member of another species, usually for food. This is so different from offensive and defensive aggression as to be better considered as feeding behavior.

In most species, offensive and defensive aggression have many features in common, but also differ in certain ways. Offensive and defensive behavior usually involve multiple or prolonged **threat displays**, and more than one attack. A defeated animal will frequently show submissive behavior, often a specific **appeasement display**, which stops further attack on the part of the victor. In this way, the confrontation is usually prevented from causing serious damage to either animal. One function of these aggressive displays is to maintain **dominance hierarchies** in group-living species. The victor gains or maintains higher status, which gives it advantages in relation to mating rights and access to food.

The patterns of aggressive behavior in male and female animals seem to be the same, and the neural circuits underlying them are probably the same. However, the nature of aggression does differ between the sexes. Most aggression is directly related to reproduction. Since male and female mammals usually have different reproductive roles, aggression serves different immediate ends for males and females. These sex-specific roles vary from species to species. In some, males maintain territories (and females), which they protect by aggressive displays towards other males. Female aggression is usually deployed in the defense of the young.

Hormones in aggression

Perinatal testosterone has organizing effects on *offensive* aggression in animals that parallel its sexual effects almost exactly (see Topic B4). In normal rats, for example, males show far more offensive behavior than do females. Male aggression is directed almost entirely at other males. Rats castrated at birth do not respond to testosterone with increased aggression, while those castrated in adult life respond rapidly. Aggressive behavior in male rats is also affected by social factors, and here too there is interaction with testosterone. A male rat housed with a female shows more aggression when tested by being placed in a cage with a strange male rat than do rats housed either alone or with another male. This aggression is increased if the animals are given testosterone. The presence of the female rat sensitizes the male to the effects of testosterone.

While offensive aggression is rare in female rodents, it does occur, and occurs to different degrees in different individuals. This is, again, related to activating and organizing effects of androgens. An activating effect is shown, for example, when adult females respond to testosterone injections with increased inter-female aggression. An organizing effect is shown in individual differences in natural response, and in response to adult testosterone injections. These result from differing degrees of exposure to androgens as fetuses. Because rats have large litters, a female fetus will be alongside different numbers of male fetuses (from none to two) (*Fig. 1*). It has been shown that those alongside two males have higher testosterone levels, and that they respond more aggressively to adult testosterone injection than those alongside none. Their exposure to higher levels of androgens before birth has a masculinizing effect.

It has been argued that environmental factors are the main influences on aggressive behavior in primates. Specifically, whether monkeys are submissive, or dominant and aggressive depends more on the level of aggression in the environment in which they are reared than on the individual's level of circulating androgens.

Human aggression

While there are some parallels between animal and human aggression, the differences are enormous. Aggression is far more frequent in men than in women. This difference is greatest in preschool children, and lowest in adults. The sex difference is largest for physical violence, and smallest when subjects believe that their behavior is unobserved. Both of these findings suggest a strong social influence on the sex difference. However, we should not rule out the possibility that perinatal testosterone influences the sex difference in humans, as it does in rodents. While evidence to assess this is hard to come by in humans, boys and girls whose mothers had been administered a synthetic androgen drug in pregnancy to prevent miscarriage are more aggressive than others whose mothers were not so treated. On the other hand, studies of girls with **adrenogenital syndrome** (prenatally overactive adrenal cortex, producing excessive androgens) have not shown a clear effect on aggression, although in other respects the girls behave like boys (they are described as 'tomboys').

Human male aggression increases at puberty, along with increased testosterone secretion, just as does that of other species. Castration of adult men not

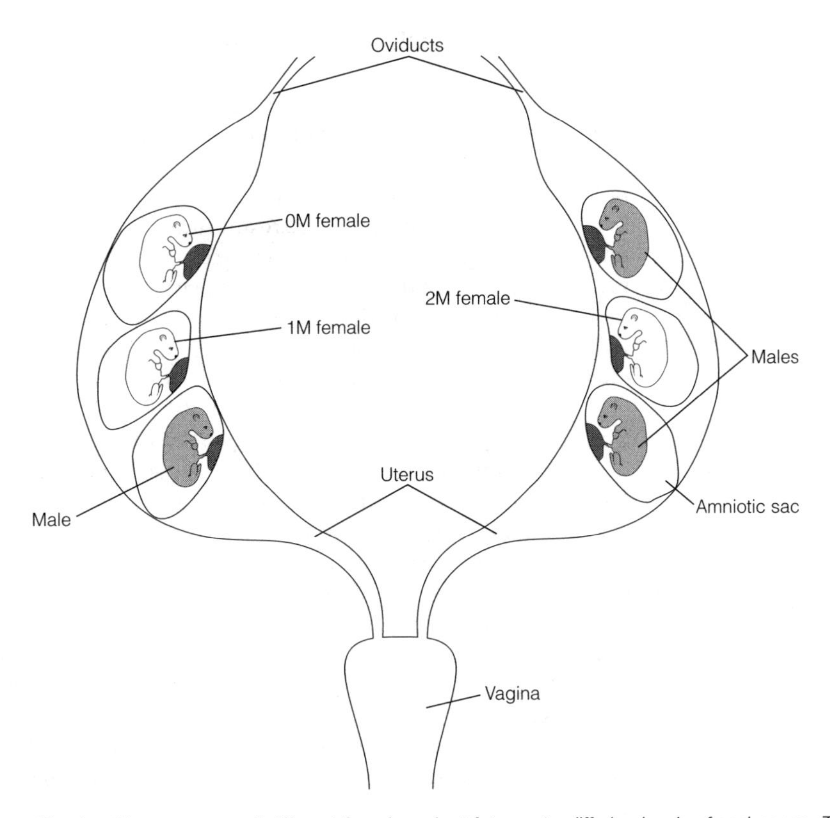

Fig. 1. The exposure of different female rodent fetuses to differing levels of androgens. The female on the right (2M) has a male on each side, and receives more androgen than the 1M female on the left, which is next to one male. The 1M female in turn receives more androgen than the 0M female, which is not next to any male fetuses. Redrawn from van Saal, F. S. (1983) Models of early hormonal effects on intrasex aggression in mice. In: Hormones and Aggressive Behavior *(ed. Svare, B. B.). By permission of Kluwer Academic/Plenum Publishers, New York.*

only frequently decreases sexual urges, but it usually also decreases aggressiveness. Correlations have been observed between individual differences in aggressiveness and amounts of circulating androgens. A number of studies have shown that men and women convicted of violent crimes have higher testosterone levels than those convicted of nonviolent crimes. However, these studies do not rule out other, covarying factors which might give the appearance of a direct relationship. A number of studies suggest that the relation is actually between *dominance* and testosterone, rather than aggression and testosterone, and further that the testosterone difference may be a *result* of changes in dominance, not a cause. For example, in one study testosterone levels were higher in men who had won a tennis match easily than in those who had won narrowly. Furthermore, testosterone levels were higher in those winning prizes as a result of their own efforts (students graduating), but not in those who won purely by chance (lottery winners). This same interpretation could apply to correlational studies in animals: animals higher up in a dominance hierarchy show higher levels of testosterone, but this may be due to the fact that they fight more, to the fact that they win more fights, or simply to their dominant position.

Other parallels with animal behavior are shown by the uses to which human males put aggression. Displays of dominance, and often aggression, by men can frequently be interpreted as serving the function of attracting and keeping desirable mates. The territorial function is less obvious at the individual level, although we do like to maintain our own private territories at home and even in public. On public transport, for example, people will usually avoid sharing a seat with a stranger as long as there are vacant seats available. People will sometimes react aggressively when they feel that their **personal space** has been invaded. So, while human aggressive behavior shows some of the relationships to hormonal function shown in animals, the relationship is much looser. Further, human aggression seems to be much less species typical than that of other species; that is, it is subject to greater individual control.

Sex and aggression

It is clear that offensive aggression is linked to sexual and reproductive activities. Many links have been noted between sex and aggression in humans. In research on pornography, for example, it has been shown that sexual and violent stimuli frequently have similar effects on viewers. One explanation is that sex and aggression are both associated with sympathetic arousal, and that both increase in intensity with greater arousal. Arousal may transfer from one to the other. Thus, unexpressed annoyance increases later sexual response, and unexpressed sexual arousal can enhance later aggression. Other common features involve cognitive processes. Cognitive control may be exerted over both aggression and sexual behavior, while, conversely, both aggression and sexual arousal may, if sufficiently intense, interfere with cognitive processes, particularly in narrowing attention to the object of the aggression or sexual behavior.

In humans, the relationship between aggression and sex goes beyond the biological role of one in achieving the other, although both may be seen to have their origins in these roles. Thus, aggression is commonly used by men in most cultures to force unwilling females into having sex (i.e. rape). Sadomasochistic sexual interactions depend on aggressive acts enhancing sexual experience. More commonly, sexual activity is accompanied by actions with an aggressive origin, such as biting and scratching.

B6 SLEEP AND DREAMING

Key Notes

Biological rhythms

The daily circadian rhythm is driven by an endogenous clock in the suprachiasmatic nucleus with a period of about 25 h, entrained to 24 h by the day–night cycle.

Stages of sleep

The electroencephalogram (EEG) shows that people cycle through alternate periods of slow-wave sleep and rapid eye-movement sleep (REM) (in which an apparently alert EEG is accompanied by sensory and motor inhibition) about every 90 min. The total time spent sleeping, and the proportion of time spent in REM sleep, are both greatest before birth, and decrease with age.

Sleep deprivation

Sleep deprivation is accompanied by little physiological change, and cognitive changes are largely limited to vigilance tasks. Enforced, prolonged sleep deprivation in animals leads to death from stress-induced changes.

Mechanisms of sleep

The onset of sleep seems to be controlled by neurons in the preoptic area. The transition to REM sleep, and its various components, are controlled by the pons.

Functions of sleep

Recuperative views regard sleep as a period during which repair processes take place, or during which processes related to learning occur. Circadian approaches suggest that sleep evolved as a way of fitting organisms to the light–dark cycle. The dynamic stabilization theory is that sleep maintains and enhances synaptic efficiency by permitting the 'off-line' activation of neural circuits.

Dreaming

Dreams occur during REM sleep. Dreams might be meaningless byproducts of the sensory activation described above (the activation–synthesis hypothesis).

Related topics

Actions of the nervous system (A2)   The cerebral cortex (A4)

Biological rhythms

Rhythmic variations in level of arousal are characteristic of animal behavior. There are rhythms with a period longer than a day, such as the monthly **menstrual cycle** (see Topic B4) and the annual **seasonal affective disorder** (see Topic L3), and others with a period shorter than a day, such as the **basic rest–activity cycle** of about 90 min (see *Stages of sleep*, below). The daily cycle of sleeping and waking is known as a **circadian rhythm** (meaning 'about daily'). Underlying the sleep/waking cycle are *continuous* variations in hormone secretion and other physiological activity, shown by a daily variation in body temperature of about 1ºC. This accompanies continuous variation in cognitive capacities, which are greatest when body temperature is highest.

Studies of animals or people in isolation from the normal cues of the day–night cycle reveal a **free-running rhythm**, which in humans has a period of about 25 h. This shows that there is an endogenous (internal) **biological clock** with a period of about 25 h which is normally **entrained** by external cues (called **Zeitgebers**), keeping it to 24 h. The most important Zeitgeber is light. Others include social interaction, feeding, exercise, and alarm clocks. If we destabilize the relation between the internal clock and the Zeitgebers, for example by jet flight to other time zones or by moving to different work shifts, we suffer disturbances of sleep and cognitive function which, in the former case, we call **jet-lag**. Recovery follows re-synchronization of the internal rhythm and the local environment. The effects of jet-lag can be minimized by gradually shifting the waking time in preparation, by exposure to intense light early in the morning, or by treatment with the hormone melatonin, which facilitates the entrainment of biological rhythms.

The biological clock controlling the circadian rhythms is in the **suprachiasmatic nucleus (SCN)** in the medial hypothalamus. Destruction of this area prevents rhythmicity, and individual neurons in the SCN show rhythmic activity. The SCN receives information directly from the **optic chiasma**, and this is the route through which light modulates the activity of the SCN, entraining the endogenous rhythm.

Stages of sleep

Sleep is not merely a state of low arousal. Recording the activity of cortical neurons through electrodes attached to the scalp (the **electroencephalogram** or **EEG**) shows that persons falling asleep and sleeping through the night pass through distinct stages (*Fig. 1*). Falling asleep is accompanied by a progressive decrease in frequency and increase in amplitude of the EEG. Stages 3 and 4 of sleep are often together called **slow-wave** (SW) **sleep**. A person cycles through these stages throughout the night, with a period of about 90 min (*Fig. 2*). The passage to SW sleep is accompanied by slowing of the heart rate and muscular relaxation. This 90-min cycle was the first evidence for the basic rest–activity cycle described earlier. After the first cycle to stage 4, occurrences of stage 1 sleep are almost always accompanied by rapid eye movements. This stage, called **rapid eye-movement sleep (REM sleep)**, is associated with the deep relaxation of the muscles of the trunk found in SW sleep (although there is likely to be twitching of the muscles of the limbs and face), but with faster breathing and heart rate. REM sleep is sometimes called **paradoxical sleep** because, amongst other paradoxes, while the EEG is most similar to the waking EEG, it is more difficult to awaken animals and people during this stage.

The total time spent sleeping, and the proportion of time spent in REM sleep, are both greatest before birth, and decrease with age. At birth, REM and SW sleep each occupy about 50%. During adulthood, the total length of sleep decreases. The proportion of REM sleep also gradually declines, as does the amount of SW sleep, which drops from about 20% at 18 years to only 2–3% at 50–60 years.

Sleep deprivation

During total sleep deprivation, the desire to sleep increases markedly for 2 or 3 days, so that it is difficult to keep people awake after the first 48 h. Sleep deprivation is accompanied by very little physiological change, and limited cognitive changes. Performance of tasks involving reasoning, spatial relations, and comprehension conducted under time pressure are usually unaffected. Performance of tasks involving vigilance or prolonged attention deteriorates.

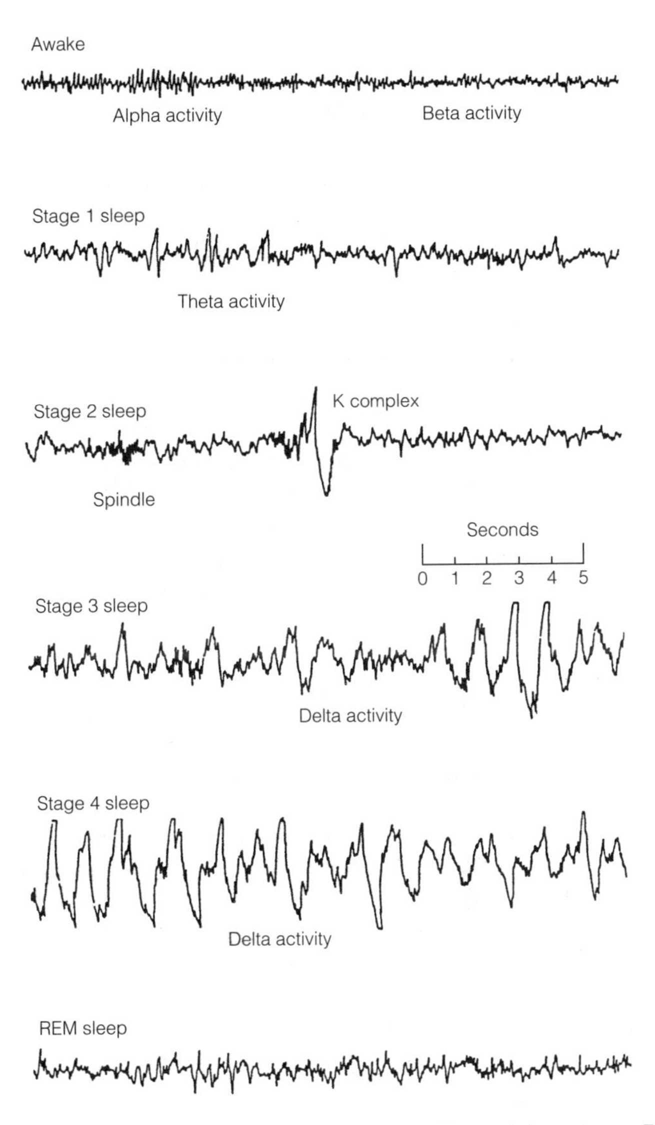

Fig. 1. The electroencephalogram during waking and sleep stages. From Horne, J. A. (1988) Why we sleep: the Functions of Sleep in Humans and other Mammals, *Oxford University Press, Oxford.*

This can be partially overcome by increasing incentives. After about 60 h of sleep deprivation, hallucinations sometimes occur. Many of these effects might be attributable to an increasing tendency to have **microsleeps** (very brief periods of REM sleep) after sleep deprivation. At the end of periods of sleep deprivation, only about 20–25% is recovered on the following 2 or 3 nights. Within this, about 70% of SW and 50% of REM sleep is recovered. Extending sleep deprivation of rats beyond what has been possible voluntarily in humans causes death after about 4 weeks. The sleep-deprived animals show the general changes seen following any prolonged stress, including enlarged adrenal glands, stomach ulcers, and internal hemorrhages.

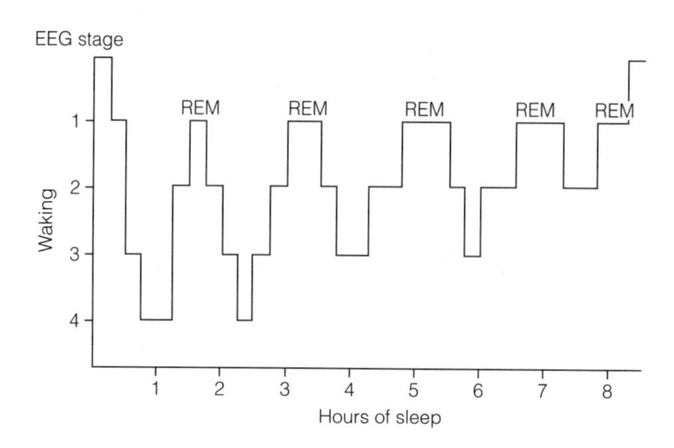

Fig. 2. The pattern of sleep during a typical night. After E. Hartmann (1967) The Biology of Dreaming. *Charles C. Thomas, Springfield, IL.*

There is a real drive to enter REM sleep. If people are awoken as soon as their EEG shows that they are entering REM sleep over successive nights, they enter REM sleep with increasing frequency, up to 50 times in one night. Subsequently, when allowed to sleep freely, they spend as much as twice the usual amount of time in REM sleep, even though the overall length of sleep is hardly increased. There are no persistent psychological effects of REM sleep deprivation that are different from those of general loss of sleep.

Mechanisms of sleep

The neural mechanisms underlying sleep are not clear. The onset of sleep follows activity in the **preoptic area** of the basal forebrain. Lesions here cause cats to stop sleeping, and stimulation induces SW sleep in cats. This activity produces a decrease in activity in the **ascending reticular activating system (RAS)** which maintains an alert cortex through cells in the thalamus. After about an hour of sleep, neurons in a center in the **pons** become active, causing increased activity in the RAS, returning us to faster EEG activity. The center in the pons connects with various other areas controlling the different aspects of the REM state, including rapid eye movements, muscular paralysis, and sensory blocking.

Functions of sleep

The two main types of theory of the functions of sleep are recuperative and circadian. Recuperative views regard sleep as a period during which repair processes take place, or during which processes related to learning occur. Circadian approaches suggest that sleep evolved as a way of fitting organisms to the light–dark cycle, conserving energy for those times of the day when they need to be active to seek food, mates and so on. The latter do not explain the different nature, and cyclical appearance of, REM sleep. **Developmental theories** of REM sleep emphasize the predominance of REM sleep early in development, and suggest that it plays a key role in brain development; perhaps by promoting synaptic connections. **Learning theories** propose that REM sleep permits or promotes the formation of long-term memories. Earlier claims that a period of sleep following learning improves memory are now treated with caution. The enhancement of memory by sleep might be *passive*, because of reduced **retroactive interference** from later stimulation, rather than an *active* process of memory formation (see Topic F3). The effect, when it occurs, is small, especially in humans, and in many studies is not shown at all.

A recent theory is that sleep maintains and enhances synaptic efficiency by permitting the 'off-line' activation of neural circuits, a process known as **dynamic stabilization**. As evolution produced greater brain complexity, the need for dynamic stabilization became greater, and necessitated the evolution of first SW sleep, and then REM sleep to isolate neural circuits from sensory and motor systems. The reason REM sleep is so prominent in the fetus is because this is when most of these circuits are being formed.

Dreaming

People woken during REM sleep almost always report vivid dreams. Waking in nonREM periods results in vague reports or no mental activity. Dreams run in real time: REM periods last up to 40 min, and people awoken at different points will give approximately accurate reports of the duration of the dream they have been awoken from. People who claim that they never dream have just as many REM sleep periods as others, and they are nearly as likely to report dreams when awoken during these periods. Long-term memory traces of the dreams are apparently not formed during sleep, so they are mostly not recalled on waking. Those that are recalled are those we awaken during, or shortly after.

Dreams might be meaningless byproducts of the sensory activation of dynamic stabilization described above (the **activation–synthesis hypothesis**), or of memory consolidation processes. Some have viewed dreams as the important component of sleep. Freud and his followers viewed dreams as a 'safe' way in which a person can express (in a disguised manner) repressed drives, and sleep could be viewed as existing for this purpose. While dreams are frequently personally relevant, this would be expected simply on the basis that they originate in the individual's memories, isolated from sensory input, and does not imply any hidden significance.

C1 Psychophysics

Key Notes

Sensation

The term sensation is used to denote an elementary experience that arises from the stimulation of a receptor system. Psychophysics is the study of the quantitative relationships between stimulus and sensation.

Thresholds

Early workers studied the absolute and differential thresholds using psychophysical methods specially developed for the task. Both types of threshold turned out to be rather variable instead of the all-or-nothing affair that had been expected. The psychophysical function describes the probabilistic relationship between stimulus magnitude and sensation, which in turn leads to a probabilistic definition of thresholds.

Scales of sensation

Although it is possible to develop a subjective scale of sensation using the absolute threshold as a zero point and the differential threshold as a unit, direct methods proved more convenient and led to the development of a hypothesized power law relationship between stimulus magnitude and sensation. A range of different sensory attributes have been shown to have relatively stable properties with a characteristic parameter, n. The scale of loudness (sone) and pitch (mel) have proved particularly useful.

Theory of signal detection

An alternative view is that the absolute threshold arises from a decision-making process rather than the basic neurological properties of sensory systems. The theory of signal detection describes a process whereby weak stimuli are observed against a background of randomly varying noise. This theory replaces the threshold with a decision criterion, and accounts for the errors which people make when faced with the task of detecting faint stimuli of a variety of types.

Related topic

Hearing (C3)

Sensation

Sensation is any elementary experience or awareness of conditions inside or outside the body produced by the stimulation of some receptor or receptor system. It results from an initial contact between the individual and the environment. **Perception** follows from the elaboration and interpretation of sensations, and leads to the formation, and subsequent use, of representations of the outside world.

Psychophysics is concerned with the relationship between the magnitude of a sensation and the magnitude of the stimulus that produces it. While there is usually a straightforward way to measure stimulus magnitude in physical units, measuring sensation is more complicated and **psychophysical methods** were devised to make this possible.

Thresholds

Early researchers were principally concerned to measure **absolute thresholds** and **differential thresholds**. The absolute threshold was defined as the magnitude of the physical stimulus that would just produce detectable sensation by 'lifting it over the threshold of consciousness'. A differential threshold was defined as the physical difference in magnitude between two values of a stimulus that produced a **just noticeable difference** in sensation.

Attempts to measure the absolute threshold for a range of sensory modalities revealed that it was more variable than had originally been expected. For any individual, the magnitude of a stimulus that can just be perceived varies a little from trial to trial, and aggregating responses from a large number of trials reveals a characteristic graph called a **psychophysical function** (*Fig. 1*). Such a function demonstrates that there is no clear transition point between not perceiving a stimulus and perceiving it, and the absolute threshold must therefore be defined in a different way. It is now generally specified as that value of the stimulus that has an *equal probability* of being perceived or not perceived; that is, the value of the stimulus that is perceived on 50% of presentations.

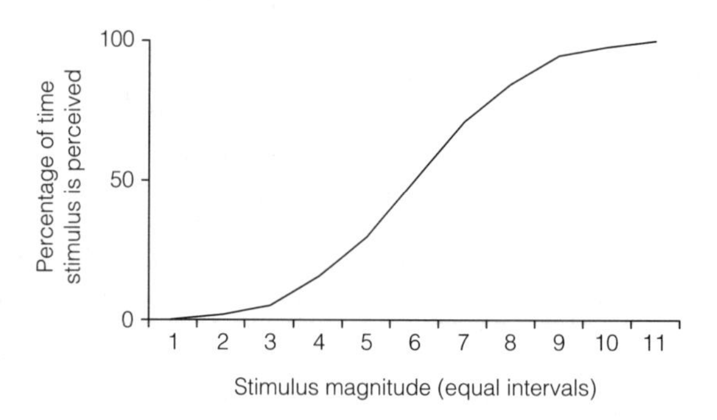

Fig. 1. Typical psychophysical function.

The **differential threshold** is estimated by comparing a fixed **standard stimulus** with a graded series of **comparison stimuli**. For example, in the case of weight, a standard 50 g weight might be compared with a 45 g weight, a 46 g weight, and so on up to, say, a 55 g weight. For each comparison, subjects say which weight they think is the heavier, the resulting judgements following a psychophysical curve. When the weights appear the same (the **point of subjective equality**), 50% of judgements will be correct. The just noticeable difference (sometimes simply called the differential threshold) is defined as the weight which is judged heavier than the standard on 75% of trials; that is, when a *difference* is perceived 50% of the time. This just noticeable difference turns out to be a constant proportion (about 4%) of the standard weight, whatever value that might take. Thus, if we call the increment in weight at this point ΔI, then $\Delta I/I$ turns out to be a constant: 0.04. This relationship is known as Weber's law and holds true for a variety of sensory attributes. A very acute sense will have a small constant and a less acute sense will have a larger one.

Scales of sensation

Gustav Fechner proposed that a sensory scale could be constructed using just noticeable differences as the unit of sensation, and this is equivalent to stating that sensation increases with the logarithm of stimulus magnitude. However,

S.S. Stevens constructed scales of sensation in a more direct manner using the methods of **magnitude estimation** or **magnitude production**. These methods were applied to a range of different stimulus attributes, and the data were found to conform to the general equation now known as Stevens' power law:

$$\text{Mean magnitude estimate} = aX^n$$

where X is the magnitude of the stimulus in physical units, a is a constant which depends on the unit of measurement used for X, and the exponent, n, is a characteristic of the sensory attribute. Examples of estimates of n are given in *Table 1*.

Table 1. Estimates of the exponent of Stevens' power law for a variety of sensory attributes

Attribute	*n*
Brightness (point source)	0.5
Smell (coffee)	0.55
Loudness (white noise)	0.6
Taste (salt)	1.3
Lifted weights	1.45
Tactile roughness (emery paper)	1.5
Electric shock through fingers (60 Hz)	3.5

These two methods of constructing scales (logarithmic versus power function) both give consistent accounts of experimental data, but are also clearly in conflict. This is probably due to their requiring people to judge their sensory experience in different, though equally valid, ways.

The concept of the absolute threshold still has practical application, most notably in the clinical assessment of hearing, where a chart of an individual's auditory thresholds over a range of frequencies is known as an **audiogram** (Topic C3). However, the current understanding of the processes involved in the detection of very small stimuli characterizes people more as decision makers than as passive receivers of information.

Theory of signal detection

The **theory of signal detection** is based on the idea that all sensory observations take place against a background of noise which varies randomly in intensity over time. When a signal of fixed intensity is presented, it is observed against whatever intensity of noise there happens to be at the time and the individual, usually called the observer, has to decide whether the signal was present or not. If the signal is large compared with the noise then this presents no difficulty. When it is small, the observer has to adopt a **decision criterion** (usually known as β) and respond according to whether the observation exceeds the criterion or not. The mistaken judgements that give rise to the psychophysical function result from this decision-making procedure. Observers may decide a signal was present when it was not (a **false alarm**) or miss a signal that was actually present. Moving the decision criterion changes the relative proportions of the two kinds of mistake, but does not improve decision making overall; that is determined by the **discriminability** of the signal (written as d', pronounced 'd-prime') from the noise. d' can be calculated for any specified signal from the relative proportions of correct and incorrect decisions taken over a few hundred observations, and takes the place of the absolute threshold. The theory has been extensively tested and provides an accurate account of how we detect very low intensity stimuli across a variety of modalities. Its essential components are illustrated in *Fig. 2*.

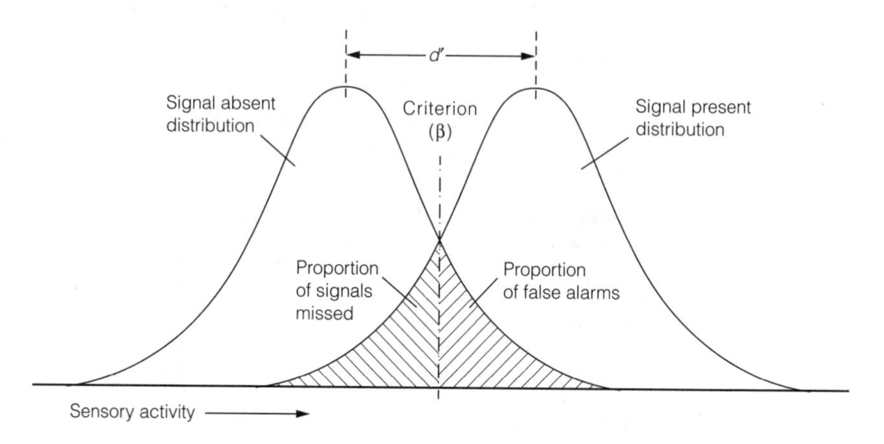

Fig. 2. The detection of low intensity signals represented as two overlapping distributions.

C2 VISION

Key Notes

The eye

The eye contains two types of light-sensitive cell, rods and cones, which are attached to the retina. Rods are most responsive at low light levels and do not contribute to color vision. Cones respond to normal daylight levels and are responsible for color vision. Rods and cones connect to layers of X and Y ganglion cells and bipolar cells in the retina, and these in turn connect to the optic nerve.

The visual pathway

Information is transmitted by the optic nerve to the visual cortex, via the optic chiasma and the lateral geniculate bodies. Cells in the pathway have specialized functions which respond to visual information in progressively more complex ways. The pattern of light that causes a neuron anywhere in the pathway to respond is called the receptive field for that neuron.

Dark adaptation

Dark adaptation occurs when we move from a bright environment into a dim one. The process takes about 30 min to complete, the delay arising from the need for visual pigments to regenerate. Detailed study of visual thresholds during adaptation reveals that cones adapt more quickly than rods.

Color vision

There are three types of cone in the retina, each of which is maximally sensitive to a different wavelength of light. Their responses effectively analyze the light that enters the eye, dividing it into different components, and it is this analysis which underpins our sensations of color. Deficiencies in color perception arise from the absence of one or more type of color receptor.

Related topics

Movement perception (D3) Form perception (D4)

The eye

The visual stimulus

Light is the name we give to that part of the spectrum of **electromagnetic radiation** to which our eyes are sensitive; visible wavelengths range from about 360 nanometers (nm) (violet) to 750 nm (red). Single wavelength light is rare, and most of what we perceive is a mixture of different wavelengths. Some objects in our environment (e.g. the sun, a torch) emit light, but most only reflect part of the light which falls upon them while absorbing the remainder.

The structure of the eye

The basic components of the eye are the **lens**, protected by the transparent **cornea**, which refracts the incoming light on to a surface of light-sensitive cells called the **retina**. The shape of the lens can be altered by the muscles that support it in order to focus on objects at different distances; this is known as **accommodation**. A smooth circular muscle called the **iris**, which is under reflex control, regulates the amount of light entering the eye (*Fig. 1*).

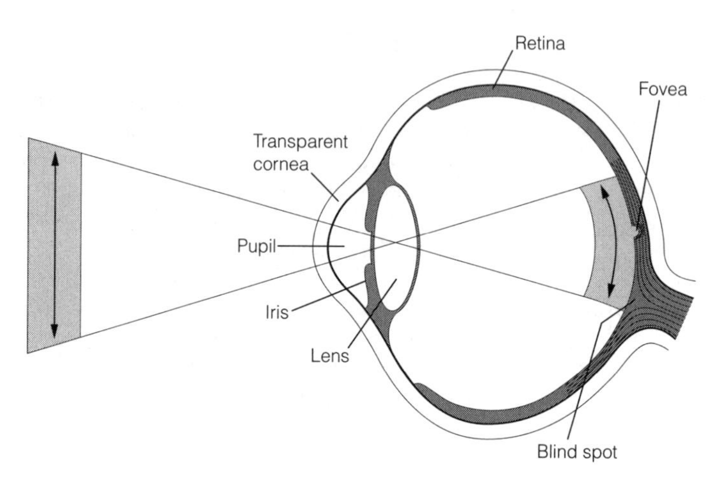

Fig. 1. Cross-section of the eye.

The retina contains two types of light-sensitive cell: **rods** and **cones**. The cones are plentiful in the small circular region directly opposite the lens, called the **fovea**, but become more sparse moving away from the centre to the periphery. The opposite is true of the rods, which are plentiful towards the edge of the retina but absent from the fovea. The two types of cell have different functions. The rods operate at low light levels, respond to the shorter wavelengths, and give rise to **achromatic** vision, while the cones operate at normal daylight levels and give rise to **color** vision. Both types of light-sensitive cell are connected to the optic nerve via a layer of **bipolar** cells and a layer of **ganglion** cells. Two other types of cell, **horizontal cells** and **amacrine cells**, make lateral connections, allowing adjacent cells in the retina to interact (*Fig. 2*).

There are three types of ganglion cell known as **X**, **Y** and **W cells**. X and Y cells have distinctive and complementary properties (*Table 1*). W cells have a much more variable distribution in the retina than X and Y cells, and also have a more variable type of receptive field and the slowest response time. At present, their function is poorly understood.

Table 1. Contrasting properties of X and Y cells

X cells	Y cells
Located in the fovea	Located in the periphery of the retina
Small receptive field	Large receptive field
Slow and sustained response	Fast and transient response
Involved in perception of fine detail	Involved in the perception of movement

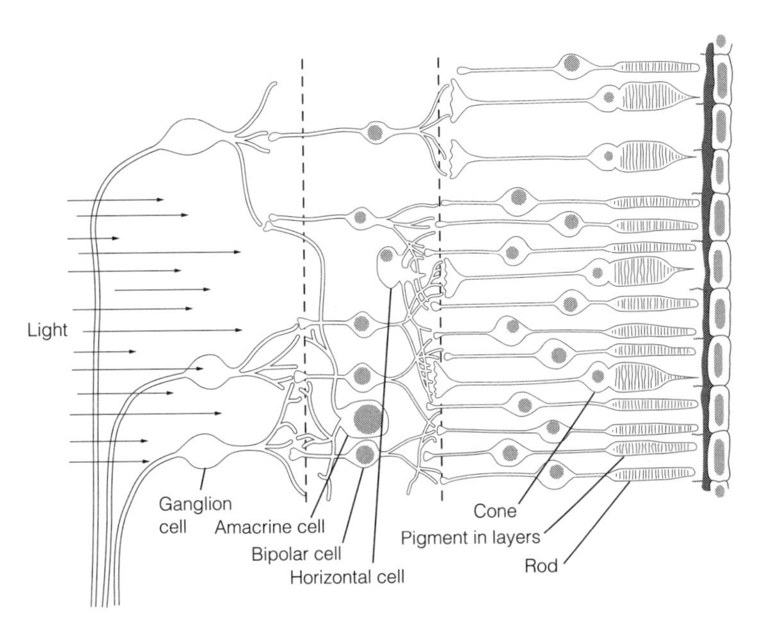

Fig. 2. The retina. Reproduced from Coren, S. et al. *(1994)* Sensation and Perception, *4th edn. Harcourt Brace College Publishers, Fort Worth.*

The point at which the **optic nerve** passes out of the eye contains no light-sensitive cells, and is known as the **blind spot**. Note that light passes through the layers of neurons *before* reaching the light-sensitive cells.

The visual pathway

The optic nerves carry information from the eyes to the **visual** (or **striate**) **cortex**, which is at the rear of the brain. The pathways from the **nasal** (inner) halves of each retina cross over at the **optic chiasma** and continue to the *opposite* cerebral hemisphere, while the pathways from the **temporal** (outer) halves continue to the same-side cerebral hemisphere. As a consequence, information from the right-hand side of the visual field is directed to the left visual cortex and *vice versa* (*Fig. 3*).

After the optic chiasma, the optic nerves continue to structures in the thalamus called the **lateral geniculate nuclei (LGN)** which contain two types of neurons, distinguished by their anatomy and function: **parvocellular** and **magnocellular** (*Table 2*). Connections from LGN cells then proceed to the visual (or striate) cortex. This area of the brain has a highly structured architecture whose primary function appears to be the decoding of detailed features of the visual input (Topic D3).

Table 2. Contrasting properties of magnocellular and parvocellular neurons

Parvocellular neurons	Magnocellular neurons
Small receptive field	Large receptive field
Slow and sustained response	Fast and transient response
Input almost entirely from X cells	60% input from X cells, 40% from Y cells
Respond to high contrast	Respond to low contrast
Respond to color	Do not respond to color
Respond to fine detail	Do not respond to fine detail
Involved in color vision, acuity	Involved in movement, depth perception

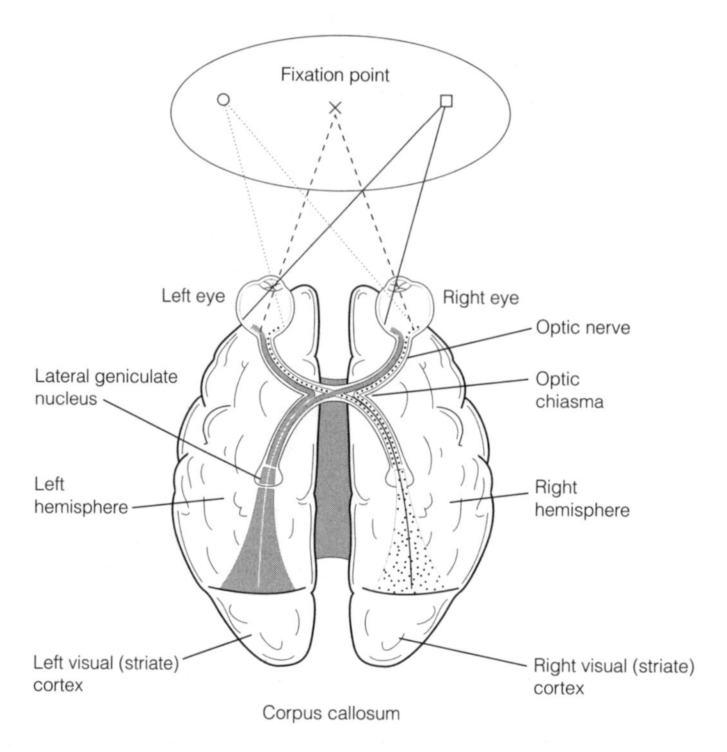

Fig. 3. The visual pathway. Reproduced from Gross, R.D. (1992) Psychology: The Science of Mind and Behaviour, *2nd Edn, p. 212. Hodder and Stoughton, London.*

Receptive fields

The receptive field of a neuron in the visual pathway is that area of the retina which, when stimulated by light, will cause activity in that neuron. Receptive fields can be very complex; for example, a cell in the visual cortex may respond only when its receptive field is stimulated by a dark line moving diagonally across it. Such fields are built up by interactions between cells lower down in the pathway (i.e. nearer the eye), and the most basic interaction is called **lateral inhibition**. This is the process whereby cells that are activated by light *inhibit* the activity of adjacent cells. The degree of inhibition depends on how strongly the cell is responding; the stronger the response the greater the degree of inhibition. Cells in the centre of an activated area therefore have their activity tempered by their neighbors. But cells at the *edge* of an activated area receive *less* inhibition because some of their neighboring cells are quiescent, and the *edges* of the activated area are thus accentuated. One example of the operation of lateral inhibition is **brightness contrast** (*Fig. 4*). Complex receptive fields are built up by the combination of more simple ones, so that when several neurons connect with some other cell at the next level in the visual system, the *sum* of their receptive fields forms the receptive field of that next cell (Topic D4).

Dark adaptation

When we move from a brightly lit environment into a dark one we are temporarily unable to see. Our visual sensitivity gradually improves over a period of about 30 min, and this is called dark adaptation. During dark adaptation, the photosensitive pigment (**rhodopsin**) in the rods and cones regenerates,

Fig. 4. An illustration of brightness contrast. The faint gray squares which appear at the intersections of the white lines are caused by lateral inhibition.

following its depletion during exposure to light, and the sensitivity of the receptors is greatly increased. A graph of the visual threshold against time reveals a two-part curve. By changing the color of the light and the area of the retina being stimulated, it is possible to show that the first part of the curve is due to adaptation of the cones, which reach their maximum sensitivity after about 10 min. The second part of the curve represents the adaptation of the rods, and these reach their maximum sensitivity after about 30 min (*Fig. 5*). Vitamin A plays an important part in the regeneration of rhodopsin and a deficiency of this vitamin can impair dark adaptation. In extreme cases, this results in a pathological loss of sensitivity to low light levels called **night blindness**.

Color vision

A person with normal color vision can discriminate more than 5 million colors. All these colors can be described in terms of three dimensions: **hue**, **brightness** and **saturation**. Hue is what in everyday language is called color. It is that attribute which enables us to distinguish between the color of an orange and the color of a lemon, which enables us to identify lots of things as different shades of green. Brightness differentiates black from white, and all the shades of gray in between, from each other. These are all achromatic colors: they have no hue. But chromatic colors differ in brightness too, in the sense that there are light and dark shades of the same hue. Saturation corresponds to the purity of the color, and we can think of it as the amount of white mixed with the hue. High saturation corresponds to a dense, pure color, while low saturation appears 'washed out'. A variety of geometric representations, called **color wheels**, have been devised to illustrate how these three dimensions account for the entire range of visible colors.

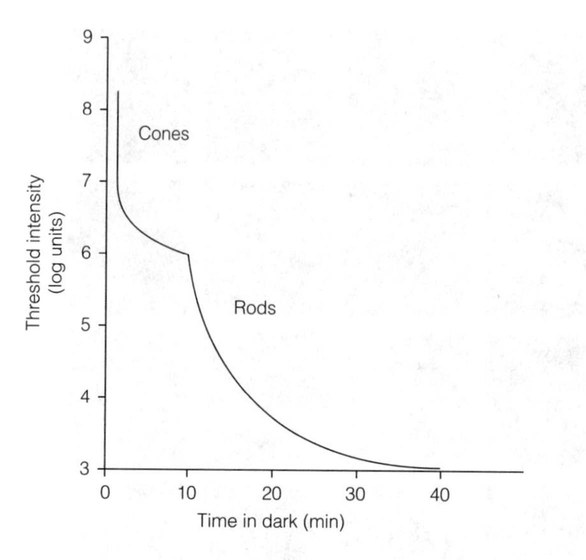

Fig. 5. The progress of dark adaptation over time.

The physiological basis for color vision is that there are three types of cone, each of which contains a different color-absorbing pigment that is maximally sensitive to different wavelengths of light (*Fig. 6*). This is the **trichromatic** theory, and it was suggested some 200 years before techniques were devised to confirm it. Early workers believed the three types of cone would be sensitive to red, green and blue light, and that is how they are labeled, even thought the sensitivity peaks turned out to be in the orange, yellow-green and violet regions of the spectrum.

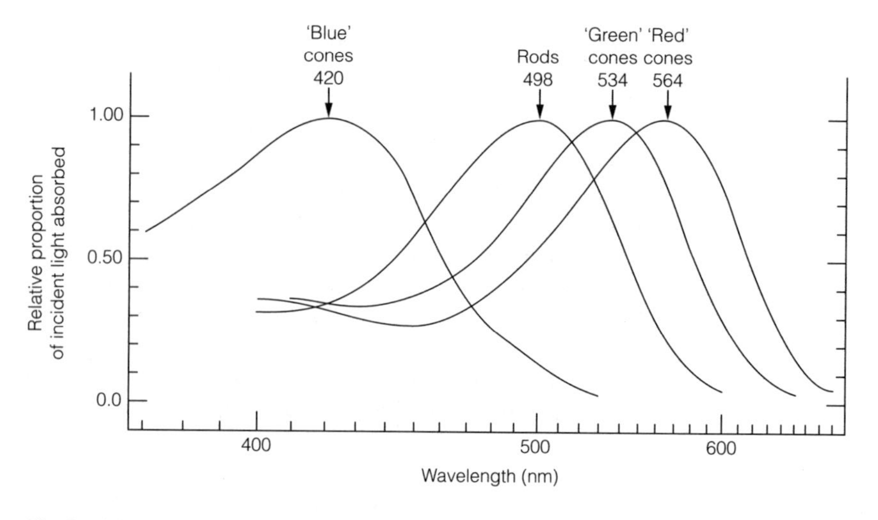

Fig. 6. Light absorption curves for rods and cones.

Color vision deficiencies

The trichromatic theory predicts five different types of color vision deficiency. People with no functioning cones cannot perceive color and, since they depend entirely on their rod cells, have rather poor overall visual acuity and find daylight quite uncomfortable. People with only one type of functioning cone cannot discriminate between colors and effectively have monochromatic vision. There are then three types of deficiency in which only two of the three cone systems are functioning. **Protanopia** arises from a lack of red-sensitive cones, and the resulting color experience is restricted to yellows, blue, and purple. **Deuteranopia** (the most common kind) arises from a lack of green-sensitive cones and, as a result, individuals cannot discriminate between green and certain shades of red and blue. **Tritanopia** is predicted to occur when the blue cone system is absent. It is exceedingly rare and results in a color experience of red and blue-green. There are less dramatic deficiencies, called **color weaknesses**, in which all cone systems are functioning but color matches require more red (**protoanomaly**) or green (**deuteranomaly**) than in normal individuals. Color deficiencies are inherited by a sex-linked mechanism; about 8% of males and 0.05% of females have some color weaknesses.

C3 Hearing

Key Notes

The ear

The ear detects sound pressure waves and is divided into three sections. The outer ear and middle ear function to transmit pressure waves to the inner ear, the cochlea. This fluid-filled spiral structure contains sound receptors supported on the basilar membrane, and these are stimulated when pressure changes give rise to waves in the fluid.

Pitch

Sound waves are coded into neural impulses in two ways. Nerve firing rates follow the frequency of the sound at low frequencies, while at higher frequencies coding is determined by the point of maximum displacement on the basilar membrane. The pitch of a sound is principally, but not entirely, determined by the frequency of the sound waveform.

Loudness

The amplitude of a sound wave is the main determinant of the sensation of loudness, but frequency is also important since the ear is not equally sensitive across the entire audible range. Amplitude is coded into the neural impulse in two ways: louder sounds cause more neurons to fire and, in addition, some neurons only fire in response to high amplitude sounds.

Localization

We are able to localize a sound in space by making a comparison between the waveforms that reach each ear. Sounds that are to one side reach the nearer ear slightly earlier and are slightly louder than those that reach the further ear, and these two cues enable us to judge where the sound came from with considerable accuracy. The shape of our outer ears can also provide a cue about whether a sound is in front of or behind us.

Hearing loss

Hearing loss can be caused by loud noise and may be permanent if exposure is regular and prolonged; it is also a common feature of the ageing process. In both circumstances onset is gradual and the condition is irreversible.

Related topic

Speech recognition and production (G2)

The ear

The auditory stimulus

Sound waves are pressure changes in a fluid medium, usually air, which are caused by the vibration or movement of an object. When an animal detects these pressure changes (usually with an ear, but there are exceptions), they give rise to neural signals which are in turn processed by that animal's brain in a way which depends on the complexity of its nervous system. There can be no sound in a vacuum (e.g. outside a planetary atmosphere) and, arguably, there can be no sound without an animal to detect it.

All sounds can be described in terms of pressure changes over time (called the sound **waveform**). However, it is often more useful to think of complex sounds as being broken down into simple components called **sine waves**, each with a specific **frequency** and **amplitude** – a process known as **Fourier analysis** after the mathematician who first demonstrated it (*Fig. 1*).

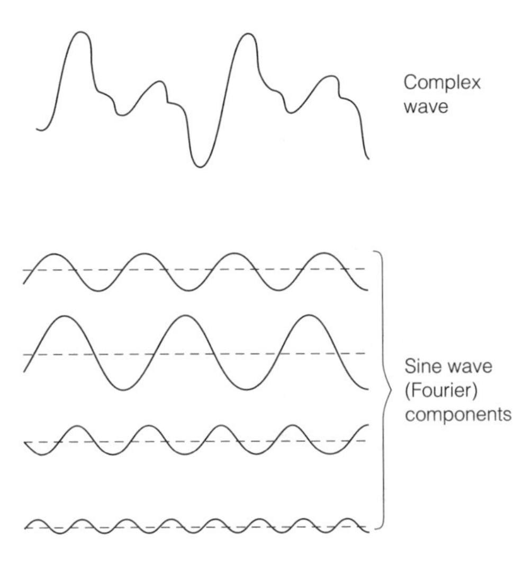

Fig. 1. Sine wave components of a complex wave.

Sine waves are often called **pure tones** and can be produced by a tuning fork or by electronic means. The **frequency** of a sine wave is measured in **hertz (Hz)** which corresponds to the number of repetitions (cycles) per second. The **amplitude** of a sine wave (expressed as sound pressure) is measured in **decibels (dB)** (*Fig. 2*). Note that the decibel is *not* a measure of loudness.

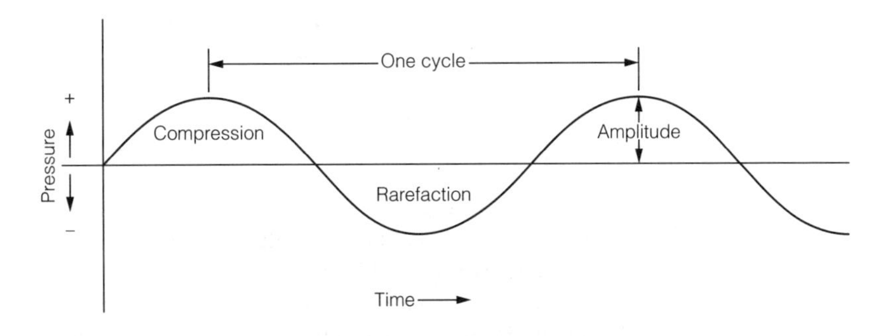

Fig. 2. A sine wave.

The structure of the ear

The ear has three regions: the **outer**, **middle** and **inner ear** (*Fig. 3*). The outer ear comprises the visible part (the **pinna**) and the **auditory canal**, which terminates at the eardrum or **tympanic membrane**. Sound pressure waves travel down the auditory canal and cause the eardrum to vibrate. These vibrations are transmitted across the middle ear by three tiny bones (the malleus, the incus and the stapes, collectively called **ossicles**) to the inner ear (the **cochlea**) by way of a membrane-covered opening (the **oval window**) in its bony wall.

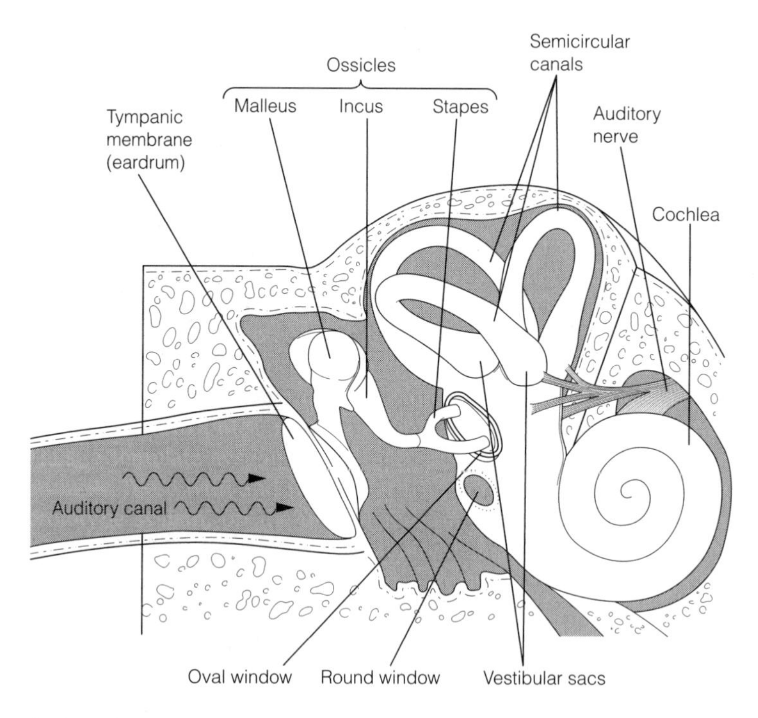

Fig. 3. The structure of the ear. Reproduced from Smith, R.E. (1993) Psychology. *Published by West Publishing Company, Minneapolis/St Paul.*

The cochlea is filled with fluid and divided along its whole length by the **basilar membrane** and the **tectorial membrane**. Between the two are the hair cells (which form part of the **organ of Corti**), which are the sound receptors. When a sound pressure wave reaches the oval window, it sets the cochlear fluid in motion, which in turn causes the hair cells to bend. This bending generates nerve impulses that are transmitted along the **auditory nerve** to the brain.

Pitch

Pitch is the sensation that corresponds to how high or low, in the musical sense, a sound appears to be. The most important determinant of pitch is the frequency of the sound stimulus, although it is also affected by the **duration** (very short sounds are perceived as clicks, regardless of frequency) and by **intensity** (above 1 kHz sounds rise in pitch when their intensity is increased; below 1 kHz they fall).

The frequency of sounds is coded by the ear in two ways. At low frequencies, the hairs cells fire at the same frequency as the sound; this is the **frequency principle** of pitch coding. However, the refractory period limits the rate at which this can occur, and above about 400 Hz a second mechanism takes over. This is based on the part of the basilar membrane that is displaced the most by the wave action in the cochlea, and is known as the **place** theory of pitch coding (*Fig. 4*).

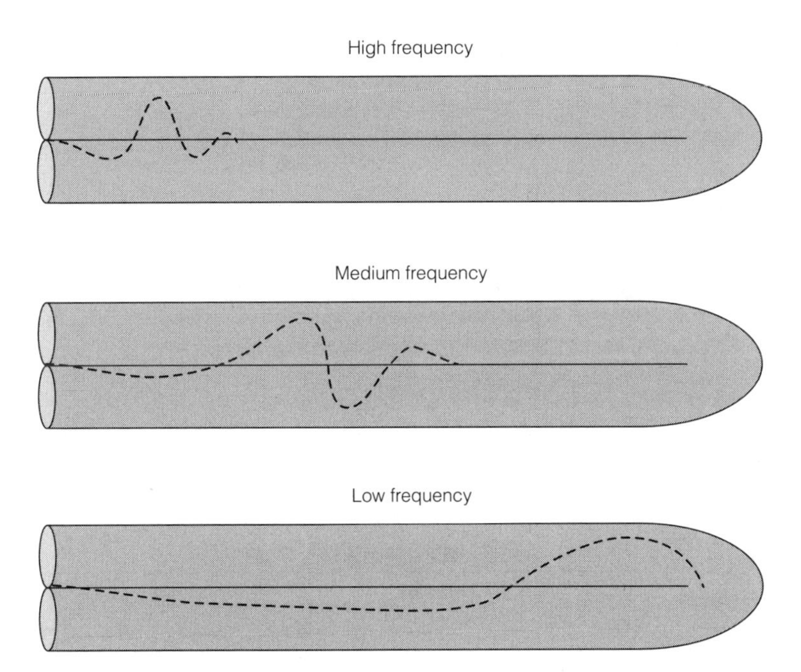

Fig. 4. Displacement of the basilar membrane (dotted line) by fluid waves of different frequencies. From Smith, R.E. (1993) Psychology. *Published by West Publishing Company, Minneapolis/St Paul.*

The pitch of a sound is usually measured on the musical scale, which is built around the octave. A note is one octave above another note of the same letter name when it has a frequency of exactly double the lower note. This results in an equal logarithmic spacing of frequencies across the entire scale, and the musical 'distance' between all octaves is regarded as the same. However, a psychological scale of pitch (the **mel** scale) has also been devised, and there are interesting discrepancies between the two. In particular, the psychological distance between octaves is *not* the same, which reflects the common opinion of musicians (*Fig. 5*).

Loudness

The sensation of loudness arises primarily from the amplitude of the sound pressure wave, and it appears to be coded by the cochlea in two ways: loud sounds cause more nerve cells to fire than quieter sounds, and some nerve cells only fire in response to intense sound. The ear is responsive to a very wide range of sound pressures (*Table 1*).

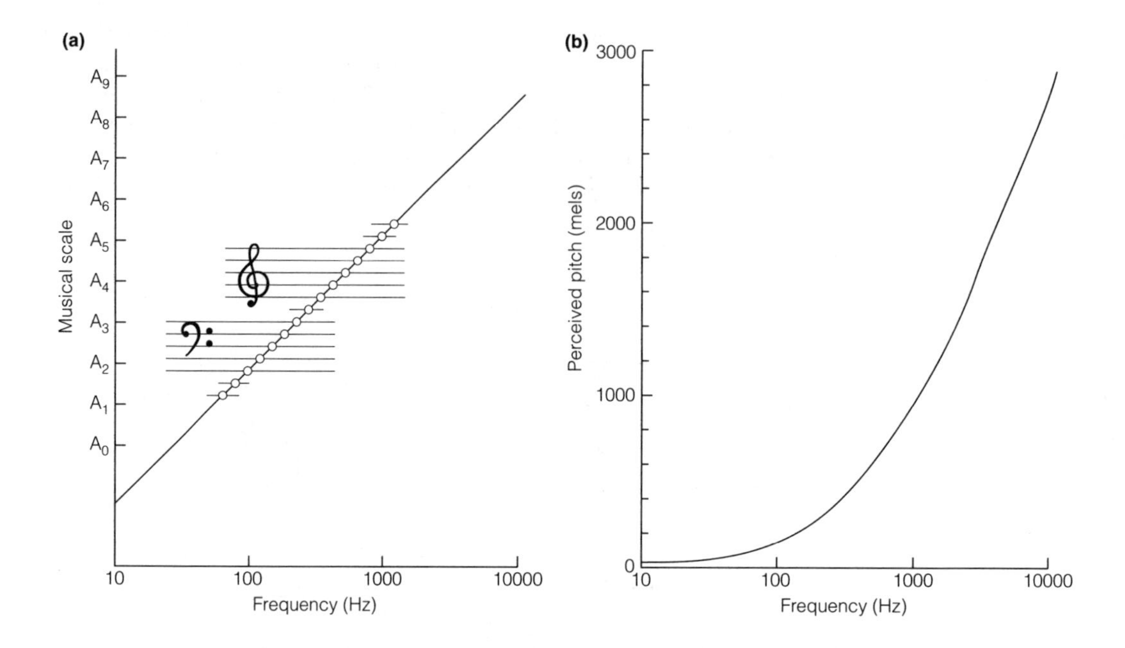

Fig. 5. Relationship between log frequency and (a) the musical scale; (b) the mel scale of pitch. Reproduced from Lindsay, P.H. and Norman, D.A. (1977) Human Information Processing, *2nd Edn, p. 163. Academic Press, New York.*

Table 1. Some comparative sound levels

Sound pressure level (dB)	Common sounds
140	Jet aircraft taking off at 30 m
120	Rock band
100	Pneumatic drill at 1 m
80	Factory noise
60	Normal conversation
40	Quiet room
20	Whisper, rustling leaves

The decibel is a logarithmic unit and an increment of 20 dB corresponds to a 10-fold increase in sound pressure level.

The loudness of a sound can be measured in a logarithmic unit called a **sone**. The relationship between loudness and the physical intensity of the stimulus that gives rise to it is approximately linear, down to about 30 dB. Below this level, the ear is rather more sensitive, and changes in loudness are more rapid with increases in sound pressure.

The loudness of a tone also depends on its frequency, and the ear is most sensitive in the region of 1 kHz to 4 kHz. By matching the loudness of any given pure tone to a standard 1 kHz tone, it is possible to construct **equal loudness contours** (*Fig. 6*).

Localization

If a sound source is directly in front of us, the sound pressure waves that reach our two ears will be identical, and we can use this as a **cue** to judge the location

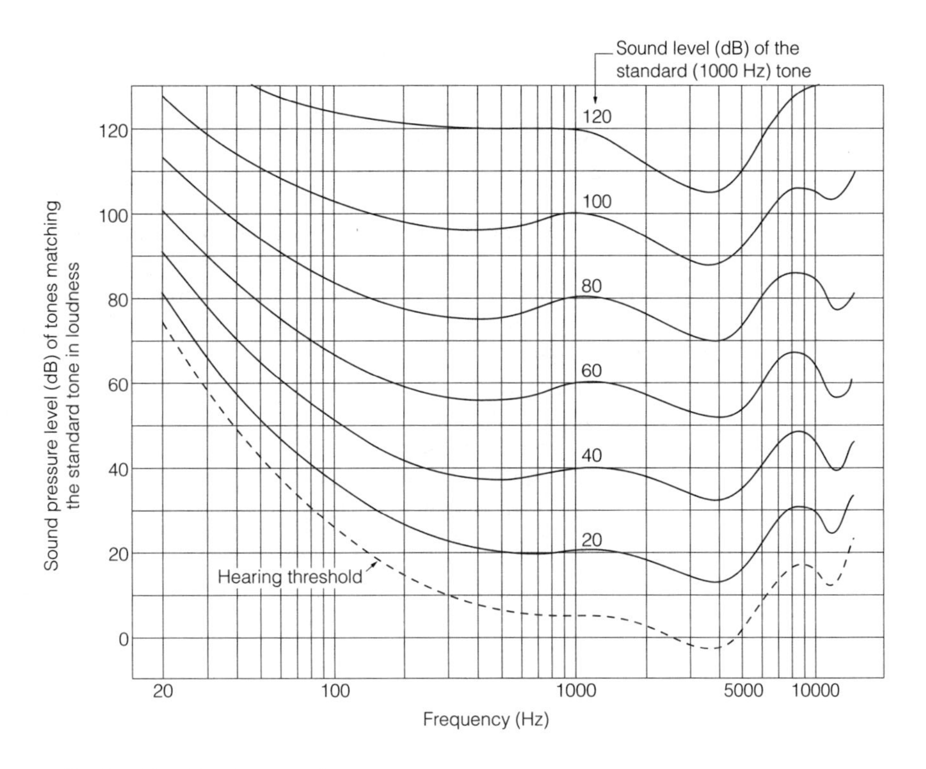

Fig. 6. Equal loudness contours.

of the source when we cannot see it. If the sound source is to one side, the differences between the sound pressure waves at each ear provide sensitive cues to localization. For example, if a sound source is to our left, the sound will arrive at the left ear first, and will be more intense. The **time of arrival** cue works best for sounds below 1.5 kHz, while the **relative intensity** cue works best for sounds above about 2.5 kHz.

In real life, sounds often reach our ears both directly and after reflection from other surfaces such as walls. We are not generally aware of these echoes, and we base our judgement on the first sound to arrive; this is known as the **precedence effect**. The pinna also provides cues for sound localization, especially in distinguishing whether a sound source is in front of or behind us. The folds in the pinna introduce subtle distortions into the sound pressure wave in a way that depends on the direction from which it comes. When sounds (such as music) are heard through headphones, they bypass these distortions and are typically heard as being inside the head.

Hearing loss

Exposure to very loud sounds, for example amplified music or workplace noise, for as little as 20 min can cause a hearing loss that persists for up to 24 h. This is measured in terms of how much louder than normal a sound must be in order to reach threshold. The condition is known as **auditory fatigue** and the degree of loss is called the **temporary threshold shift**. Prolonged and regular exposure will result in a **noise-induced** hearing loss that is progressive and irreversible, since it is due to the destruction of hair cells in the cochlea. Onset is very gradual and often not noticed.

Tinnitus is a persistent humming or ringing in the ears in the absence of any external stimulus. It frequently follows exposure to loud noise, and is a symptom of several ear disorders including cochlear damage, and disorders due to mechanical injury or certain drugs.

Hearing loss that occurs as a result of the ageing process is called **presbyacusis**. It is generally the cumulative effect of small changes in such things as the elasticity of the eardrum and basilar membrane, and the loss of sensorineural elements in the cochlear. Presbyacusis initially affects sensitivity to high frequency sounds, but losses at progressively lower frequencies occur with advancing years.

Hearing loss due to cochlear damage is often accompanied by **loudness recruitment**. Although *thresholds* are raised, the increase in perceived loudness with increased intensity is more rapid than usual, with the result that high intensity sounds may be perceived as normal. Hence raising one's voice when talking to someone with a hearing loss may result in the admonition that 'there's no need to shout'.

C4 The cutaneous senses

Key Notes

Touch and pressure

There are three main types of touch receptor that contribute to the sensation of touch and pressure, although their separate functions are not clearly understood. Their distribution is uneven across the surface of the body, giving rise to widely differing sensitivities in different regions. All touch receptors adapt to steady stimulation, which is quickly reversed when changes occur.

Pain

Pain differs from the other cutaneous senses in that there is a much less straightforward relationship between the stimulation of receptors and the resulting sensation. Sensations of pain may be altered by circumstances and by psychological states, and it appears that cognitive processes may play an important part in the experience, although the mechanisms of this are as yet unclear.

Temperature

The skin contains receptors for both warmth and cold that are irregularly distributed. Sensations of warmth and cold are relative to a narrow range of temperatures called physiological zero, which feels neither warm nor cold. Humans adapt readily to temperature changes, although outside of the range 16ºC–42ºC there will generally be a sensation of cold or warmth.

Kinesthesis

A range of receptors situated in muscles, joints, and tendons provide information about the relative positions of our limbs and the load on our skeletal structure. These receptors do not give rise to conscious sensations, but their proper functioning is essential for the control of movement and bodily position.

Vestibular system

The semicircular canals and vestibular sacs form part of the inner ear and provide information about our orientation in space and about rotational movement. This information is essential for keeping our balance.

Related topic

Movement perception (D3)

Touch and pressure

There are three main types of receptor cell that are sensitive to touch: **basket cells**, which are found at the base of hair shafts; **Pacinian corpuscles**, which are embedded in hairless (**glabrous**) skin; and **free nerve endings**, which have no specialized receptors and are found in all types of skin. Although structurally distinct, these three cells types do not appear to be differentially sensitive to particular types of stimulus, and they do not give rise to different types of sensation.

The sensitivity of the skin to pressure varies across different regions of the body in a way that is closely related to the distribution of receptor cells.

Sensitivity is measured by the **two-point threshold**, which is the minimum separation that can be detected between two sharp points applied to the skin. It is lowest on the fingers and highest on the backs of the calves (*Fig. 1*). Continued steady pressure on the skin results in **adaptation** and the reduction, or even complete disappearance, of sensation. Adaptation is quickly reversed when the stimulus moves or changes intensity.

We can learn to extract complex and spatially detailed information from changes in pressure on the skin; for example, an experienced **Braille** reader can achieve around 100 words per minute, and we all readily learn to identify three-dimensional shapes by touch alone. Experiments in **tactile–visual substitution** demonstrate that, with some effort, we can also learn to retrieve spatial information from a pattern of vibratory stimuli on the skin surface that have been derived from a transformed video image.

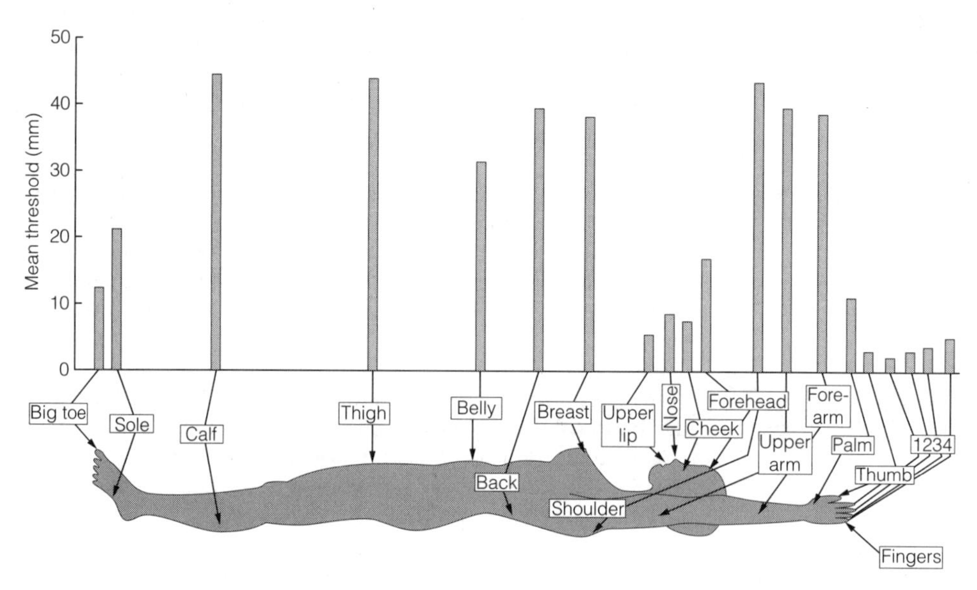

Fig. 1. Two-point thresholds for the body surface. Reproduced from Atkinson, R.L., Atkinson, R.C., Smith, E.E. and Bem, D.J. (1993) Introduction to Psychology, *11th Edn. Harcourt Inc., Orlando, FL.*

Pain

Pain differs from other sensations in that it can be located in almost any region of the body, including internal organs, and, rarely, in parts of the body which have been amputated. Pain may also be experienced at a site in the body when it in fact arises from somewhere else; this is known as **referred pain**. The receptors for pain (called **nociceptors**) are believed to be free nerve endings that occur throughout the skin, as well as in joints, muscles, and much of our internal anatomy. A wide range of stimuli, including mechanical and chemical tissue damage, and extremes of temperature and pressure, can trigger these receptors. However, the relationship between the intensity of pain and the extent of an injury can vary from minor injury with intense pain to severe injury with little pain.

A range of psychological factors including emotional state, expectations, and social context can modulate the experience of pain, and different stimuli can produce pain of differing quality. The experience of pain can also be masked by an intense but nonpainful cutaneous stimulus (such as rubbing or scratching) at the injury site. It is difficult to make general statements about the psychophysical properties of pain, since thresholds vary considerably with individual circumstances. The concept of specific pain receptors cannot readily account for this variability. However, the **gate control theory** proposes a comprehensive mechanism whereby signals generated by cognitive mechanisms may control the onward transmission of neural signals from nociceptors through 'neural gates' in the spinal cord.

Temperature

The surface of the skin contains temperature-sensitive spots that are irregularly distributed; some spots are more sensitive to warmth and some are more sensitive to cold. It appears that the receptors are free nerve endings, but the mechanism whereby the temperature stimulus is transduced into nerve firing is not known.

Sensations of temperature are relative, except at extremes. Under normal circumstances, the body is adapted to a narrow range of temperatures (about 33ºC ± 2ºC) called **physiological zero**, and temperatures close to 33ºC will feel neither warm nor cold when applied to the skin. However, adaptation takes place very rapidly. Raising the temperature of a region of skin by several degrees will result in a temporary sensation of warmth that will soon be replaced by the neutral sensation again. This relativity is easily demonstrated by putting one hand in a bowl of warm water and the other in a bowl of cold water until both have adapted to the water temperature. Then place both hands in a bowl of water at room temperature. The hand that had adapted to the warm water will now feel cool, and the one that had adapted to the cold water will feel warm.

In general, the time taken to adapt is longer the more extreme the temperature. Complete adaptation takes place within a limited range of temperatures (about 16ºC–42ºC) outside of which there will always be a sensation of cold or warmth, respectively. Under certain conditions, a very warm stimulus can produce a sensation of cold, called **paradoxical cold**. This is thought to demonstrate that although the nerve endings in some cold spots respond to both hot and cold stimuli, cold spots only give rise to *sensations* of cold.

Kinesthesis

Kinesthesis (sometimes termed **proprioception**) refers to the sensory system that provides information about the relative position of the components of our jointed skeleton. There appear to be three types of receptor involved. **Pacinian corpuscles** (similar to those found in the skin) are located in the joints, where they are mechanically stimulated by the relative movement of their surfaces, and in the muscles, where they are stimulated by changes in tension. **Ruffini cylinders**, also located in the joints, respond to changes in the angles at which bones are held. **Golgi tendon organs** are located at the junction between tendon and muscle and are also responsive to changes in muscle tension. Almost all the information about posture comes from the receptors in the joints. The other receptors provide information about strain, which enables us to adjust posture and muscle effort in response to varying skeletal load.

The kinesthetic system does not give rise to clearly identifiable sensations but, nevertheless, we have no difficulty in knowing the position and orientation of our limbs in space. For example, we can accurately point to parts of our bodies without using vision and walk up stairs without looking at our feet. The combination of tactile and kinesthetic information forms the basis for **haptic** perception. The active exploration of objects by touch (usually, but not necessarily, using the hands) provides unitary experiences about their shape, weight, surface texture, and consistency. Our ability to recognize common objects using this information is known as **tactual stereognosis**.

Vestibular system

The vestibular system lies within the inner ear and provides information about the body's motion and its position relative to gravity. It has two components: the **vestibular sacs** and the **semicircular canals**. The two fluid-filled vestibular sacs (the utricle and the saccule) are lined with **ciliated hair cells**, and contain tiny crystals of calcium carbonate called **otoliths** that are relatively free to move within the cavity. The system operates by the otoliths bending the hair cells with which they come into contact. When we are upright and at rest, the otoliths sink to the bottom of the sacs and stimulate the hairs cells there. When we move, the mass of the otoliths lags slightly, because of their inertia, and they bend the cilia around the side or top of the sac depending on the direction of movement. In this way, the cilia function as detectors of both gravity and of *changes* in movement. Nerve fibers from the vestibular sacs extend to various regions of the brain, notably the cerebellum.

The primary function of the three semicircular canals is to detect the direction and extent of circular movement. They lie approximately at right angles to each other so that each is related to one of the major planes of the body: vertical, horizontal, and front/back. This arrangement allows them to detect **angular acceleration** of the head in three-dimensional space, but they are not sensitive to movement at a *constant* rate. The semicircular canals are filled with fluid and lined with receptor cells. This fluid moves relative to the receptor cells when the head moves, and the cells generate signals about the nature of the movement (Topic D3).

Stimulation of the semicircular canals causes the eyes to move slowly in the *opposite* direction to the rotation then rapidly back, in a rhythmic fashion. This reflex is called **vestibular nystagmus**, and it helps us to maintain a stable visual image as our head moves.

C5 THE CHEMICAL SENSES

Key Notes

Taste	The taste of a substance is related to its chemical composition, but the details of that relationship remain elusive. Taste buds in the oral cavity, and especially on the tongue, response to all four basic tastes, but there are localized differences in sensitivity.
Smell	The smell of a subtance is related to its chemical composition, but the sensation of smell is more complex than taste and the question of whether there are primary odors and, if so, what they are is unresolved. Only very small quantities of odorants are needed for detection, and women are much more sensitive to some odors than are men. Smell combines with taste to provide sensations of flavor.

Taste

Substances that we can taste (called sapid) must be soluble in water. Four **primary tastes** have been distinguished: sour, sweet, salty, and bitter, although others have been suggested, notably *umami* (loosely translated as 'savoriness'), which is considered a basic taste in Japan.

The chemical composition of a substance is a critical determinant of its taste, although there are too many exceptions for us to account for all tastes on this basis. Substances that taste sour are generally acid compounds, but some acids (e.g. amino acids) taste sweet. Sweet tastes (generally linked with nutrients) are associated with carbohydrates, although nonnutritive compounds such as saccharine are also intensely sweet. Inorganic salts generally taste salty while bitter tastes result from alkaloids such as quinine and nicotine.

The basic receptors for taste are called **taste buds**. These are located in tiny pits and grooves throughout the oral cavity and, particularly, in clusters on the visible elevations on the tongue called **papillae**. All areas of the tongue respond to almost all of the basic tastes, but there are localized differences in sensitivity.

Smell

Substances that we can smell (called odorous) must be volatile, and must be water- and fat-soluble in order to reach the olfactory receptors. Odorants are usually organic, rather than inorganic, and are often mixtures of extremely complex compounds.

Odor-sensitive cells, called **olfactory cells**, are situated on the **olfactory mucosa** located in the mucus membrane high in the nasal cavity. These cells are connected by the olfactory nerve to the **olfactory bulbs** at the base of the brain. When air is inhaled, it is carried to the mucosa where odorants dissolve and are trapped by **olfactory binding proteins**, molecules which concentrate odorant molecules at receptor sites. Odor stimulation can also occur while *exhaling*, especially while eating.

Odor *intensity* depends on the concentration of odorant molecules reaching the receptors. The question of odor *quality* is more complex, and there have been several attempts to determine whether there are such things as 'primary' odors.

For example, the **smell prism** is based on six hypothesized attributes arranged in a geometrical configuration (*Fig. 1*). Specific odors are represented by points within the triangular prism that indicate the relative contribution of each of the primary odors.

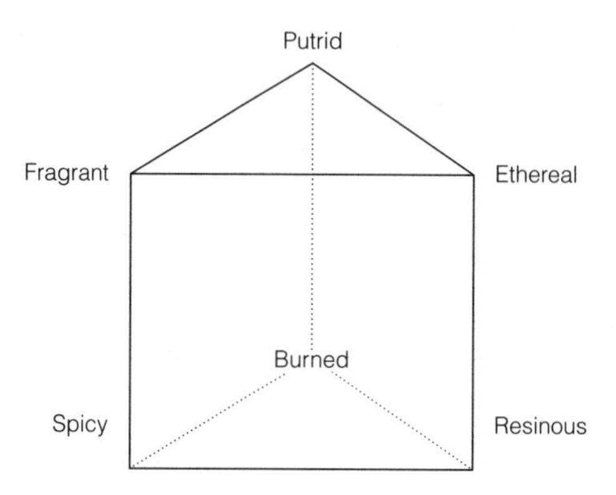

Fig. 1. The smell prism.

Another scheme makes use of terms found to be used most frequently to describe the smell of organic compounds, and attempts to link them with the chemical properties of substances which have these characteristic smells (*Table 1*).

Table 1. Examples of chemicals which give rise to particular odors

Odor	Chemical example	Common example
Camphoraceous	Hexachloroethane	Moth balls
Musky	Butylbenzene	Musk
Floral	Ethyl carbinol	Rose, lavender
Minty	Menthol	Peppermint
Ethereal	Diethyl ether	Dry-cleaning fluid
Pungent	Formic acid	Vinegar
Putrid	Ethyl mercaptan	Decaying cabbage

The **stereochemical** or **lock and key** hypothesis was based on the observation that many of the molecules that shared a 'primary odor' had similar geometric properties. The suggestion was that the shapes of the odorant molecules acted as keys that allowed them to bind to specific receptor sites on the surface of the olfactory membrane and hence give rise to specific odors. Unfortunately, research has since demonstrated that molecules with similar shapes can in fact give rise to very different odors, and it appears likely that a range of chemical properties are involved in the determination of odor.

Absolute thresholds have been calculated for a wide range of odorants, and it is known that only very tiny concentrations are needed for detection. Some examples are given in *Table 2*; however, making these measurements is complex, and estimates vary depending on the precise methodology used.

Table 2. Examples of threshold concentration for sample odorants

Substance	Odor	Threshold concentration (mg/l of air)
Amyl acetate	Banana oil	0.039
Butyric acid	Perspiration	0.009
Hydrogen sulfide	Rotten eggs	0.00018
Citral	Lemon	0.000003
Ethyl mercaptan	Decaying cabbage	0.00000066

Females are generally more *sensitive* to odors than males. A well-documented example is the case of Exaltolide, a synthetic odorant used in the manufacture of perfume. The threshold for this substance is 1000 times lower in sexually mature females than in males, and 50% of males cannot smell it at all. There is evidence that sensitivity for this and many other odorants interacts with hormonal variation during the menstrual cycle. Females are also better at *identifying* odors than males and, since this superiority is evident before puberty, the gender differences cannot be entirely due to hormonal differences.

Humans have very good memories for the sources of odors and laboratory studies show that smells associated with real-life experiences (called **episodic odors**) are particularly long lasting. Odors can also create emotional arousal, which is mediated by the direct link between the olfactory bulbs and the limbic system. Even neutral odors that are paired with stressful events can later elicit mood and attitude changes, and these links can occur without awareness. **Conditioned taste aversion** can occur when tastes are paired with intensely unpleasant conditions.

The overall appeal of food and drink is due to a combination of sensory effects called **flavor**. Contributory attributes include not only taste and smell but also temperature, texture, and irritation of the oral and nasal cavities such as may be caused by carbonated drinks and spices.

D1 THEORETICAL PERSPECTIVES

Key Notes

Nativism and empiricism

One of the central questions in the study of perception concerns the origins of our perceptual abilities. One extreme view (nativism) is that they are all inherited and another (empiricism) is that they are all learned. It is not useful to regard these positions as mutually exclusive since we certainly inherit those basic neurophysiological structures that determine much of what we can sense, but then make extensive use of stored past experiences to interpret it.

Direct perception

The empiricist view generally describes perception in terms of information processing. In contrast, direct perception is the view that animals can directly pick up information that is of significance to them without the need for intermediate processing. This is a controversial view, although there is some experimental evidence to support it.

The computational approach

The computational approach to perception is concerned with specifying the exact nature of that information processing which must take place if clearly defined perceptual tasks are to be accomplished.

Related topics

Movement perception (D3)
Form perception (D4)
Selective attention (D6)

Nativism and empiricism

Our perceptual abilities enable us to organize sensations into representations of the world around us. We take these abilities for granted but it has proved very difficult to provide satisfactory explanations of the processes involved. Much of the research has centered on vision because it is arguably our most important sense, and also because visual stimuli are easier to specify precisely and to produce in the laboratory. Consequently, many important ideas about perception have been stated in terms of vision although they may often prove to be useful in understanding audition as well.

A very early debate in the development of psychology concerned whether our perceptual abilities are inherited (the **nativist** view) or learned (the **empiricist** view), an argument which has surfaced in areas other than perception (see Topic K8). We will not attempt to discuss the merits of each case because:

> "No simple exposition of this great and largely fruitless controversy can, however, be adequate to its complexities. For one thing, almost every protagonist turns out, whatever he was called, to have been both nativist and empiricist. Everyone believed that the organism brought something congenitally to the problem of space; everyone believed that the organisation of space may be altered or developed in experience."
> (from Boring, E.G. (1942) *Sensation and Perception in the History of Experimental Psychology*. Appleton-Century-Crofts, New York, p. 233)

The modern view of the issue is that we certainly inherit the neurological structures that underpin our perceptual abilities. Some of these structures also develop to a greater or lesser degree as a result of our experience, especially early on in life. However, a very substantial component of our perceptual ability depends upon things that have been learned.

One school of thought that belonged to the nativist tradition, however, left an enduring legacy. The **Gestalt** psychologists were especially interested in the way in which the visual system organizes perception so that we can effortlessly separate objects from their background. The fundamental tenet of the Gestalt project was that the 'whole is greater than the sum of its parts'. This summarized the view that perception could not profitably be understood by studying perceptual elements and then trying to synthesize the whole from them, because the qualities of the whole *determined* the nature of the parts. The Gestalt psychologists made an important contribution to the study of visual form (see Topic D4).

The **empiricist** view of perception is that learning and memory play an integral part in perception and that, therefore, intermediate processes must exist that mediate between our sensations and our conscious perception. These processes resemble inferential thinking in that they enable us to go beyond the evidence of our senses, which is often incomplete and distorted. They permit us to *anticipate future events* so that we can, for example, catch a ball, and they make use of material stored in memory to take advantage of past experience. The term **constructivism** is sometimes used to express this idea that perception is an *active* process through which we *construct* our conception of the world from the evidence of all our senses together with information held in memory. The empiricist approach dominated perceptual research for some decades, and the term **information processing** characterized the way in which the processes of perception were often studied in relative isolation from their content (see Topic D6).

Direct perception

The empiricist view is that perception is an **indirect** affair, that sensory evidence must always be elaborated by cognitive processes. In contrast, the ecological view (primarily associated with the work of J.J. Gibson) is based on the premise that animals learn to analyze their environment in terms of its meaning *for them* and to differentiate these meanings (called **affordances)** *directly* without the need for intermediate cognitive processes.

Gibson was clear that perceptual theory should be concerned with the real world, and believed (in direct contrast to the empirical approach) that it is not useful to analyze perceptual problems in terms of stimuli and retinal images. Instead, he began from the idea that the visual world is made up of illuminated surfaces, and that light which reaches our eyes after reflection off these surfaces carries information about their nature and position; that is, *the light itself is structured.* Gibson called this structured light the **ambient optic array**. As we move through our environment, the **flow** of light moving across our retina contains patterns of regularities which derive from this structure, and such patterns can be detected directly rather than derived from some unspecified form of information processing. One of Gibson's examples may help make this clear. As a pilot prepares to land an aircraft, the point toward which the plane is heading appears stationary while everything else in the visual field 'flows' away from it. If the pilot changes direction, that stationary point moves and the flow pattern changes. As the plane gets nearer to the ground there is an increase in the speed of flow in some parts of the environment. Thus, the pattern of flow contains detailed information about the plane's speed, direction, and distance from the ground which, according to Gibson, can then be perceived directly rather than as the result of some kind of intermediate calculation (see Topic D2, *Fig. 7*).

The ecological approach is centrally concerned with the relationship between perception and action, and movement is an integral part of this relationship. Some properties of the ambient optic array remain constant as we move around and provide important information about the structure of the environment. Gibson called these properties **invariants** and believed that the sensory systems of animals evolved to detect those which were significant to them. Invariants can be quite complex relationships but, according to Gibson, their perception does not involve any sort of computation or comparison of input with previously stored memories. The animal simply **resonates** to invariants; that is, it picks them up directly (see Topic D3).

Gibson's position is radically different from that of most other perceptual theorists and has been regarded by many as highly controversial, or just plain wrong. Its emphasis on the relationship between visual processing and action has proved valuable, but it denies a distinction between this form of perception and the processes that lead to an *understanding* of the visual world, which relies to some extent on representation in memory.

The computational approach

The computational approach to the analysis of perception takes as its starting point a set of questions about the kind of information processing a perceptual system must carry out in order to accomplish tasks such as perceptual discrimination and visually guided action. Furthermore, the problems are addressed in terms of the way in which a computer might process such information, and this requires formal, precise and unambiguous specification of both the information and the computational processes.

The most influential computational approach to vision is that of David Marr, who suggested that descriptions of perceptual processing must be addressed at three levels: the **computational theory**, **representations and algorithms**, and **hardware implementation**. At the level of computational theory, we are concerned with the formal statement of the problem: precisely what it is that is to be recognized, the information that is present in the retinal image, and the information needed to achieve a correct solution. At the second level, we must specify the ways in which information is represented in the system and the kinds of calculations that must be accomplished to achieve the solution. Finally, at the third level we must examine the specification of the device that will implement the solution. Generally, this will be a computer but research into ways in which the brain might accomplish the same tasks is increasingly fruitful.

D2 Depth perception

Key Notes

Cues for the perception of depth and distance

Although we can readily perceive depth and distance, it is not entirely clear how we mange to do so since there is no depth stimulus. The most likely mechanisms involve our detecting sources of information called depth cues from which we then draw inferences.

Oculomotor cues

Oculomotor cues are sources of information which originate in the muscles which control the direction in which our eyes are pointing. They are thought to be of limited use on their own.

Pictorial cues

Pictorial cues are sources of information within the visual scene that enable us to make inferences about depth and distance. In general, a visual scene will yield several cues and these are used in combination to yield the best judgment.

Stereopsis

The use of two eyes provides a stereoscopic, or three-dimensional, view of the world by fusing the two, slightly disparate, images. The exact mechanism underlying this is not fully understood, although a computational approach to the problem has yielded some insights.

Related topic

Constancies and illusions (D5)

Cues for the perception of depth and distance

The perception of depth and distance involves both judging the distance of an object *from yourself* and judging the distance *between* other objects in the environment. Both tasks require us to solve the problem of making inferences about three-dimensional space from the two-dimensional image which falls on the retina. The data for these inferences come from a number of possible sources known as **cues**.

Oculomotor cues

Accommodation is the process of adjusting the shape of the lens in order to focus an image on the retina. Since objects at different distances require different degrees of accommodation, the tension in the **ciliary muscles** may provide a cue for depth. However, accommodation in humans operates across a limited range of distances (about 20 cm to 3 m) and the process is relatively slow, it is thought to offer rather imprecise information.

Convergence refers to the process whereby the two eyes are turned slightly inward towards the nose so that the image will fall on the fovea and thus achieve optimum visual acuity. The resulting angle between the eyes is called the **convergence angle**, and for distances up to about 6 m the convergence angle can provide distance information. There is still some controversy about just how useful this is.

Pictorial cues

Interposition is the appearance of *relative* depth that occurs when one object is partly obscured by another; the fully visible object is seen as being closer to the observer than the occluded one (*Fig. 1*). An important feature of occlusion is that we are often not aware of the missing parts of objects because the visual system rapidly fills in the gaps. In *Fig. 1* for example, the background shape will be perceived as a square even though more than half of it is obscured.

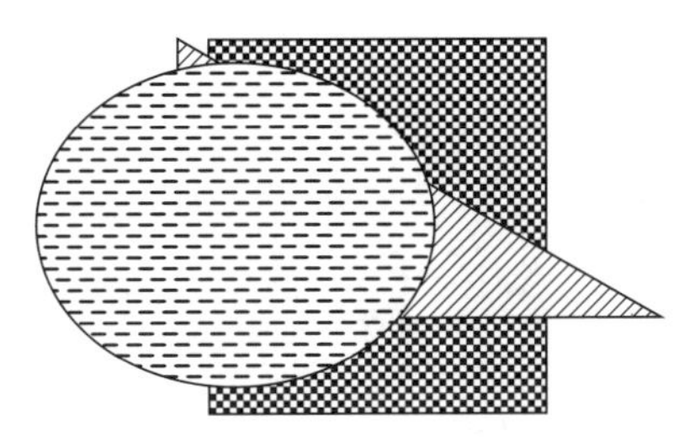

Fig. 1. Interposition as a cue for relative depth.

Elevation, or height in the visual field, refers to the fact that objects which are further away tend to be higher in the visual field than objects which are close by. This cue also helps us perceive depth in pictures. When we are outdoors, objects at a distance are generally seen less clearly than objects close by due to the scattering of light by tiny particles of dust and water vapour. The further away an object, the more of the atmosphere the reflected light must pass through and hence the hazier the object will seem. This effect is called **aerial perspective**.

The **relative size** of objects also furnishes a cue to their relative distance. When identical shapes of different sizes are viewed together, or in quick succession, the larger ones are seen as nearer than the smaller ones (*Fig. 2*). This effect arises from the diminishing retinal image sizes and does not require prior experience of the objects. However, **familiar size** can contribute to spatial perception. For example, if we see a very small car on the road we use our experience to infer that it is far away, rather than a toy. A familiar extension of the retinal image cue is provided by **linear perspective** (*Fig. 3*) in which the relative size of objects, the relative distance between objects, and their height in the visual field all combine to give a two-dimensional representation of spatial relationships.

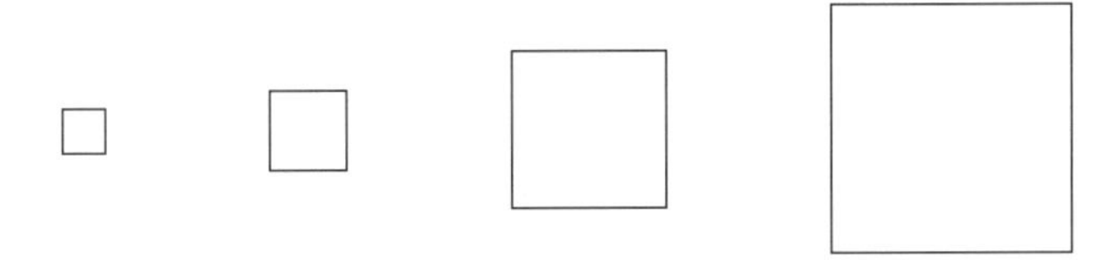

Fig. 2. Relative size as a cue to relative distance.

As we scan an outdoor visual field from nearby to far away, we can see that objects in that field appear more closely packed together the further away we look (*Fig. 4*). This is true whether we look at large objects such as trees and buildings, or small objects such as blades of grass or pebbles on a beach. This gives rise to a **texture gradient** which provides a powerful distance cue. Furthermore, discontinuities in the texture gradient can give additional information about the surfaces we are looking at (*Fig. 5*).

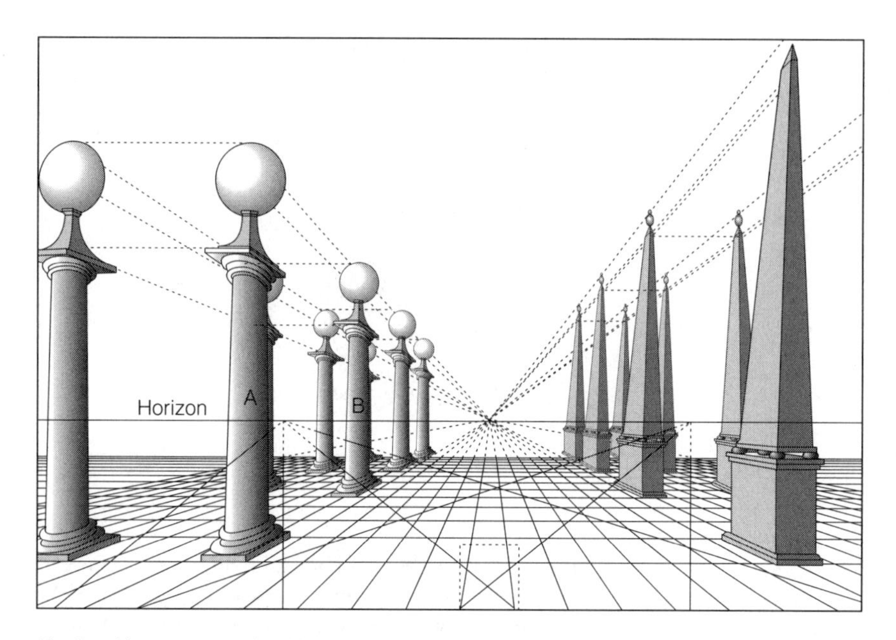

Fig. 3. Linear perspective. Note that the horizon bisects each column at the same point regardless of its distance from the observer.

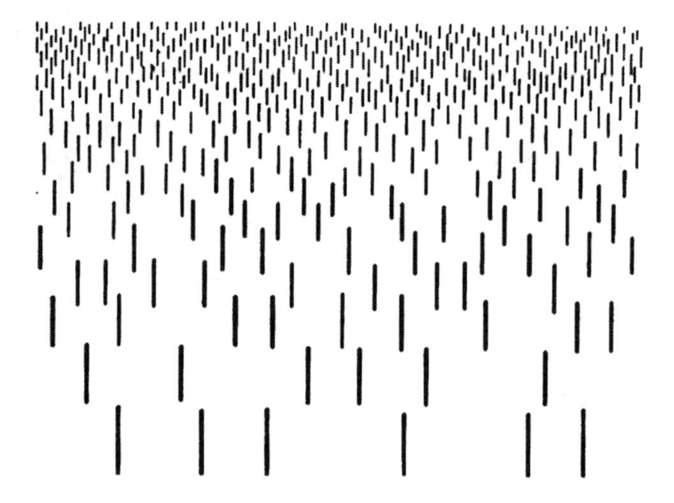

Fig. 4. An example of a texture gradient which gives cues to distance.

In everyday life, much of our perception occurs when we are moving and when objects in our environment are moving as well. Any movement of one object relative to another provides a cue to depth due to **motion parallax**. If, for example, a moving observer views stationary objects then those objects which are closer seem to move faster than those which are further away. If the observer fixates on a point some distance away, objects which are nearer than this fixation point appear to move in the *opposite* direction to the observer and those which are further away move in the *same* direction. This effect can easily be seen when looking out of the window of a car or railway carriage, and is due to the relative movement of images on the retina (*Fig. 6*).

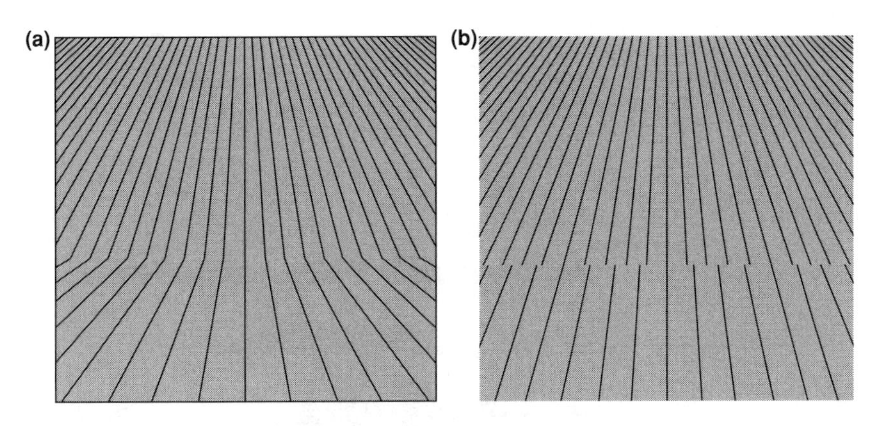

Fig. 5. Changes in texture gradients can provide information about (a) changes in elevation and (b) discontinuities in surfaces. Reproduced with permission from Gleitman, H., Fridlund, A.J. and Reisberg, D. (1999) Psychology, *4th Edn. W.W. Norton, New York, p. 220.*

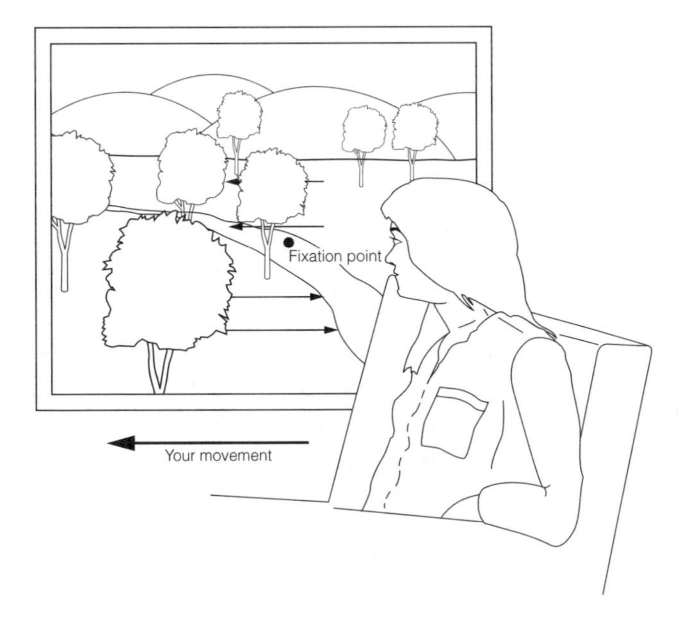

Fig. 6. Motion parallax. Objects at varying distances from the observer move at different speeds, which provides information about their distance from the observer.

When we move, we can also gain information from the way in which texture gradients change. Objects that are close by appear to move more rapidly than those that are more distant and, in the same way, texture gradients appear to expand as you move across a surface. Moving *towards* a surface causes a continuous change in perspective due to the changing position from which objects are viewed. These cues, known respectively as **optic flow** and **motion perspective**, provide powerful sources of information about relative distance (*Fig. 7*).

Stereopsis

Stereopsis (or **binocular depth perception)** refers to the three-dimensional nature of vision that is possible when we use two eyes rather than just one. The

Fig. 7. Optic flow patterns as sources of information about relative distance.

Fig. 8. Random dot stereogram. (a) In monocular vision, the pictures have no depth characteristics. (b) When stereoscopically fused, a centre square emerges from the background. Reproduced from Julesz, B. (1971) Foundations of Cyclopean Perception. *University of Chicago Press, Chicago.*

cues for this arise from **binocular disparity**; that is, the small differences between the images at each eye which occur because of their horizontal separation. A process called **fusion** merges these two images and, although fusion does not occur over the entire visual image, we tend not to notice the unfused, or double, parts. The way in which the visual system derives the perception of depth from the fusion of disparate images is only partially understood. Psychophysiological research has demonstrated the existence of **disparity-tuned** detectors in the visual cortex of the cat that respond to small differences in the relative horizontal displacement of images. However, the random dot stereograms developed by Julesz permit the perception of depth without any *monocular* shapes (*Fig. 8*).

These pictures are created from large numbers of *identical* elements without any contours, and to fuse such images by means of disparity-tuned detectors

Fig. 9. A 1754 engraving by Hogarth entitled False Perspective. *The contradictory cues are perspective, interposition, and relative size.*

would require the system to know which dot had to be paired with which. Computational theorists have devised algorithms for solving this **correspondence problem**. The basis of the solutions requires detectors tuned to the same disparity to facilitate each other while those tuned to different disparities inhibit each other, and there is some evidence from psychophysiological work with monkeys that such activity does in fact occur.

The accuracy of our perception of distance and depth generally depends on the interaction of a number of these cues, and our sensitivity to differences in distance using multiple cues is roughly equal to the sum of the sensitivity for each cue acting separately. However, it is possible to construct artificial situations where cues contradict each other, and therefore cannot be summed. Under these circumstances, subjects will *select* a cue on which to base their judgment, and this may give rise to an illusory perception (*Fig. 9*).

D3 Movement perception

Key Notes

Mechanisms of motion perception

There are two visual mechanisms for detecting movement. Movement of an image across the retina when the eye is stationary is detected by parvocellular and magnocellular ganglion cells. When the eyes move across a stationary scene, or move to track a moving object, corollary discharge signals from the eye muscles enable the visual system to differentiate the relative movements involved.

Apparent movement

Apparent movement is an illusion induced by switching stationary stimuli on and off in alternating sequences; its most familiar application is in motion pictures. Small apparent movements probably stimulate motion detectors in the visual pathway, but larger ones appear to arise from cognitive processes.

Biological motion

We need only very small amounts of information to recognize human movement, to discriminate the movement of a man from a woman, and to discriminate our own movement from that of others. This phenomenon provides a good example of direct perception.

Movement after-effects

Movement after-effects occur after watching a continuously moving scene, such as a waterfall, for some minutes. It is thought to be due to adaptation of the retinal motion detectors.

Related topics

Vision (C2)
Theoretical perspectives (D1)
Form perception (D4)

Mechanisms of motion perception

Movement of the retinal image

If the eye is held still by **fixating** on a point and the image of a moving object then falls on the retina, movement is registered by the stimulation of a succession of adjacent retinal units. Our ability to detect movement in this way depends on the area of the retina on which the image falls. Moderate and fast velocities are better detected towards the periphery while slower velocities are better detected towards the fovea. This difference arises from the distribution of magnocellular and parvocellular ganglion cells in the retina (see Topic C2).

Movement of the eye and head

When we **track** a moving object, the movement of our eyes results in the image of the object staying more or less stationary on the retina. If the object is moving against a textured background, the image of the background moves on the retina and the retinal movement system can detect the changes. However, this background stimulation is not *essential* for the perception of movement (for example, we can detect the movement of a spot of light in an otherwise darkened room) since the **efferent neural signals** which control the voluntary movement of the eyes provide information about movement. When the eyes move they send back corresponding **afferent** signals (called **corollary discharge signals**) which are then related to the movement of an image on the retina. Thus, when we scan a *stationary* visual scene the retinal movement is canceled out and we perceive it as stationary. However, when we track a *moving* object the corollary discharge signals do not cancel the stationary image on the retina, which is then perceived as moving (*Table 1*).

Table 1. Perceptual outcomes in the perception of movement

System	Activity	Target	Retinal image	Corollary discharge	Perception
Retinal image	Fixation	Moving	Changing	No	Movement
Eye movement	Scanning	Stationary	Changing	Yes	No movement
Eye movement	Tracking	Moving	Stationary	Yes	Movement

After Shiffman, H.R. (1996), Sensation and Perception: An Integrated Approach. 4th Edn. New York: John Wiley Inc.

Apparent movement

When two lights, separated in space, are switched on and off in sequence, it is possible to induce an illusion of movement whereby the light is seen to move from one location to another. This is known as **apparent movement** although the Gestalt psychologists called it the **phi phenomenon**.

It appears that apparent movement is mediated by two separate mechanisms. The first (the **short-range process**) interprets small target jumps of up to about 15 min of visual angle as motion, and is probably mediated by the activity of motion detectors in the visual pathway. The second (the **long-range process**) deals with larger jumps and appears to be mediated by more complex cognitive factors of an interpretative nature. This corresponds to the Gestalt psychologists' observation that apparent movement obeys the law of Pragnänz (see Topic D4); that is, what is perceived is the simplest interpretation of the stimulus array (*Fig. 1*).

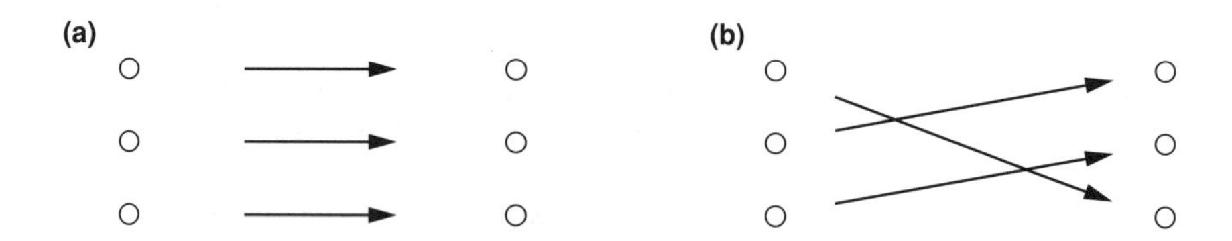

Fig. 1. The law of Prägnanz applied to apparent movement. If three pairs of lights are arranged to induce apparent movement, the perceived pattern will always be as illustrated in (a), never as in (b), or other variations.

Motion pictures are produced by projecting a series of still pictures at a rate of 24 per second. The smooth transition of the features of one frame into those of the next relies on two factors. Firstly, the visual response to a stimulus **persists** for a brief period after the stimulus ends. Secondly, there is a close correspondence between the structural features of successive frames so that the interpretation of movement is the simplest interpretation of the visual stimulus. If this correspondence is interrupted during editing, the resulting sequence may prove a challenge to the system and the perception of smooth movement will be reduced.

Autokinetic movement occurs when we try to fixate a stationary point of light in an otherwise completely dark room that provides no fixed reference points. After some minutes, the light will typically appear to wander, often considerably. The most likely explanation is that this kind of abnormal fixation creates fatigue in the eye muscles causing the *eye* to wander. Since this is not a voluntary movement (and hence does not give rise to corollary discharge signals), the cancellation process involved in scanning is not instigated and the experience is of the light moving.

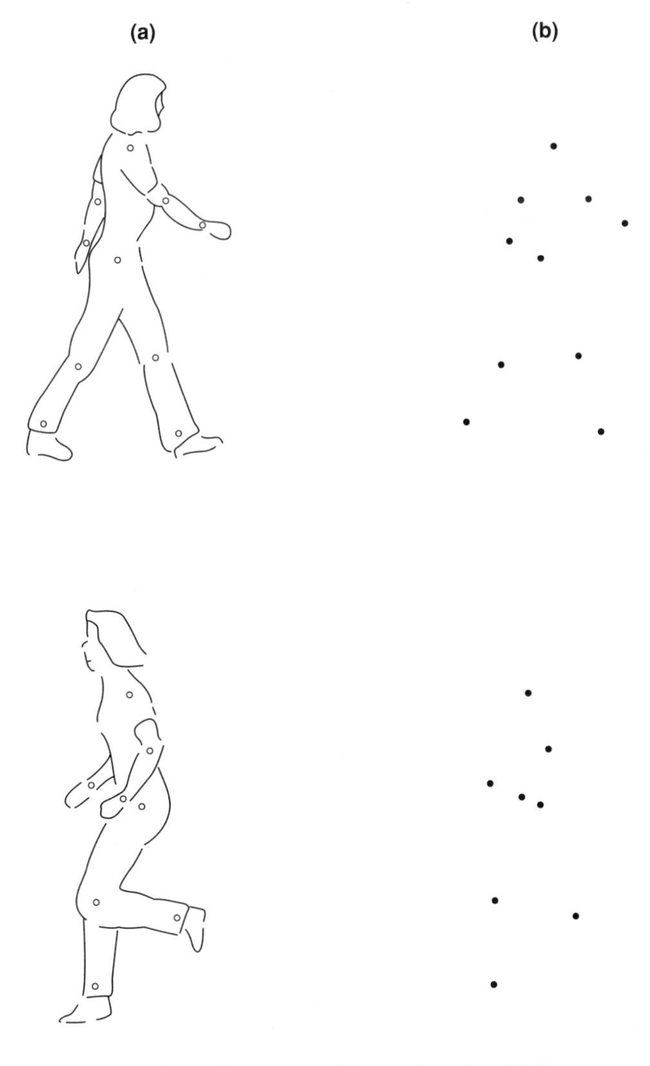

Fig. 2. (a) Outlines of a person walking and running. (b) The corresponding dot configurations. Subjects observing only the dots perceive a walking or running person after just one or two steps. Reproduced from Johansson, G. (1973) Visual perception of biological motion and a model for its analysis. Perception and Psychophysics, *Vol. 14, p. 202.*

Biological motion

The perception of movements made by other human beings is known as biological motion. This has been extensively studied using film of people dressed in black and moving in complete darkness with tiny lights fixed to body joints. Such films record the movement of parts of the body relative to each other but contain no other visual information (*Fig. 2*).

People viewing these films are quite confident about what they are seeing after the actor has taken only one or two steps, and familiar movements can be identified from clips as brief as 100 ms. However, when stationary, the patterns of light do not resemble a human form, demonstrating that it is the *movement* that is typical of human gait. Much work has been done in the context of direct perception (see Topic D1) to discover the nature of the information employed in this phenomenon, with some success. For example, we can distinguish between men and women using these films, the basis for this being that males walk with larger movements at the shoulder than at the hip while the opposite is true for females. It is important to note that these are immediate experiences requiring no complex processing on the part of the observer.

Movement after-effects

If we spend some minutes gazing at something moving then look away, we may get a sensation of movement in the opposite direction. This is known as a **movement after-effect**. A well-known example is the **waterfall effect**; if we stare at a waterfall for some time then shift our gaze to a stationary scene, the scene will appear to move upwards. This effect is probably caused by the selective adaptation of movement detectors during the prolonged gaze. When the gaze is shifted, the activity of those detectors is suddenly reduced giving the impression of movement in the opposite direction. If we adapt to movement using only one eye then shift our gaze to look at a stationary scene with the *other* eye, we still experience the after-effect. This demonstrates that movement after effects are likely to have origins in a central region of the visual system.

D4 FORM PERCEPTION

Key Notes

Structuralism

The view that perceptions are built up from elementary sensations is called structuralism. Attempts to discover and catalogue elementary sensations proved unworkable, but a more modern version of structuralism based on the analysis of visual features is still influential.

Gestalt

Gestalt theorists held that the perception of form is primary and cannot be reduced to smaller elements. Their principles of perceptual organization are still useful today although the neurological mechanisms that were though to underpin them have been proved to be wrong.

Feature analysis

Feature analysis is a form of information processing by which the retinal image is analyzed into fundamental components using specialized cells, or groups of cells, which have evolved for the purpose. One shortcoming of the concept is that it has proved difficult to determine exactly what these components (or features) are. It has, however, been successfully used in machine-based visual recognition systems where the feature set can be precisely specified.

Pandemonium

Pandemonium was a particular recognition system that used feature analysis as its first stage, and built in further stages which completed a recognition system. The final stage is characterized as a decision-making process and, as a consequence, the system can make mistakes when the sensory evidence is ambiguous or of poor quality.

Model-based theories

Template and prototype theories are both referred to as model based because they characterize recognition as a process of comparing the retinal image with stored representations, or models, of objects. Neither system can satisfactorily deal with the recognition of novel objects.

Computational theories

Marr's computational theory addresses the need to specify the nature of analytical and computational processes involved in image analysis and object identification. It can be thought of as the most rigorous version of the information processing approach to vision.

Component-based theories

Component-based theories explain object identification in terms of fundamental geometric shapes rather than elementary features. These shapes, or geons, can be used to build large numbers of object prototypes, and the system can deal with the recognition of unfamiliar objects.

Related topics

Vision (C2)

Selective Attention (D6)

Structuralism

Structuralism was formulated in the early years of the 20th century. It is based on the fundamental assumption that all psychological experience arises from the combination of simple components in the same way that chemical compounds are built up from elements. Perceived form was thought to be constructed from elementary sensations and the task of identifying these sensations was approached through the method of **analytical introspection**. Observers were trained to attend to the immediate and primary qualities of an object, and to report these sensations while ignoring all other aspects. Thus, for example, a person looking at a small box might report edges and corners, shadows, colors shading into each other, and so on; they must however ignore the fact that it contained breakfast cereal. After some years of use, this method had produced a catalogue of some 40 000 elementary sensations, which seemed a rather large number. Furthermore, there were often big differences between the data produced by different individuals and it became apparent that the method was flawed. However, the underlying idea of structuralism has remained influential, as we shall see later.

Gestalt

The Gestalt theorists took a view that was opposed to structuralism, holding that the natural units of perception were forms and shapes rather than any more fundamental class of sensations. Thus they argued that the visual array will always be perceived in terms of the most simple and stable forms possible (the **Law of Pragnänz**) and their **principles of perceptual organization** describe how we respond to the relationship between elements of the perceptual field rather than to the elements themselves. The most important of these principles are as follows (see also *Fig. 1*).

1. The **principle of proximity** (*Fig. 1a–c*). Elements that are closer together are more likely to be perceived as belonging together. In (*a*) we see *rows* of elements because the horizontal spacing is narrower than the vertical. In (*b*) the organization disappears because horizontal and vertical spacing are equal, and in (*c*) we see *columns* of elements because the vertical spacing is now narrower.
2. The **principle of symmetry** (*Fig. 1d*). Elements that form symmetrical units are more likely to be perceived as belonging together than those which do not. In the left-hand panel we see white patterns against a black background, and it is almost impossible to see black patterns against a white background for more than a few seconds. In the right-hand panel the reverse is true. The stable pattern is determined by the symmetry of the bars, not the color.
3. The **principle of good continuation** (*Fig. 1e*). Elements group together to form uninterrupted smooth lines wherever possible. In this illustration we see the dots forming two smooth lines which cross. Although other interpretations are possible they are not readily perceived.
4. The **principle of similarity** (*Fig. 1f*). Elements that are similar to each other are more likely to be seen as belonging to the same figure. In this illustration we see two triangles more readily than a six-pointed star.
5. The **principle of closure** (*Fig. 1g*). Elements group together in a way that favors the more enclosed or complete figures. In this illustration there are three distinct shapes, but the principles of good continuation and closure favor the perception of one bar lying behind another.

Comparable patterns of organization are found in audition, particularly in the perception of melody. Notes that are close together in pitch are perceived as being part of the same melody (an example of the law of proximity), and separa-

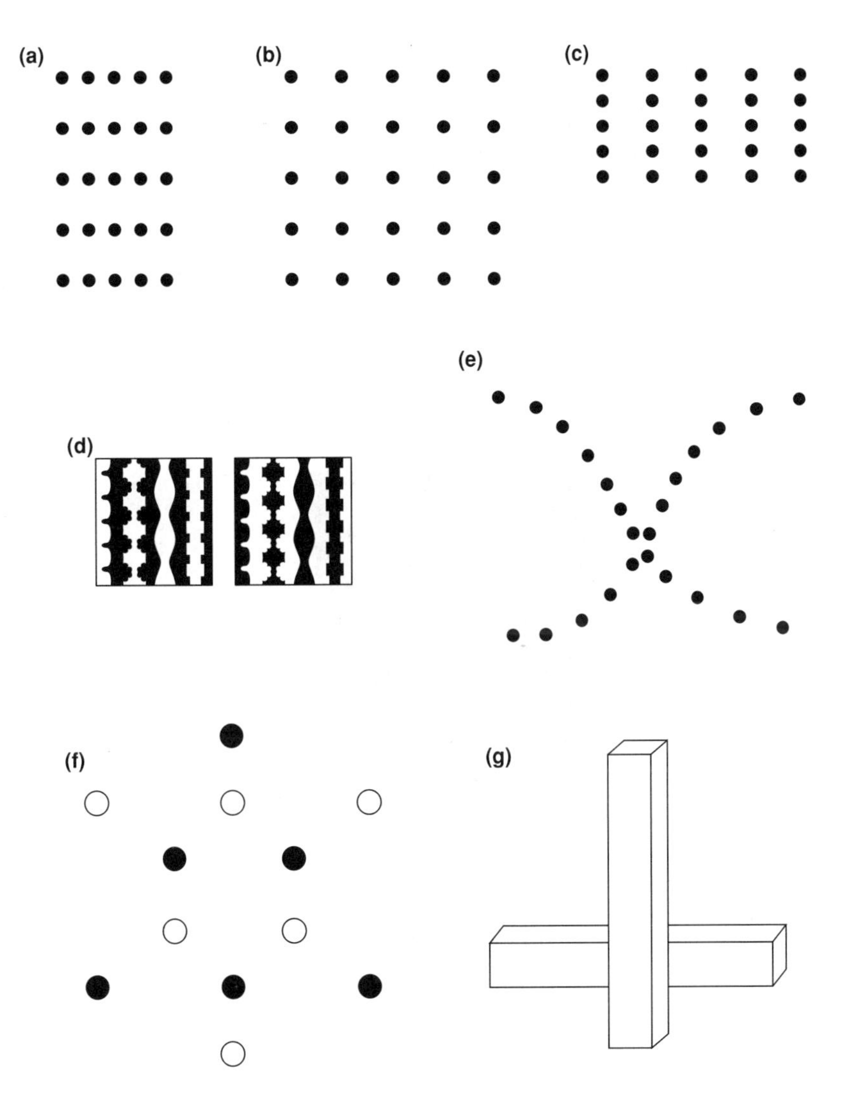

Fig. 1. The Gestalt principles of perceptual organization.

tion between figure and ground becomes the distinction between melody and accompaniment. Similarly, when several instruments play simultaneously we tend to group those of similar timbre into units; an example of the law of similarity. Also, sequences of notes going in the same direction tend to be perceived as part of the same melody even when they are being played as part of a different sequence, an example of the law of good continuation (*Fig. 2*).

Some figure–ground distinctions are ambiguous. The Gestalt school used the fact that one stimulus could give rise to more than one distinct perception to illustrate the point that the behavior of the whole is not determined by that of its individual elements (*Fig. 3*).

The Gestalt psychologists provided good *descriptions* of some examples of the perception of form, and many of their demonstrations were very convincing, but their *explanations* were much less useful. These were based on the principle

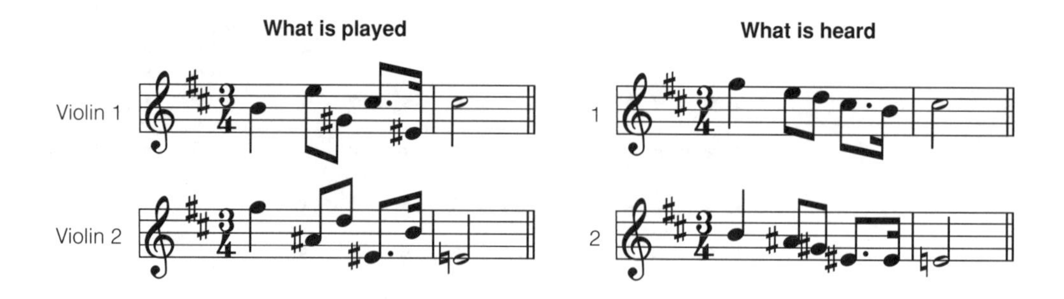

Fig. 2. An auditory example of the law of good continuation. From Tchaikovsky's Sixth Symphony.

Fig. 3. A reversible figure–ground stimulus which can be seen as a pair of faces or a vase. Reversals occur spontaneously if you fixate on the figure.

of **isomorphism**, which stated that there was a correspondence between the organizational process of perception and the organization of the electrical activity of the brain. However, attempts to provide experimental evidence for this principle have failed.

Feature analysis

We can regard shapes as regions of the image on the retina that are surrounded by **contours**. These regions have qualities of spatial extent, texture, color, shading, and so on. We can call these qualities **features** and they enable us to discriminate objects from each other. Sometimes such discrimination only requires a small number of features, for example distinguishing between a dog and a sheep, and sometimes it requires rather more. Note that *perceptual* discrimination does not depend upon prior knowledge of the objects in question.

The structuralist assumption can now be rephrased in terms of not sensations, but of *features*, and a range of experimental tasks has led to the identification of a relatively small number of fundamental features which include color, brightness, orientation, length, curvature, line endings, line intersections, and line closures. This approach avoids one of the major difficulties with analytical introspection because features can be *seen directly* rather than *inferred* from verbal report. However, the group of features that emerges from a particular experiment does depend to some extent on the kind of task that is used and, consequently, the total feature set is still open to revision. When visual features are characterized in this way we can see the possibility of tracing them to their source in the receptive fields of cells in the visual pathway (see Topic C2).

This type of processing is called **bottom-up** (or data-driven) because the incoming information is processed according to rules which are determined by the structure of the sensory pathways. Data-driven processing analyzes and registers the presence of features and patterns in the sensory input, and might be considered to be concerned more with the **local** than the **global** details of an object. Under some circumstances, however, global details may be recognized more readily than local ones (*Fig. 4*), and there is evidence that observers may be able to switch their attention voluntarily between local and global aspects of a figure (see Topic D6).

E E E E   D D D D
E   D
E   D
E   D
E   D
E E E   D D D
E   D
E   D
E   D
E   D
E E E E   D D D D

Fig. 4. These figures have different local features, but the same global feature. The global feature is generally recognized first, but you will have no difficulty in switching your attention to the local features if you need to.

Identifying objects may begin with feature analysis but it also employs **top-down** (or concept-driven) processing, which makes use of our past experience and knowledge of the world around us. Concept-driven processing is responsible for integrating features into perceptual objects, often in conjunction with expectations about what is probable in a given situation. Context is a very important factor in the identification process; for example, briefly presented objects are more readily recognized when they are depicted in familiar rather than unfamiliar contexts, and the perception of ambiguous objects can be determined by their context (*Fig. 5*).

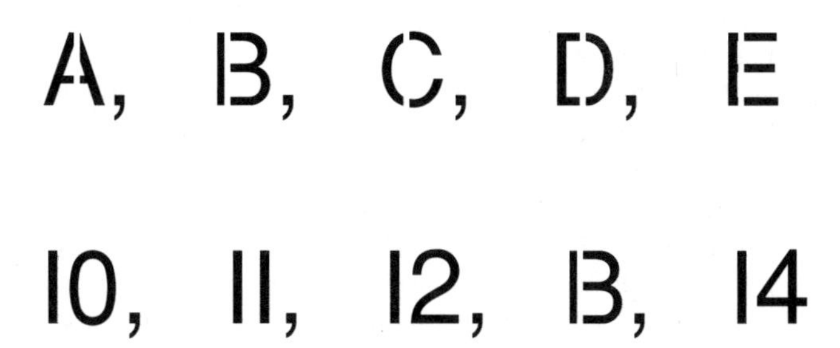

Fig. 5. The effect of context on recognition: the 'B' and the '13' are the same characters.

Pandemonium

Selfridge's **pandemonium** theory emphasizes the feature extraction element of identification. Each stage in the process is accomplished by 'demons' with specific duties. Image demons record the external signal and pass it on to feature demons, which each look for their particular feature within the image and respond when they find it. At the next level, cognitive demons watch the responses of the feature demons because they in turn must respond when particular combinations of features are detected. They react enthusiastically when this happens, but also show a little activity when something similar to their pattern is signaled. The resulting activity, described as pandemonium, is interpreted by the decision demon who must determine which of the cognitive demons is showing the most activity. It is possible for the decision demon to make mistakes, leading to errors of identification, when several cognitive demons show similar patterns of activity. Pandemonium models have been incorporated into machines that can recognize typefaces, such as those used on bank cheques.

Model-based theories

Model-based theories rely on our detailed conceptual knowledge of the world rather than on features. **Template theory** is based on the idea that we compare an incoming image with representations stored in memory (the template), which can be scaled for size and rotated to match any orientation; successive comparisons are made until a match is found. This system has been used to design robots that can identify objects and manipulate them, using video camera input. However, a template system cannot identify novel objects, and the theory does not address the question of how templates arise in the first place. The robots, of course, need the assistance of humans to acquire them. **Prototype** theories differ from template theories in that the stored representation is not an image but a collection of attributes, making a more flexible identification process possible. Thus, a template system would need several images of a chair including, for example, those with and without arms, in order to identify all chairs reliably. A prototype system, on the other hand, only needs a set of defining attributes, which introduces the realistic possibility that some objects might be identified as chairs when they were not intended as such. The identification of novel objects, however, remains problematic.

Computational theories

The most influential of the computational theories is that of David Marr (see Topic D1). His work emphasized the need to understand the *structure* of the visual image, a matter generally neglected by other theorists, with the notable exception of Gibson. Marr's account describes object recognition in terms of three levels of processing, each of which are expressed as computations performed on the visual image. The first level is the computation of a **primal sketch**, which is a description of the major features of the image in terms of brightness, contours, and areas where there are sharp changes in light intensity. These are critical for the location of edges and boundaries that in turn convey information about shape and orientation. The primal sketch is then used to construct the **2½-D sketch**. This is a description of the orientation and depth relationships between surfaces in the image *from the viewpoint of the observer.* We need this image to guide actions, such as moving about in our surroundings. Finally the 2½-D sketch is used to construct a **3-D representation** that is *independent* of the viewpoint of the observer. At this stage, we can store descriptions of objects and later use them for identification (see below).

Research based on this approach has a natural affinity with work in artificial intelligence and computer science. However, one of its more important consequences for psychology has been the demonstration that object recognition is much more complicated than had been assumed.

Component-based theories

Component-based theories are an extension of the computational approach and represent an attempt to deal with the problem of recognizing objects (and, particularly, novel objects) without relying on sets of features which have been arbitrarily defined. Biederman suggested that one could specify a set of primitive *geometric* elements from which, with rotation and scaling, object prototypes could be constructed. Examples of these geometric icons (or **geons**) and objects which can be constructed from them are illustrated in *Fig. 6*. A relatively small number of geons (about 36) could generate an extremely large number of objects.

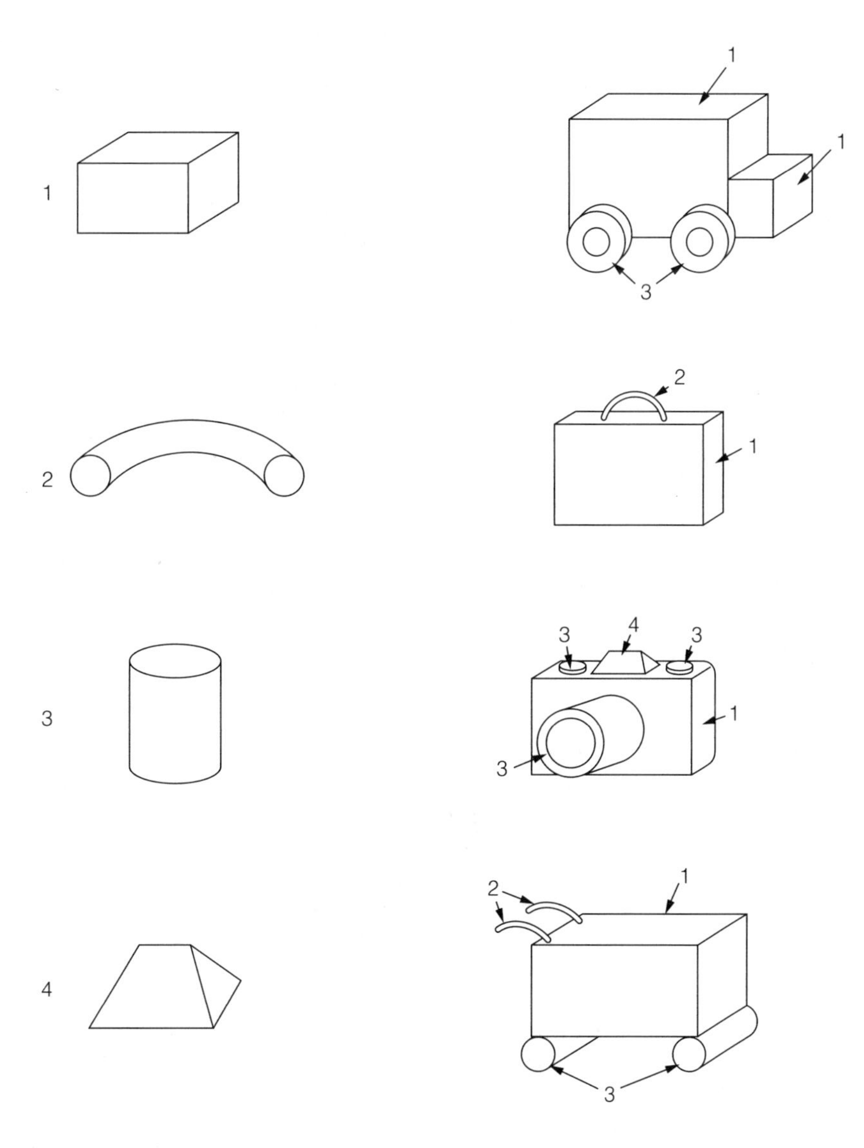

Fig. 6. A selection of four geons and examples of prototype objects that can be constructed from them.

D5 CONSTANCIES AND ILLUSIONS

Key Notes

Lightness constancy Lightness remains constant despite changes in illumination because our perception of it depends on the proportion of incident light reflected from an object rather than the absolute amount. If an object is illuminated independently of its surroundings, lightness constancy cannot operate and the perception of lightness depends on absolute illumination.

Size constancy Our perception of the size of objects remains roughly constant at different distances despite the fact that the size of the retinal image changes. The effectiveness of this mechanism depends on a number of factors including familiarity with the objects concerned and the availability of distance cues.

Shape constancy Shape constancy operates in a similar fashion to size constancy, ensuring that we perceive objects to be of the same shape in a variety of orientations. Shape constancy can operate on unfamiliar objects if they can be broken down into familiar components.

Visual illusions A range of visual illusions can occur when cues to depth and size constancy are misinterpreted, or when the contrast between neighboring objects makes the judgment of relative size and orientation difficult. No one type of explanation for them has proved satisfactory, and it is likely that illusions arise from processes at several levels in the visual system.

Related topic Form perception (D4)

Lightness constancy

The lightness of an object remains relatively constant even though its illumination changes. Thus, a piece of white paper in a shadow is still seen as white and a black cat in bright sunlight is still seen as black, even though there may be more light reflected from the cat than from the sheet of paper. The basis of this constancy is the **albedo**; that is, the *proportion* of incident light that is reflected from a surface. Reducing the illumination of a piece of paper from bright to dim will reduce the *absolute* amount of light reflected off it, but will not change the albedo, and any surrounding objects or background will be affected in the same way. The visual system thus compensates for the *overall* level of illumination. Lightness constancy can break down when an object is specially illuminated so that the light source is concealed and no light at all falls on the background. The visual system cannot then judge the albedo and lightness will depend on absolute illumination.

Size constancy

As we move towards or away from an object, the size of the retinal image changes considerably: doubling the distance at which you view something will halve the size of its retinal image. Within broad limits, however, the perceived size of objects does not change in a corresponding fashion. **Size constancy** involves a process of adjusting the perceived sizes of objects to take account of their distance from the observer. Cues for apparent distance are therefore important in maintaining size constancy (*Fig. 1*) and *reducing* the effectiveness of distance cues will lead to less effective size constancy. Size constancy is also affected by the interaction of learned factors, expectations, and even motivational states. Familiarity with the size of an object, for example, will enable us to judge distance based on retinal size and known size, and the range of distances over which size constancy operates is larger in adults than in children since, presumably, their experience is less extensive. However, looking *down* from a high building we do tend to see tiny people and tiny vehicles since our lack of familiarity with *vertical* distance cues hampers the constancy mechanism.

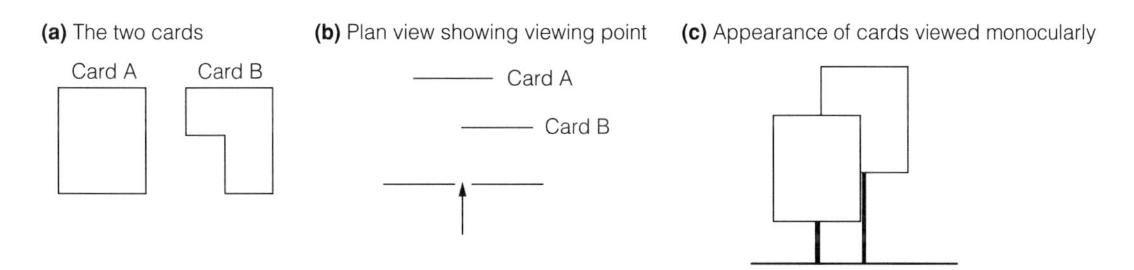

Fig. 1. Illusory distance with conflicting cues. Playing cards A and B are arranged on stands so that the cut-out in card B gives the appearance of occlusion, indicating that it is behind *card A. Given that they are familiar objects and are expected to be the same size, the relative size of the retinal image gives the veridical cue, but it is ignored. The usual perception is that card B is behind card A, but bigger.*

Shape constancy

Shape constancy is the relative stability of perceived shape under variations in orientation, and it seems to occur as a result of a compensatory mechanism much as size constancy does. A door, for example, will 'look' rectangular when it is open and the retinal image is a trapezoid, because the visual system appears to compensate for the changes in the retinal image (*Fig. 2*). As a consequence, familiar shapes are more subject to shape constancy than unfamiliar ones, although an unfamiliar object made up of familiar shapes may exhibit elements of constancy (see Topic D4). Thus, the shape in *Fig. 3* appears to be made up of rectangular blocks but, as drawn, contains no right angles.

Size and shape constancy are closely related, and the addition of depth cues to a drawing can radically alter the perception of relative size (*Fig 4*).

Visual illusions

Visual constancies have been used to provide explanations for a range of visual illusions. The Müller-Lyer, Ponzo and Poggendorf illusions, for example (*Fig. 5*), are often cited as instances of misplaced **perspective constancy**. The basis for this is the suggestion that the figures contain elements that are automatically recognized as perspective cues and which *inappropriately* trigger the size constancy mechanism. The two components of the Müller-Lyer illusion, for

Fig. 2. Shape constancy. Our familiarity with doors leads us to perceive them as rectangular, even when the image is trapezoid.

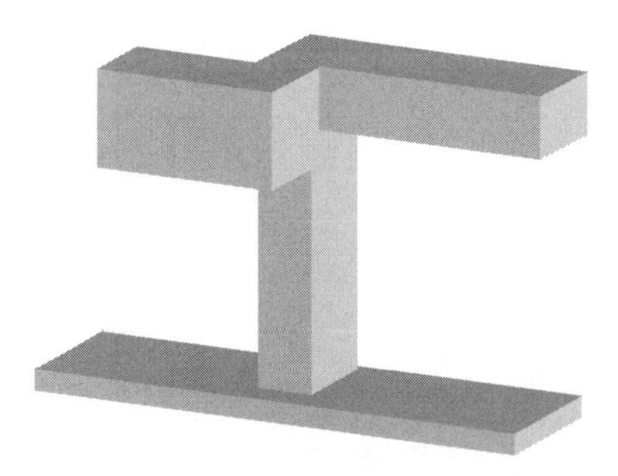

Fig. 3. An unfamiliar object, which appears to be made up of rectangular blocks, but as drawn, contains no right angles.

Fig. 4. Depth cues and the perception of relative size.

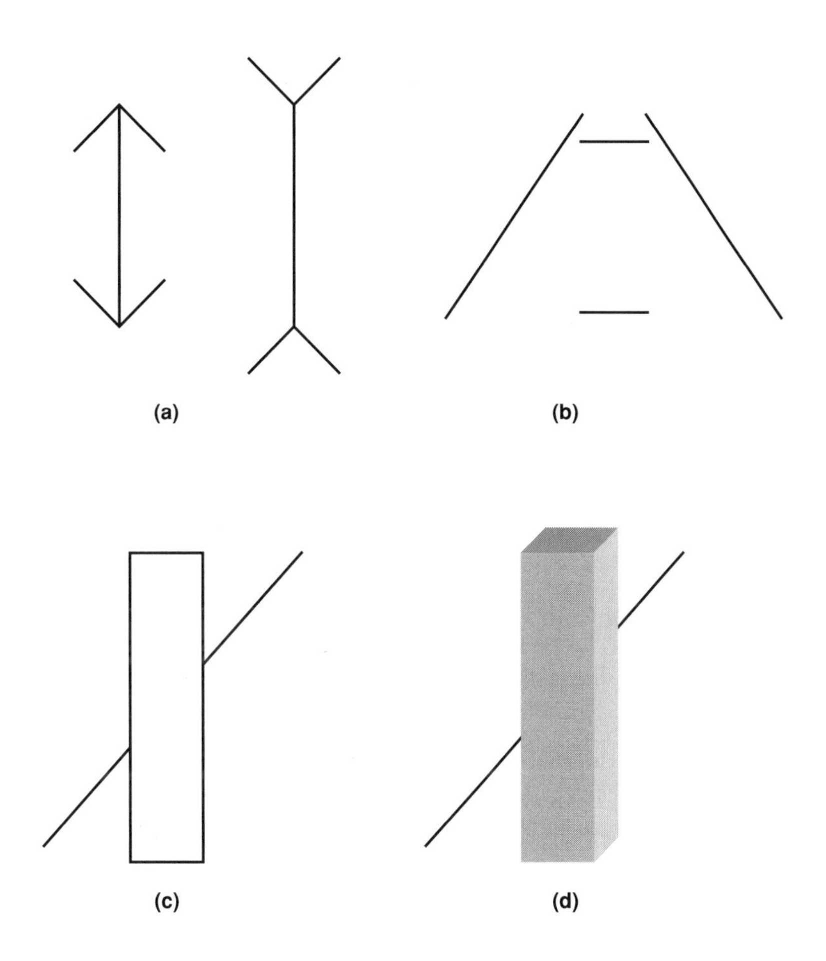

Fig. 5. Visual illusions. (a) Müller–Lyer: the two vertical lines are of equal length, but the right-hand one appears longer. (b) Ponzo: the two horizontal lines are of equal length, but the upper one appears longer. (c) Poggendorf: the two parts of the diagonal line appear misaligned, but they are not. When depth cues are added to the Poggendorf (d), the illusion of misalignment is considerably weakened.

example, have been likened to perspective drawings of the inside and outside of buildings. The inside corner is perceived as the further of the two from the observer, which automatically triggers a size constancy correction thus making that component of the figure seem *longer* than its retinal image would suggest and, therefore, *longer* than the other line. The upper bar in the Ponzo illusion is seen as further away and the same explanation accounts for why it is perceived to be longer than the lower bar. The perspective explanation of the Poggendorf illusion is that the central bar is perceived as lying in the plane at right angles to the viewer while the oblique lines are seen as the edges of surfaces on *different* planes that are receding into the distance. If perspective cues are added so that the central bar appears to recede in the same direction as the oblique lines, the illusion of misalignment is weakened or disappears altogether.

If the depth information in a figure is ambiguous, an observer may experience a spontaneous *reversal* of perspective, and the rate of this reversal will increase with the period of inspection (*Fig. 6*).

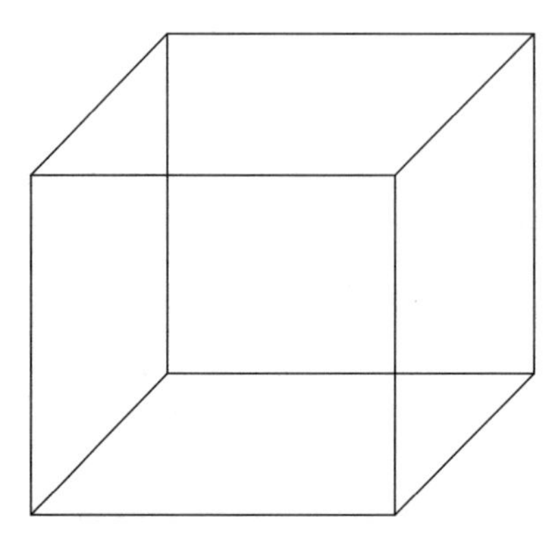

Fig. 6. Reversal of perspective illustrated by the Necker cube. Look closely at the figure for a few seconds and the reversal will happen spontaneously.

Illusory perceptions of size and alignments can occur under conditions where perspective is not a factor. These **contrast** illusions are caused by the effects of context on the judgement of size (*Fig. 7*) and orientation (*Fig. 8*).

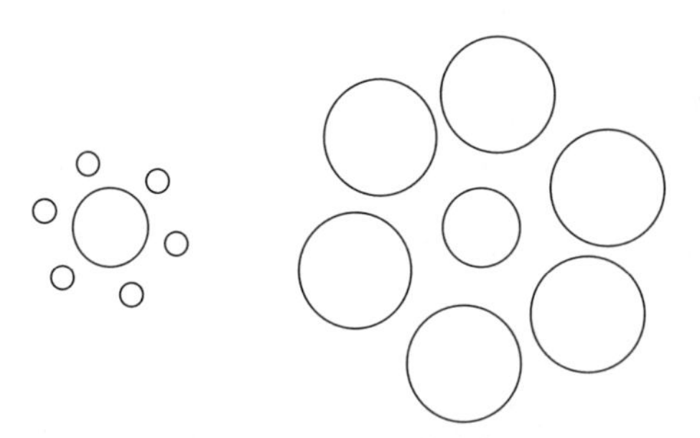

Fig. 7. A size contrast illusion. The central circles of the two groups are the same size, but the one on the left seems larger due to the smaller circles surrounding it.

While the perspective theory gives an intuitively appealing account of some visual illusions, it has been challenged on the grounds that, for example, the Müller-Lyer illusion occurs using figures which do not incorporate depth cues (*Fig. 9*). It has also been demonstrated in lower animals such as the pigeon using discrimination learning (see Topic E1). In fact, no single perceptual mechanism can satisfactorily explain visual illusions, and it is likely that they arise from the interaction of a variety of processes at different levels of the visual system.

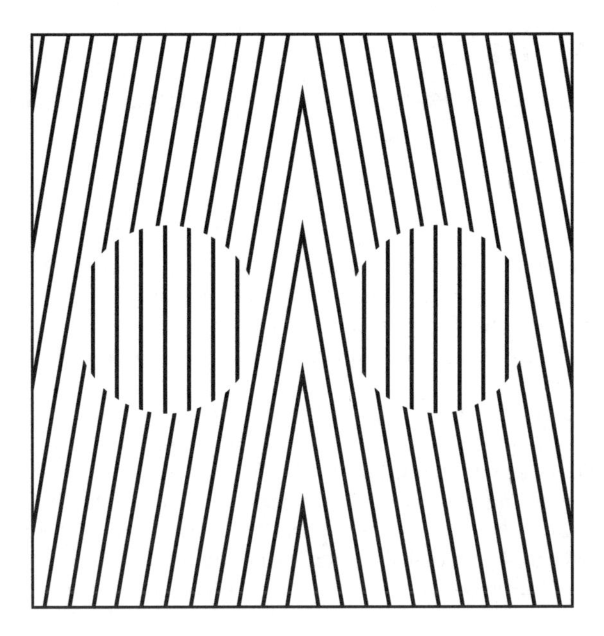

Fig. 8. *An orientation contrast illusion. The circular areas contain vertical lines, but they appear tilted in the direction opposite to the lines which surround them. Reproduced from Schiffman, H.R. (1996)* Sensation and Perception: An Integrated Approach, *4th Edn. John Wiley & Sons, Inc., New York, p. 276.*

Fig. 9. *An example of the Müller–Lyer illusion that does not incorporate depth cues. The distance between the dots on the edges of circles 1 and 2 appears smaller than the distance between the dots on circles 2 and 3.*

D6 SELECTIVE ATTENTION

Key Notes

Orienting

Orienting processes are evident in humans from birth. Their function is to turn our attention to potentially significant sources of information. They may involve the movement of just the eyes or the head, or the whole body. Eye movements associated with orienting are called saccades.

Selective listening

We can select one source of information to listen to while ignoring others, and it is suggested that this is accomplished by a process of filtering. It is not clear whether filtering takes place early or late during information processing. The fact that some meaning is apparently extracted from unattended messages supports the late filtering hypothesis, but there is psychophysiological evidence of early filtering too. There is some evidence that visual filtering takes place in a similar fashion.

Selective looking

Searching for information in the visual field involves controlled scanning and a process of feature identification. These two types of activity are linked so that scanning can be guided by the expectation of where useful information is to be found. Searches for simple features are fast and relatively unaffected by the amount of irrelevant information. Searches for the conjunction of several features are slower and more prone to error.

Divided attention

We can divide our attention between two simultaneous tasks, but our success will depend on the difficulty of the tasks and how practiced we are. With high levels of practice we can acquire a degree of automatic processing which then enables us to deal with more demanding simultaneous activity.

Theories of attention

Structural theories of attention focus on filtering processes and the problem of early or late selection. Attentional resource theories concentrate on divided attention and the mechanisms that determine the limits on our performance of simultaneous tasks. Neither approach has yet afforded a comprehensive account of our abilities and limitations.

Related topic

Encoding in memory (F1)

Orienting

Attention refers to a group of processes that enable us to **select** those sensory inputs which are important to us and, to some extent, exclude those which are not. The orienting response occurs when attention is drawn to a sudden change in the environment such as a noise, movement in the peripheral visual field, or a change in illumination. It involves moving the sense organs so that they are in the best position to monitor the source of the changes, and this generally involves moving the eyes, the head, or sometimes the whole body. Small eye and head movements, known as the **orienting reflex**, occur in newborn infants, and their occurrence in response to sudden noises can be used as an indication that a baby's hearing is functioning.

The eye movements associated with the orienting response are short, jerky movements known as **saccades**, or saccadic eye movements. When we are *voluntarily* scanning the visual scene it takes about 200 ms to initiate a saccade in response to a new stimulus. Reflexes are typically faster than voluntary activity, and the orienting reflex initiates saccades in about 75 ms; these are called **express saccades**.

Orienting is not necessarily accompanied by an overt response. For example, our attention may be *captured* by a sudden change in the visual field or by a particular voice amongst many in a crowd; this is sometimes termed **covert orienting**.

Selective listening

Orienting processes *prepare* us to attend to the source of some interesting or useful information, but then we may wish to attend closely to some things and exclude others. The processes that are employed to do this depend to some extent on whether the information is auditory or visual. Auditory information appears to be selected by **filtering**. Imagine that you are in a room with a lot of people who are all talking to each other: if you want to attend to one conversation you will somehow have to ignore everything else. This problem (originally called the **cocktail party phenomenon**) has been studied using a procedure called **dichotic listening**. An observer listens to two unrelated messages that are delivered, one to each ear, through headphones. The task, known as **shadowing**, is to repeat one of the two messages (the target message) aloud as it is presented, while ignoring the other one. Data show that observers can attend to one message and filter out the other, but that not all messages are equally easy to shadow and not all unattended messages are completely ignored. Prose is more readily shadowed than poetry or disconnected words, and nonsense syllables (nonword fragments such as *dop, vil* or *pab*) are quite difficult. This demonstrates that meaning and grammar help the shadowing process. Target messages which differ from the unattended message in some physical characteristic, such as rate of speech or the pitch of the speaker's voice, are also easier to shadow. Equally interesting is what happens to the messages that the observer was supposed to ignore. In general, it appears that they can remember very little of this message, although if the shadowing task is *interrupted* then a few words of the unattended message will be remembered, which indicates that it was held in short-term memory (see Topic F2). However, this does not mean that the unattended message is not processed at all, since there is evidence that some changes (such as a switch from a male to a female voice) or the presentation of certain words (such as the observer's name) will capture attention away from the target message.

An analogous process, visual filtering, can be demonstrated using two visual scenes that are superimposed on a video screen. Observers who have to shadow the activity in one scene are generally unable to report anything specific about the activity in the unattended scene.

Selective looking

The most common form of selective looking involves some kind of **visual search**, and studies of saccadic eye movements reveal how we scan a scene in order to detect a target or to extract other kinds of information. Newborn infants tend only to fixate on small portions of the visual scene, but scanning develops markedly over the first few months of life to take in progressively more extensive areas. Development of the scanning mechanism is complete by about 7 years, but can become much more difficult in old age.

Scanning is not a random process. The way we scan depends upon what kind of information we hope to capture (*Fig. 1*), and we use our past experience to search for information in the most likely parts of the visual scene. Successful searching involves the identification of **features** within the scene, and if the target differs from all the other elements by a single feature (e.g. one person in a white coat when all others are wearing black) then this is all you have to search for. Reasonably enough, this is called a **feature search**. When targets possess a specific *combination* of attributes then a **conjunction** search is necessary. This principle is illustrated in *Fig. 2*. Conjunction searches take longer than feature searches and are more prone to error. When a feature search is possible, the number and type of other elements (known as **distracters**) does not affect the searching speed. This suggests that the searching process is a **parallel** one; that is, all the items are effectively scanned at the same time. When a conjunction search is necessary, however, the number of distracters has a direct effect on search speed. This suggests that conjunction searching is a **serial** process in which each item has to be inspected individually.

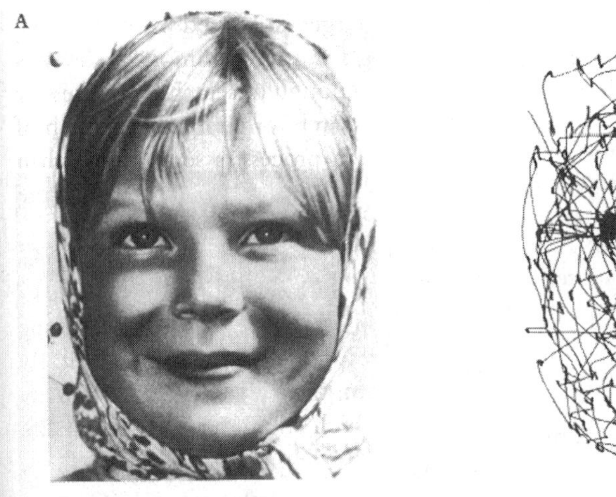

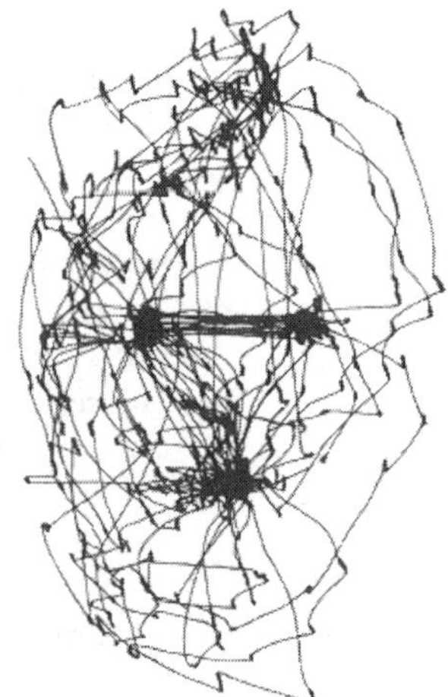

Fig. 1. Eye-movement records of a subject inspecting the photograph. The record shows that eye movements are directed towards the eyes and the mouth, which are the visually more informative regions. Reproduced from Yarbus, A.L. (1967) Eye Movements and Vision. *Plenum Press, New York.*

Divided attention

When observers are asked to *divide* their attention between two verbal or visual tasks, their performance depends on several factors. Tasks that use the same sensory modality are likely to interfere with each other. However, practice may bring about substantial improvement which may be due to increasing **automaticity**, practice in alternating rapidly between the attentional demands of the two tasks, or the development of new strategies for reducing the attentional load. This type of experiment has led to the idea that there is a limited **attentional resource** available for dual processing.

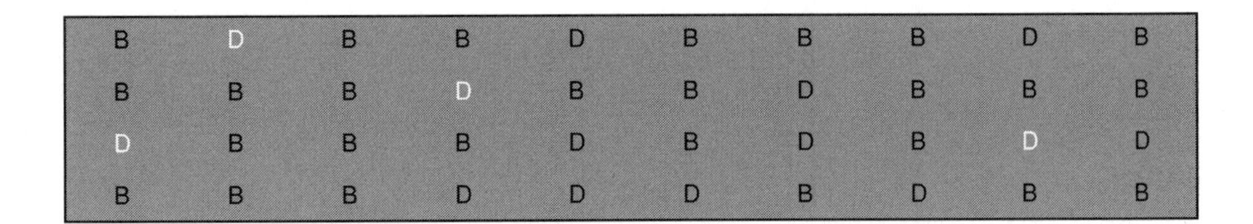

B D B B B B B B D B
B B B B B B B B B B
B B B B B B B B B D
B D B B B B B B B B

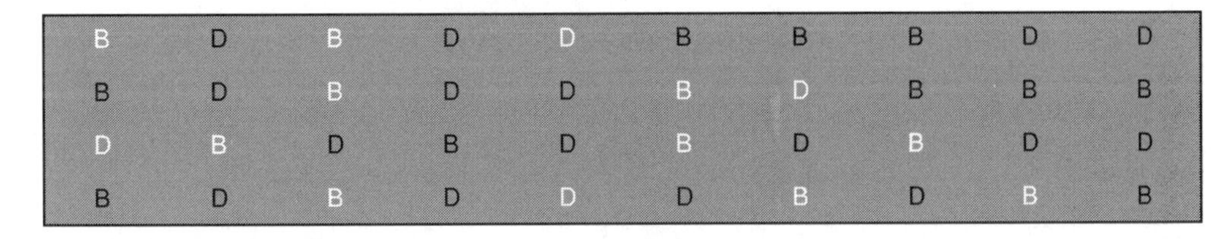

Fig. 2. An illustration of feature and conjunction searching. In each of the three blocks there are target letters to find: the target is a white D. The top and middle blocks require only feature searching, for brightness and shape respectively, to accomplish the search, while the bottom block requires a conjunction search of both features. A successful search will identify four targets in each block.

Theories of attention

As we have seen, attention operates to enable some information to reach consciousness while the rest is filtered out. Two types of theory have been proposed to explain how this is achieved, and neither of them is entirely satisfactory. The work on shadowing led to the **structural theory** that uses the core idea of a filter through which a restricted number of sensory inputs (perhaps only one) can pass at a time. The filter may operate very early on in the perceptual process (the **early selection** model) so that selection is made on the basis of physical parameters such as voice and speech rate. There are problems with this version of the theory in explaining how any information from the unattended inputs is ever processed. The **late selection** model holds that all inputs get some degree of processing, and that the final selection is made on the basis of meaning. This version of events does not account for electrophysiological evidence of early attentive processes that are insensitive to differences in meaning.

The work on divided attention has given rise to theories that use the notion of **attentional resource**. The basic concept here is of a central attentional capacity that can be deployed across tasks so that two concurrent tasks can be performed successfully only if their combined demands do not exceed the available resource. Furthermore, if the attentional demands of one task increase then the resource available for the other task will decrease. Experiments that manipulate the difficulty of concurrent tasks often, but not always, produce the predicted results. Unfortunately, the measurement of attentional resource and task difficulty are interdependent. The explanatory power of this idea is therefore limited, because it is not possible to define two tasks which are independently calculated to exceed an observer's attentional resource *in advance of the experiment* and then test the degree to which performance is affected.

E1 Associative learning

Key Notes

The biological function of learning
Although animals may have built-in instinctive adaptations it is also essential that they can learn about the environment. Associative learning is a basic form of learning, typically studied in lower animals; it is concerned with learning about regularities in the external world.

Habituation
Habituation is the simplest form of associative learning and involves learning to ignore uninformative stimuli.

Classical conditioning
Classical conditioning, originally studied by Pavlov, involves the association of an initially neutral stimulus (conditioned stimulus) with a biologically significant event (unconditioned stimulus). Such associations allow an animal to predict significant events.

The conditioned response
Classical conditioning is measured by the response (conditioned response) made to the conditioned stimulus. This is often similar to the response made to the unconditioned stimulus but is not usually identical to it. It indicates that the animal has made the association.

Extinction
If a conditioned stimulus is no longer followed by the unconditioned stimulus the animal ceases to produce the conditioned response and the association is said to have extinguished.

Generalization and discrimination
If an animal has learnt to respond to a conditioned stimulus, it will also respond, but to a lesser degree, to other similar stimuli; this is known as generalization. Alternatively, animals can be conditioned to one stimulus and not to another. In this case, they discriminate between them by responding to one and not the other.

Inhibition
Stimuli which have been associated with the absence of an unconditioned stimulus are not neutral. They become inhibitory stimuli and reduce the level of response to other conditioned stimuli.

Related topics
Instrumental learning (E2)
Mechanisms of learning (E3)

The biological function of learning

Animals live in a changing environment so they need to be able to modify their behavior to adapt to these changes. Some creatures, such as social insects, depend largely on built-in '**instinctive**' associations; but even they alter their behavior in response to environmental changes. An ant may bring food back to the nest by instinct, but it needs to learn topographical details in order find its way. Mammals, and in particular primates, have specializd in this type of adaptive behavior.

The scientific study of learning began about 100 years ago in Russia with Pavlov and in America with Thorndike. Although their approaches to the subject were very different, they shared the belief that there were basic laws which applied to all forms of learning in all species. The objective was to discover these laws, and the best way of doing so was to work with lower animals; dogs, rats or pigeons being the preferred species. Today, it is generally agreed that there are different types of learning process and that learning in different species may vary greatly. In particular, learning in humans is fundamentally affected by our conceptual and linguistic abilities. Nevertheless, much can be learnt about the fundamental processes of association by the study of lower animals.

It is clearly adaptive for an animal to learn from experience that a specific sort of rustle in the bushes signifies a meal, or that a berry of a particular appearance is unpleasant to eat. **Associative learning** is concerned with the way in which environmental events are related to one another in this fashion.

Habituation

If an animal is repeatedly presented with a stimulus which is not followed by any biologically significant event it learns to ignore the stimulus. This simple form of associative learning is called **habituation**. It is observed in virtually all species from sea anemones to primates. For example, most mammals respond with a startle reaction to a sudden loud tone, but this response wanes with repeated presentation if nothing follows the tone. Habituation enables an animal to ignore a previously novel stimulus which tells it nothing about the world. If the stimulus that has been habituated is changed (e.g., if the tone changes in frequency or duration) the original response may reappear; this is known as **dishabituation**. In higher mammals at least, habituation involves the formation of an internal representation of the habituated stimulus. Dishabituation results when the event in the external world fails to match this representation; dishabituation can therefore inform us about the nature of an animal's event representation. This has been used most effectively in studying preverbal human infants (Topic H2).

Classical conditioning

Classical conditioning is a form of associative learning originally studied by Pavlov. It is based on unlearnt reflexes. In such reflexes an **unconditioned stimulus** (UCS) reliably produces a response (the **unconditioned response** or UCR). An initially neutral stimulus, the **conditioned stimulus** (CS), is presented shortly before the UCS. In Pavlov's classic experiments with dogs, a buzzer (CS) was paired with presentation of food in the mouth (UCS) which reliably produced salivation (UCR). (This process of following the CS with the UCS is often termed *reinforcement*; this usage will be avoided here to avoid confusion with the different process of reinforcing instrumental learning; see Topic E2.) After a number of such pairings the CS alone came to elicit the response. Classical conditioning has been carried out using a wide variety of other UCSs and a large number of species from invertebrates to humans.

UCSs are nearly always biologically significant events; the function of classical conditioning therefore is to enable the organism to predict such important events as food or pain. This explains why some reflexes which have little biological significance, such as the knee jerk, are very hard to condition. Effective conditioning also occurs only when the CS *precedes* the UCS (the optimum interval depends on the particular situation). This is to be expected if classical conditioning reflects the *predictive* value of the CS. A CS which arrives at the same time as or after the UCS is not much of a predictor!

Since classical conditioning is concerned with the prediction of significant environmental events it works in much the same way whether the UCS is beneficial or painful for the animal; it is adaptive to be able to predict the event in either case.

The conditioned response

Pavlov assumed that conditioning was simply **stimulus substitution**. He believed that the conditioned response (CR) was the *same* response that had previously been elicited by the UCS and was now elicited by the CS. In this view, classical conditioning cannot result in any *new* responses, it can only allow existing responses to become attached to new stimuli.

Many CRs are, indeed, similar to the UCR. Generally, however, detailed analysis of apparently similar CRs and UCRs reveals subtle differences. For example, the salivation in preparation for food is very like that for food in the mouth; however, it is somewhat different in chemical composition. Sometimes the CR and UCR are quite different from one another. A rat's UCR to the UCS of electric shock to the feet is to jump around and squeal; the CR to a stimulus predicting shock (CS) is to cower and become immobile.

The main function of classical conditioning is to detect regularities in the world. The critical events are internal associations; the responses that are measured are merely indices of these central events. If this is the case there is no reason to expect the CR to be the same as the UCR, though since it is a preparatory response it is often similar to it.

The strength of the CR may be measured by:

- amplitude of response, e.g. drops of saliva;
- probability of response: the likelihood that a CS will elicit a CR;
- latency: the longer the delay in occurrence of the CR the weaker it is.

Extinction

If, following conditioning, the CS is presented and is not followed by the UCS, the CR will decrease and eventually cease. The association is said to have **extinguished**. There are regularities in the environment, but the environment also changes over time; extinction is an indication of the way in which an organism tracks these changes.

Extinction does not entirely eliminate the association. This is shown by two results:

- retraining of a response after extinction is more rapid than the original training so some residual association must remain;
- on retesting following a rest period an extinguished CR will typically reappear. This is known as **spontaneous recovery** (*Fig. 1*).

Generalization and discrimination

If an animal is conditioned to respond to a particular CS, say a tone of 1000 Hz, and is then tested with tones of other frequencies, it will also respond to them, but to a lesser degree than to the original CS. This is known as **stimulus generalization.** The larger the difference between the CS and the test stimulus the less the response will be. This produces a **generalization gradient** (*Fig. 2*). Generalization is not the result of the animal failing to distinguish between the CS and the test stimulus, it is rather a reflection of the fact that similar stimuli in the environment will usually have similar consequences. There would be little value in learning only about stimuli *exactly* the same as the CS since two natural events are rarely precisely the same.

There are occasions when two similar stimuli have *different* consequences; one berry may be good to eat, a similar looking one may not. Here, it is important

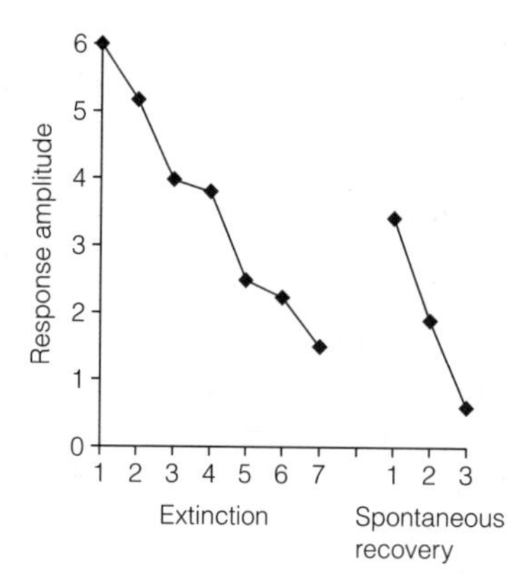

Fig. 1. Extinction and spontaneous recovery of a conditioned response.

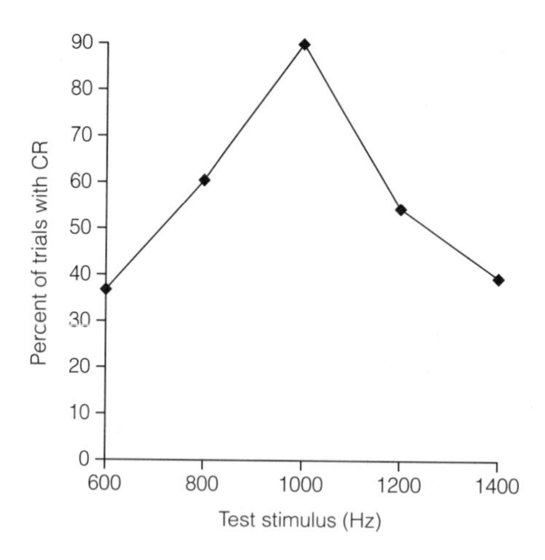

Fig. 2. Generalization gradient of a response conditioned to a tone of 1000 Hz.

that the animal can **discriminate** between the two. In discrimination training the animal is initially conditioned to respond to a CS, say a tone of 1000 Hz. Once the CR is well established, another similar stimulus (e.g., a tone of 850 Hz) is presented but not followed by the UCS. After a number of trials, the animal produces a CR in the presence of the CS but none to the other stimulus; it is said to have discriminated between the two stimuli.

Inhibition

Following discrimination training an animal will respond to the CS but not to the other, negative, stimulus. However, this stimulus is not neutral. If it is presented at the same time as a third independent CS (which has previously been conditioned using the same UCS) the response to this third stimulus will be markedly reduced. The animal has learnt that the negative stimulus predicts no reinforcement; it therefore tends to inhibit the response to other CSs. The stimulus has become a **conditioned inhibitor**. The characteristics of such inhibitory stimuli are not fully understood, but it is clear that inhibitory learning is an important part of associative learning.

E2 INSTRUMENTAL LEARNING

Key Notes

Instrumental learning

Whereas classical conditioning deals with the detection of regularities in the environment, instrumental learning is concerned with the effects of an animal's responses on that environment.

The Law of Effect

The Law of Effect states that responses which are followed by positive consequences are more likely to be performed in the future. It provides a simple mechanism for trial-and-error learning.

Reward and reinforcement

Positive reinforcement consists of the presentation of a reward *or* the removal of an aversive stimulus. Negative reinforcement involves removal of a reward or presentation of an aversive stimulus. Stimuli consistently associated with a primary reinforcer become conditioned reinforcers capable of reinforcing behavior themselves.

Schedules of reinforcement

Instrumental behavior can be maintained when only a proportion of responses are reinforced; extinction may be slower following such training than following continuous reinforcement. The four best known schedules of reinforcement are: fixed ratio, fixed interval, variable ratio, and variable interval. Instrumental behavior can be shaped by a process of selective reinforcement so as to produce novel responses. Animals can also learn to make different instrumental responses to different stimuli.

Punishment, escape, and avoidance

Responses may be punished by following them with a negative reinforcer such as electric shock. Alternatively, animals may be trained to escape or avoid negative reinforcers. Avoidance behavior is highly resistant to extinction; it may provide a model for human phobias.

Related topics

Approaches to motivation (B1)
Associative learning (E1)
Mechanisms of learning (E3)
Complex learning (E4)

Instrumental learning

Associative learning is concerned with what leads to what in the environment. However, in order to deal effectively with that environment, an animal needs to act on it. The way in which it learns to do this is known as **instrumental learning**. When a dog is trained to 'beg' for food it is showing instrumental learning. The essential feature of such learning is that the outcome depends on the animal's behavior. If it begs it will get fed, but no begging and no food.

By contrast, in classical conditioning, the animal is learning the connection between the conditioned and unconditioned stimuli (CS and UCS), two environmental events. Here, the presentation of the UCS will occur whether or not the animal makes a conditioned response. (See Topic E1.)

The Law of Effect

Instrumental learning was first systematically studied by Thorndike in the USA about 100 years ago. Thorndike proposed **The Law of Effect** to explain this form of learning. This states that responses which are followed by 'satisfying states of affairs' are subsequently more likely to occur than those which are not. In an unfamiliar situation, an animal will initially produce a wide range of responses, some based on innate reflexes, some on previous learning. If one of these responses eventually results in a 'satisfying state of affairs' it will be more likely to occur on future occasions. In this way, experience will ensure that the effective response will be strengthened until it is the dominant one in the animal's repertoire. This provides a purely mechanical procedure accounting for instrumental learning. The analogy with Darwinian natural selection is obvious.

Thorndike studied the behavior of hungry cats getting out of a fairly complex piece of apparatus known as a puzzle box. Later experimenters simplified the situation, using rats in mazes to study various aspects of instrumental learning. The simplest apparatus of all is the **Skinner box** (named for its inventor B.F. Skinner), in which the animal, typically a rat or a pigeon, presses a bar or pecks a key in order to obtain food. In this radically simplified situation, Skinner was able to demonstrate a number of impressive regularities in what he termed **operant behavior.**

Instrumental learning results in many phenomena parallel to those found in classical conditioning (Topic E1). These include extinction, spontaneous recovery, and stimulus generalization.

Reward and reinforcement

Food for a hungry animal and water for a thirsty one are examples of **appetitive stimuli**; such stimuli are ones which an animal will generally work to obtain. Other stimuli, such as electric shock or other painful events, are **aversive stimuli**; the animal will work to avoid or remove such stimuli. **Positive reinforcement** of a response occurs when it is followed by the presentation of an appetitive stimulus *or* removal of an aversive one. **Negative reinforcement** involves the presentation of an aversive stimulus or removal of an appetitive one.

It is not necessary to make these distinctions in classical conditioning since the UCS occurs whether or not the animal responds. The animal is learning about regularities in the environment, not how to act on that environment.

Food and water are primary positive reinforcers, but stimuli which are regularly associated with primary reinforcers become reinforcers in their own right. They are known as **conditioned** or **secondary reinforcers**. Conditioned reinforcers have been demonstrated in numerous experiments with a variety of species; from these experiments it is evident that they are established by classical conditioning. Money and social approval are standard examples of conditioned reinforcers for human beings.

Schedules of reinforcement

If every response, say a bar press in a Skinner box, is rewarded, the animal is on a **continuous reinforcement** schedule (CRF). However, behavior can be very effectively maintained if only some responses are rewarded. It is generally found that a partially rewarded instrumental response is *more* resistant to extinction than responses which have been continuously rewarded: the **partial reinforcement effect (PRE)**. This is quite different from classical conditioning, where intermittent presentation of the UCS results in poor learning and fast extinction.

Skinner studied four major **schedules of reinforcement**:

- **Fixed ratio (FR)**: the animal is rewarded only after a fixed number of responses. This produces high rates of responding with postreinforcement pauses when the animal stops responding altogether. Extinction is relatively rapid.
- **Fixed interval (FI)**: the first response after a fixed period of time is rewarded. Eventually, the animal learns the length of the interval and responds only towards the end of it. Extinction is relatively rapid.
- **Variable ratio (VR)**: the animal is rewarded *on average* after a certain number of responses, say 40 (VR40); however, which particular response will be reinforced is unpredictable. This results in high rates of responding and much resistance to extinction.
- **Variable interval (VI)**: Reinforcement is given, say, every 30 sec (VI30), but the actual intervals may vary from a few seconds to over a minute. This produces fairly fast and very stable rates of responding and is very resistant to extinction.

Responses (operants) other than bar pressing are subject to the same laws of learning. One-armed bandits, for example, are programmed on a variable ratio schedule, and produce high and prolonged rates of response in human gamblers. Skinner claimed that any response, however complex, could be explained in terms of the same basic principles of operant conditioning. It is unlikely that this is so, particularly for complex human behavior such as language learning.

Shaping involves selectively rewarding behaviors in the animal's repertoire which approximate to the desired response. Thus, to get a pigeon to turn in circles you would initially reward it *only* when it turned 10° to the right; once it was regularly turning this far the criterion for reward would be increased to 20° and so on, until the pigeon was turning round in circles. By this technique, pigeons have been trained to play ping-pong or direct guided missiles; shaping demonstrates that instrumental learning permits the emergence of *new* responses.

Instrumental responses are not directly produced by an external stimulus, in the way that a CR is produced by the CS. The animal makes the instrumental response 'spontaneously'. The CS is said to **elicit** the CR, but an instrumental response is **emitted** by the animal. Of course, the animal's surroundings affect which responses are emitted; the cat in the puzzle box or the rat in the maze behave quite differently from the way that they do in their home cages. Using instrumental procedures, different responses can be trained to occur according to the stimulus situation. A pigeon in a Skinner box may be reinforced for pecking a key lit by a green light, but not when the light is red. After training, the bird will respond only in the presence of the green key. The response has come under **stimulus control**. Many of our everyday responses are under stimulus control in this way; consider your behavior at a road crossing. This approach to instrumental learning is known as S–R theory because it assumes that what is learnt is a mechanical association between a stimulus and a response.

Punishment, escape, and avoidance

If a response is followed by the presentation of an aversive stimulus it is less likely to occur in the future. **Punishment** training of this sort specifies what response the animal should *not* make, but it does not identify what response it *should* make. Punishment may be effective in removing unwanted behavior; it is not useful in training alternative behaviors. The applications to everyday life are obvious.

If a response is followed by the *removal* of an aversive stimulus it is positively reinforced; this is **escape** learning. For example, a dog may readily be trained to jump over a hurdle to escape a shock to its feet.

If the shock is preceded by a warning signal, say a buzzer, the dog will learn to respond during the buzzer and so avoid the shock (*Fig. 1*). During the escape phase of the training, the buzzer has been regularly associated with the shock, and has therefore become a conditioned aversive stimulus. The avoidance response removes the buzzer (and prevents the shock); removal of an aversive stimulus is positively reinforcing, and so the avoidance response continues to be strengthened even in the absence of shock. But, once the dog is regularly avoiding shock, the buzzer is no longer paired with shock and should lose its conditioned aversive properties. It will then no longer maintain the avoidance response which should extinguish for lack of reinforcement.

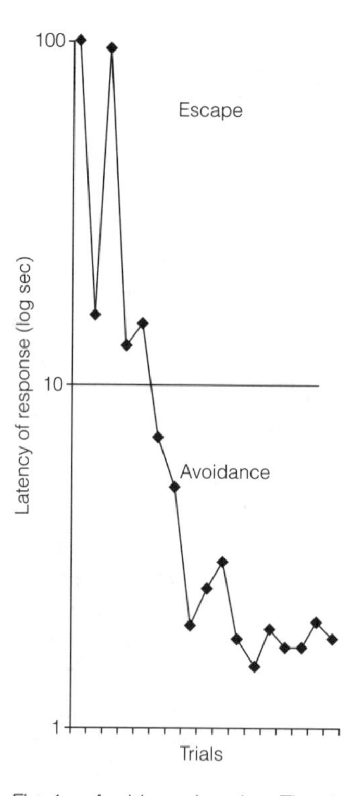

Fig. 1. Avoidance learning. The shock is switched on 10 sec after the warning signal. Responses with a latency of greater than 10 sec are therefore escape responses; those with shorter latencies are avoidance responses.

In fact, avoidance responding is very resistant to extinction. A dog will continue to avoid shock for many hundreds of trials. This can best be explained in terms of **expectancies** (Topic E3) rather than S–R learning. Following the warning signal, the dog expects the shock *if it has not made the avoidance response.*

This expectancy is never disconfirmed so long as it continues to avoid, and so the avoidance behavior is maintained indefinitely. If, however, the dog is prevented from making the avoidance response and no shock is delivered, the response can be rapidly extinguished.

It has been suggested that human phobias may be the result of maladaptive avoidance learning (Topic L4). Once learned, such behaviors are self-perpetuating, since the person never exposes himself to the feared situation and therefore never has the chance to extinguish the fear. Two forms of behavior therapy have been used to cure such phobias. In **flooding**, the patient is forced to remain in the presence of the feared stimulus until the fear extinguishes; this is obviously a potentially dangerous procedure. An alternative is **systematic desensitization** in which the feared stimulus is presented in a very mild form until the patient feels no fear of it; this is repeated with increasingly intense examples of the stimulus until eventually the patient experiences no fear (Topic L7).

E3 Mechanisms of learning

Key Notes

The generality of laws of learning

In both classical conditioning and instrumental learning, some associations can be made more easily than others. Studies of taste aversion show that taste can readily be associated with feeling ill, but not with a light or sound. Instrumental responses which can easily be learned to obtain food cannot be learned as avoidance responses.

Associative learning and contingency

Classical conditioning does not depend on simple pairing of conditioned and unconditioned stimuli (CS and UCS); it requires a contingency between them. The CS must predict the occurrence of the UCS. Contingency learning is mediated by a simpler process (blocking) which determines that CSs must be informative stimuli.

S–R theory and instrumental learning

The simple view that instrumental learning involves the stamping in of stimulus–response (S–R) connections by reinforcement is challenged by a number of findings. Responses cannot be defined in physical terms, but rather by their effects on the environment. Learning can, on occasion, occur in the absence of reinforcement. What is learned can best be explained in terms of act–outcome expectations rather than S–R bonds.

Learned helplessness

Animals exposed to unavoidable shock learn that their behavior has no consequences; they are subsequently unable to acquire avoidance responses. This learned helplessness may be a model for some forms of human depression.

Related topics

Associative learning (E1)
Instrumental learning (E2)
Complex learning (E4)

The generality of laws of learning

Early investigators assumed that any conditioned stimulus (CS) could be associated with any unconditioned stimulus (UCS), and that any instrumental response could be strengthened by any reinforcer. This simplifying assumption is known as the **equipotentiality principle**. It is now known that not all associations are equally easy to learn.

The clearest demonstration of this in classical conditioning comes from studies of **learned taste aversion**. In a typical experiment, thirsty rats were placed in a box where they could lick a spout to obtain water distinctively flavored with saccharine (*Table 1*). As the water was delivered, a light flashed

Table 1. Procedure for experiment on learned taste aversion

	Shock		X-ray treatment	
Test	Light + buzzer	Saccharine	Light + buzzer	Saccharine
Outcome	Avoidance	No avoidance	No avoidance	Avoidance

and a buzzer sounded ('bright, noisy' water). One group of rats were then given brief electric shocks immediately after they had started licking. A second group were given doses of X-ray radiation sufficient to make them ill. In both groups the CSs of light, sound and taste had been followed by an aversive UCS so, according to the equipotentiality principle, conditioning could be expected to take place. The rats were tested by giving them *either* unflavored 'bright, noisy' water *or* saccharine-tasting water without the light and sound. The rats for which the X-ray had been the UCS avoided licking under the saccharine condition but not the 'bright, noisy' condition; they had associated feeling ill with the taste of the water, but not with the external stimuli of light and sound. This is known as taste aversion. By contrast, the shocked group were undeterred by the taste of saccharine but avoided the 'bright, noisy' water. They had learned that the light and sound predicted shock but had made no association with the taste.

Thus, a novel taste can readily be associated with feeling ill, but not with external events such as shock to the feet. External events such as sounds and lights can also be associated with shock, but not with feeling ill. Furthermore, while in most forms of conditioning the optimum CS–UCS interval is a second or two, taste aversion conditioning is most effective when X-ray exposure follows the CS by about an hour (Topic B3).

This makes good ecological sense. Aversive events in the outside world are often preceded by visual or auditory stimuli, but not by distinctive tastes. On the other hand, tastes may well be connected to food which makes us feel ill an hour or two later. The equipotentiality principle does not apply to classical conditioning. Some CS–US associations are more easily learned than others.

The position is similar for instrumental learning. For example, pigeons can easily be trained to peck a lighted key for food, but it is virtually impossible to train them to do so to escape or avoid shock, although they can learn other responses to do this. This is because pecking is a bird's natural response to food, whereas it is very remote from the natural responses to shock to the feet. Where the lighted key signals water, pigeons will learn to peck, but the response resembles the normal drinking response. There are numerous other examples which show that, just as in classical conditioning, the ease of instrumental learning depends on the particular stimuli, responses and reinforcers involved.

Associative learning and contingency

Pavlov and other early investigators believed that *pairing* of the CS and the UCS was sufficient to establish a conditioned reflex. However, whenever a UCS is presented, many stimuli other than the CS will also be present; for example, the appearance of the room in which the experiment is taking place, internal stimuli, such as the beating of the heart, and many others. These background stimuli do not become CSs. This is because they do not *predict* the occurrence of the UCS; they are present whether the UCS is about to occur or not. The CS is uniquely predictive because the UCS occurs following the CS but *does not occur* in the absence of the CS. The UCS is said to be **contingent** on the CS, but it is merely paired with the background stimuli. Classical conditioning can be shown to depend on contingency, not pairing.

The mechanism by which contingency learning occurs depends on a simpler phenomenon known as **blocking**. In a typical experiment, a group of rats (the experimental group) are given a number of pairings of a buzzer with shock; in this stage, a control group receive no pairings (*Table 2*). In the next stage, both groups receive pairings of a compound stimulus of buzzer + light with shock. Both groups are then tested for their response to the light alone. The control

Table 2. Procedure for experiment on blocking

Group	Stage 1	Stage 2	Test	Outcome
Blocking	Buzzer with shock	Buzzer + light with shock	Light only	No CR
Control	–	Buzzer + light with shock	Light only	CR

group give strong CRs to the light, showing that, during stage 2, they learnt the association with shock, but the experimental group show little if any response. If conditioning depended only on *pairing* of CS (light) and UCS (shock), both groups should have learnt equally well during the second stage. However, the experimental group had already learnt that the buzzer predicted the shock. In the second stage, they were already expecting the shock because of the buzzer; the light was therefore a redundant stimulus providing no information. Such studies suggest that an association is only learnt where a CS is informative about a UCS. Background stimuli are not informative and so do not become effective CSs. Classical conditioning, therefore, is not a process of stimulus substitution (Topic E1), nor does it depend on the simple pairing of CS and UCS.

S–R theory and instrumental learning

According to stimulus–response (S–R) theory, the animal produces a response in the presence of a stimulus, and this connection is strengthened if it is followed by reinforcement. There are a series of problems with this simple mechanical account.

Definition of a response

If a rat is rewarded for pressing the bar in a Skinner box with its left paw, this particular response is reinforced. But if it is prevented from using its left paw, it will use its right one or even its nose; it seems that reinforcement strengthens any response which results in the bar being depressed. The response cannot therefore be defined purely by its physical characteristics but rather by its effect on the environment; this is not in line with the simple assumptions of S–R theory. The problem of response definition is particularly acute in the case of complex behaviors such as human speech.

Need for reinforcement

Rats given a chance to explore a maze in the absence of reinforcement will, when food is introduced, learn the maze more rapidly than rats without previous experience. This type of **latent learning** shows that, although obviously important, reinforcement is not essential for learning.

Expectancies

S–R theory assumes that the S–R bond is purely mechanical. When a rat learns to press a bar in a Skinner box it is simply learning to make a response in the presence of certain stimuli. An alternative view would be that the animal has learnt an **act–outcome association**; that a particular act, bar pressing, leads to a particular reward. One way of testing which approach is correct is to change the value of a reward *after* learning has taken place. Rats were trained to press a bar for one distinctive tasting food or pull a chain for another. A taste aversion for one of these foods was then established by giving it to the rats and then injecting them with a toxin which made them ill. When they were returned to the original learning situation, they avoided making the response that had produced the food which was now distasteful to them. Their responding had been affected by their changed expectation of the reward. They had acquired act–outcome associations rather than S–R bonds.

These results combine to show that the S–R explanation of instrumental learning is inadequate. What an animal learns about responses, the effect of its actions on the world, is not wholly dependent on reinforcement and seems to be best described in terms of act–outcome associations (internal representations) rather than S–R bonds.

Learned helplessness

In instrumental learning the response produces the reinforcement, so there is bound to be a contingency between the two. However, animals can also learn that there is *no* contingency between their behavior and events; that what they do makes no difference. This is known as **learned helplessness.** In a typical experiment, two groups of dogs were restrained and given unavoidable shocks. The dogs in one group were able to terminate a shock by pressing a pedal. The dogs in the other group had no control over the shocks they received, although the number and duration of the shocks was exactly the same as for one of the dogs in the other group. Following this training, both groups were placed in an avoidance situation: a buzzer signalled shock and, by jumping over a hurdle, the dogs could escape or avoid the shock. The dogs who had been able to escape the shock in the first phase of the experiment learned to avoid with ease, but those who had experienced unavoidable shock failed to escape or avoid the shocks, instead they were passive, lay down, and whined. They had learned that their behavior was ineffective in controlling shock, and this learning had carried over to the avoidance situation.

It has been suggested that learned helplessness provides a model for some forms of human depression (Topic L3). Be this as it may, it is clear that animals can learn that there is no contingency between their actions and aversive events, and that this can have a major effect on behavior.

E4 Complex learning

Key Notes

Limitations of trial-and-error learning
Trial-and-error learning is an important mechanism in animals and humans, but there are a variety of more complex forms of learning, even in animals, which cannot be accounted for in this way.

Learning to learn
Animals can transfer learning from one situation to another. A simple example is serial reversal learning in which successive reversals of a discrimination are learnt with fewer and fewer errors. More complex cases are learning sets and matching to sample. Such behavior can only be explained if animals are responding, not merely to external stimuli, but also to internal representations of events.

Cognitive maps
Both in natural situations and in the laboratory a variety of species, from insects to apes, show behavior that can best be explained by assuming that it is controlled by an internal representation of their environment.

Insight
Many standard experimental situations provide little opportunity for anything other than trial-and-error learning. In some circumstances, however, apes may solve problems by what has been described as insight; an internal re-structuring of the problem leading to a solution.

'Language' learning
Several studies have shown that chimpanzees can learn over 100 arbitrary symbols to represent object and actions. Whether or not this constitutes language learning, it is a form of complex learning going well beyond simple trial and error.

Related topics
Associative learning (E1)
Instrumental learning (E2)
Mechanisms of learning (E3)
Reading (G3)
Learning to talk (G5)

Limitations of trial-and-error learning

Instrumental learning, as described in Topics E2 and E3, depends on trial-and-error learning. Most psychologists, while accepting that such an account may be adequate to explain some types of behavior, believe that animals are capable of more complex forms of learning. A variety of forms of learning in animals seem to be inexplicable in terms of trial and error; these types of learning appear to depend on learning rules, and on internal representations of abstract characteristics of events.

Learning to learn

Humans are certainly able to transfer learned skills from one situation to another. This implies that behavior can be controlled by internal representations, for example memories of what has already been learned, as well as by the external stimuli acting at the time. There is also good evidence that animals can learn to learn.

The simplest example of this is **serial reversal learning**. Rats are initially trained to go to a white rather than a black box for food. When they have learned to do this without error the food is switched to the black box. The animals make a large number of errors before they learn this first reversal. Once they have done so, the reward is switched once again to the white box, and so on. They are required to switch preference repeatedly from one box to the other. After about 10 such reversals, they make only one or two errors on each occasion. In analogous experiments, monkeys learn to make only a single error on each reversal. Simple trial-and-error learning cannot explain why successive reversals become easier. The animals must be responding on the basis of their memory of previous trials.

Learning sets provide a more complex example. Here, a monkey is trained to select one of two stimuli in order to get food. After a fixed number of trials, these stimuli are set aside and a new pair is introduced. This continues over many days for several hundred different sets of novel stimuli. Eventually, the monkeys reach perfect performance where, for a new pair of stimuli, they pick one at random; if it is correct they continue to select it, if it is wrong they immediately switch to the other stimulus. It appears that they have learnt a **rule** of the form *win–stay, lose–shift*. This must involve remembering the outcome of previous trials and allowing this to control choice on the current trial. Once again internal representations, not external stimuli, are controlling behavior.

Another example is **matching to sample**. Here, the animal is presented with three stimuli, a sample (target) and two discriminative stimuli, one of which is the same as the sample (*Fig. 1*). The task is to select the discriminative stimulus which matches the target. In this procedure, the correct response changes from trial to trial depending on which target is presented. A wide range of animals, from pigeons to primates, can learn to match to sample. It seems that the animals appreciate that the target and the positive stimulus are the same, and this relationship controls their choice. To test whether animals are really responding to similarity, the procedure can be modified by introducing new sets of stimuli. There is good evidence that primates, though probably not rats or pigeons, can learn to transfer in this way; given enough training and enough different sets of stimuli they can respond correctly on the first trial when presented with a completely novel set. This can only be done if they can represent and respond to 'sameness' as an abstract concept.

These types of experiments demonstrate that, in varying degrees, animals can respond not merely to the stimuli present at the time, but also to abstract characteristics of a situation.

Fig. 1. Matching to sample. In this procedure, either stimulus can be correct in either position. The animal has to learn to respond on the basis of same/different.

Cognitive maps

The S–R approach assumes that an animal in a maze simply responds to particular stimuli at the choice points; it has no internal representation of its environment. Many investigators have suggested that animals when learning a maze may represent it internally as a **cognitive map**.

In the wild, many animals such as nest-building insects, migrating birds, and a variety of mammals are adept at finding their way around their environment. These abilities suggest the existence of internal representations of the environment which have some of the characteristics of a plan or aerial view of the space being represented. There is good evidence that human beings have this capacity, but it is difficult to obtain convincing experimental evidence in animals. One example is the radial maze (*Fig. 2*). The rat is placed at the center: there is a food pellet at the end of each arm, and the rat is left in the maze until it has eaten all the pellets. Even in the early trials, rats rarely revisit an arm (as S–R theory would predict that they should after reinforcement); in due course, they learn to visit each arm once and once only. This cannot be trial-and-error learning; it is most easily explained by the rats having learnt a cognitive map of the maze.

This sort of experimental finding, combined with observations of animals in the wild, provides good evidence that some species at least can guide their behavior by means of cognitive maps.

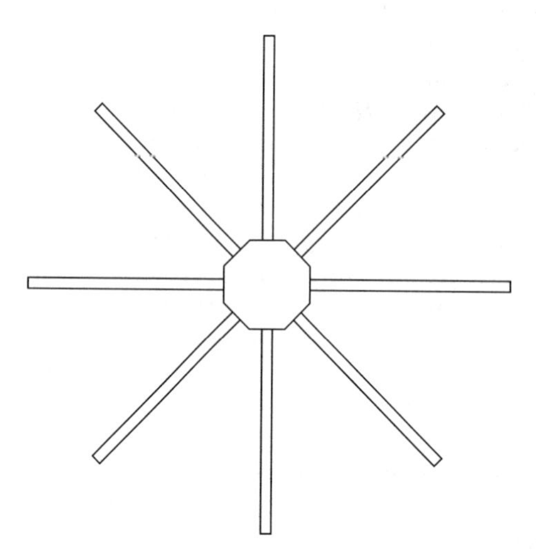

Fig. 2. Radial maze, used to test for the use of cognitive maps in rats.

Insight

Trial and error may be the only way of learning to escape from a puzzle box or negotiate a maze. Such situations provide little opportunity for real problem solving. In other situations, animals may show more complex behavior.

Köhler, a Gestalt psychologist, believed that apes at least were able to solve problems by **insight**. He believed that they would do so where all the relevant features of the situation were present to the animal. He suggested that, as in human problem solving, solutions might be reached by an internal re-structuring of the problem. He found that chimpanzees would use sticks as rakes to pull in food which was out of reach, or would move boxes in order to climb up to food suspended from the roof of the cage. He reported that, in solving these problems, there was typically a period when the animal responded incorrectly; this would often be followed by a period of inactivity. Then, the chimpanzee would produce the correct response in a smooth and organized fashion, and would make few errors on similar problems in the future. This is certainly unlike trial-and-error learning and is similar to human problem solving. It has been shown that chimpanzees will only solve problems in this way if they have had some previous experience of playing with sticks, but this does not undermine the insightful character of the problem solution.

It has proved difficult to provide scientifically convincing evidence for insightful behavior, although few with experience of the behavior of higher animals will doubt that it does occur.

'Language' learning

There have been many attempts to teach animals to use human language. The most convincing have been with chimpanzees who have been taught versions of human sign language or to use abstract shapes as 'words'. (The vocal tracts of the great apes are not suited to producing the sounds of human language.)

For example, Washoe, a chimpanzee who was taught sign language from infancy, had learned about 130 signs by the age of four; these included words for familiar objects and actions. Washoe produced these signs spontaneously and sometimes in combination. Another chimpanzee, Sarah, was taught to use metal-backed plastic shapes which could be attached to a magnetic board. Sarah also learnt about 130 words, and could use them consistently to answer questions presented in the form of a sequence of these symbols. It is clear that she understood the relation between the symbol and what it stood for. For example, she learnt the symbol for 'brown' by being given the sentence 'brown, color, chocolate', without any chocolate being present.

The interpretation of these and other similar studies has been very controversial. The chimpanzees' vocabularies were tiny compared to those of human children and it is doubtful whether they showed any understanding of syntax. However, it is clear that, given sufficient training, apes can learn the use of at least 100 symbols representing a variety of concepts. This is clearly a form of complex learning which cannot readily be accounted for by the principles discussed in Topics E1–3.

Conclusion

There are many situations in which animals, particularly primates, show learning abilities which cannot be accounted for in terms of trial-and-error learning. The degree to which they will display these abilities will depend on the species of animal, the nature of the learning situation and the animal's previous experience.

F1 ENCODING IN MEMORY

Key Notes

Approaches to memory

Memory concerns the processes that record and allow later access to events, people, and information. Ebbinghaus studied memory as a simple cognitive function, using meaningless nonsense syllables. Bartlett used stories to examine the role of meaning in memory. The information processing approach likens memory and other cognitive functions to a computer. A number of theorists have suggested that memory has two or more stages, at core a short-term store and a long-term store.

Short-term memory

Short-term memory (STM) has a limited capacity of about seven items, coded either by acoustic features or articulatory activities. Information is rapidly lost from STM, either through decay (the gradual deterioration of the memory trace with time) or displacement (items are 'pushed out' of STM by new material). Rehearsal maintains information in STM. The information capacity of STM may be increased by recoding information into larger chunks, usually by assigning meaning to sequences of items. The working memory model suggests a number of independent subsystems: a general purpose central executive, a visuospatial scratch pad, an acoustic store, an input register, and an articulatory loop. Anterograde amnesia, in the absence of retrograde amnesia, demonstrates the separate existence of STM and LTM.

Long-term memory

It has been proposed that rehearsal transfers information to the long-term memory (LTM), where it is encoded semantically. Transfer to LTM requires processing for meaning. The level of processing approach argues that memory depends on how deeply material is processed, although it might be that establishing the distinctiveness of information or organizing it (for example with mnemonics) causes retention.

Related topics

Retrieval from memory (F2)
Memory loss (F3)
Disorders of personality, identity, and memory (L6)

Approaches to memory

Memory concerns the processes that record and allow later access to events, people, and information. Memory involves acquisition (or **encoding**) of information, its **storage**, and its **retrieval**. Each of these essential aspects of memory has been studied for well over a century, and we can distinguish two broad approaches. In the late 19th century, Herman Ebbinghaus set out to study memory as a simple cognitive function, uncomplicated by the meanings of the material to be remembered, or the expectations or characteristics of the person remembering. To this end, he used lists of **nonsense syllables** which were pronounceable sequences of letters; specifically, **consonant–vowel–consonant trigrams**, such as MAB, LOD. The idea was that these items, being meaningless, could not be associated with each other, or with anything the person already knew. Ebbinghaus's approach influenced research for many decades and

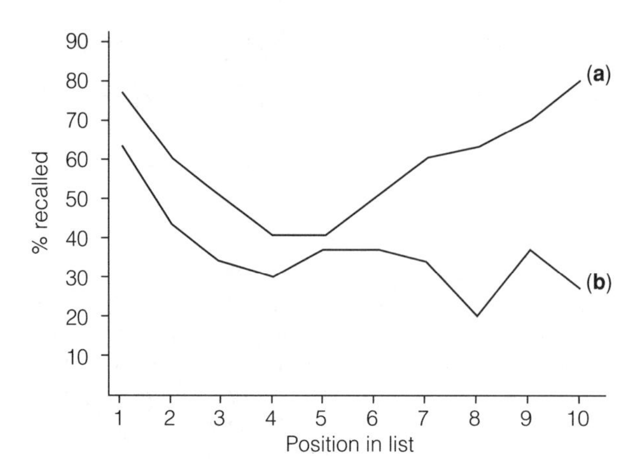

Fig. 1. (a) The serial position curve. Memory is best for items early in a list (the primacy *effect) and late in the list (the* recency *effect). (b) Delaying recall by 30 s after the end of the list selectively reduces the recency effect. Data from Postman, L. and Phillips, L.W. (1965) Short-term temporal changes in free recall.* Quarterly Journal of Experimental Psychology, ***17:** 132–138.*

revealed some basic properties of memory. One of these is the **serial position effect** (*Fig. 1*). When a list of items is presented for learning, we are more likely to recall items early in the list (**primacy effect**) and items at the end of the list (**recency effect**) than items in the middle of the list.

The other broad approach to memory views it not as a passive recording process but as an active, organizing process. This approach is usually taken as deriving from the work of Frederick Bartlett in the 1930s. Bartlett used material rich in meaning, typically stories, and examined the nature of the omissions and errors made when people retold the story. For Bartlett, active memory was characterized by **effort after meaning**, and Ebbinghaus's approach could not tell us much about real memory in everyday life. In the second half of the 20th century, memory came to be viewed within the general cognitive framework of **information processing**, which combined aspects of the two earlier approaches. Cognition was likened to the operations of a computer that receives, manipulates, stores, and outputs information. At the same time, it allows for previous knowledge and other sensory inputs to affect what is in the memory in various ways.

The idea that memory operates in different stages goes back at least to William James in the 19th century. He proposed a primary memory that stores transient memories, and a secondary memory that stores information over long periods. An elaborated extension of this was the **modal model** proposed in the 1960s by Atkinson and Shiffrin (*Fig. 2*). Information is briefly saved in 'raw' form in an appropriate **sensory register**. Research on the visual sensory register (the **icon**), in which people were cued to make partial reports of presented stimulus arrays, showed that much more information is available from the sensory register immediately after stimulus presentation than is available a few seconds later.

Short-term memory

Acts of attention transfer some of this information to a **short-term store** (or **short-term memory, STM**). This is viewed as a having a limited capacity: presented with a list of items, most people can recall around seven of them (the

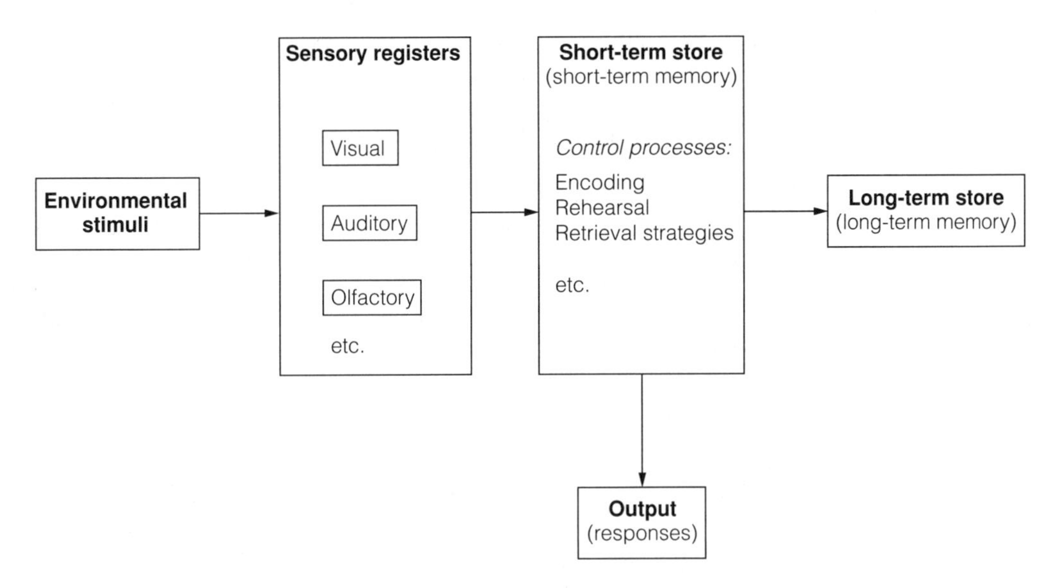

Fig. 2. The modal model of memory. After Atkinson, R.C. and Shiffrin, R.M. (1968) Human memory: A proposed system and its control processes. In: The Psychology of Learning and Motivation: Advances in Research and Theory *(ed. Spence, K.W.), Vol. 2. With permission from Academic Press, New York.*

memory span, or **digit span** if the items are digits). Items are stored in STM in verbal form, coded either by sounds (**acoustic features**) or by the movements required to voice the sounds (**articulatory activities**), and *not* by meaning. Information is rapidly lost from STM. Two processes have been proposed for this. **Decay** is the gradual deterioration of the **memory trace** with time. **Displacement** occurs when an item is 'pushed out' of STM by a new item. Material can be kept in STM by the process of **rehearsal**: the repetition (aloud or silently) of the item. This allows us, for example, to remember a telephone number between looking it up and dialing it. Preventing rehearsal causes loss from STM. The recency effect is reduced or removed by delaying recall after the presentation of the list, especially when the person is given a task such as backward counting to perform (*Fig. 1*, lower curve). This suggests that the recency effect is due to recall from STM, and that it can be maintained by rehearsal unless this is prevented by performing an interfering task.

The limited capacity of STM is largely independent of how much information is in each item. We can recall about seven digits (each selected from 10), seven letters (each selected from 26), or seven animals (from hundreds). We can increase the information capacity of STM by changing (**recoding**) the incoming information into larger **chunks**. Such recoding can be structural, like the practice in some European countries of writing and saying telephone numbers as a series of two-digit numbers (e.g. 24.18.37.87, giving four items for recall, rather than eight). But recoding usually involves assigning meaning to or recognizing meaning in sequences of items. Thus, for example, we can recall sentences of more than seven words because the words occur in predictable orders and form meaningful groups.

Alan Baddeley and Graham Hitch, in 1974, proposed a more complex view of STM: the **working memory model**. This model suggests a number of independent subsystems. Giving people a six-digit recall task, together with one of a variety of cognitive tasks such as problem solving, causes a deterioration in the cognitive performance. This suggests a core, general purpose **central executive** which performs cognitive tasks requiring the person's attention. The central executive has a limited capacity, but can operate on most types of information. Similar research suggests the existence of other, independent components. These include a **visuospatial scratch pad** and an **acoustic store**, responsible for the encoding of information. These communicate with the central executive through an **input register**, which performs some analysis of the input, perhaps separating it into meaningful units. Rehearsal takes place through an **articulatory loop**, which may have two distinct components: a **phonological store**, concerned with speech perception, and an **articulatory control process**, involved in speech production.

The separate existence of a short-term storage system is clearly revealed by the study of people with particular types of brain damage. Sufferers from **anterograde amnesia** (Topic F3) perform normally in tests of digit span, and show the usual recency effect. However, they cannot remember anything they saw or heard a few minutes earlier. They usually have a good memory for events in the more distant past. This suggests normal STM and long-term memory (LTM), but a failure of transfer to LTM.

Long-term memory

It has been proposed that rehearsal has a second function, that of transferring information to the **long-term memory (LTM)**. The primacy effect (*Fig. 1*) is usually explained as being due to rehearsal of items as they are presented, causing their transfer to LTM. Presenting items more slowly, thus giving more time for rehearsal of each item, enhances the primacy effect, but has no effect on the recency effect. However, mere rehearsal, in the sense of passive repetition, is not an effective way of consolidating information into the LTM. This is shown, for example, by the ineffectiveness of campaigns in which the same information is presented repeatedly to the public. Furthermore, while information is stored in STM in acoustic or articulatory form, studies of retrieval from memory (Topic F2) show that **semantic encoding** (on the basis of meaning) is used in LTM.

Effective transfer to LTM requires a process akin to Bartlett's effort after meaning. In the 1970s, Craik and his colleagues proposed a **level of processing** approach to encoding in LTM. In one study, people were asked to make decisions about presented words such as *Is the word in capital letters?* (visual); *Does the word rhyme with wait?* (phonemic); *Is the word a type of food?* (category); or *Would the word fit in the following sentence: ...?* (usage). They were then given a memory test. The number of words remembered was least in the visual, increased through phonemic and category judgments, and was highest in sentence usage. They argued that memory depended on how *deeply* material was processed. It is not necessary, they proposed, to postulate a separate STM; rather, material is lost from memory simply because it is not processed deeply enough.

There are, however, problems with the notion of depth of processing. In particular, it is difficult to define without involving circular argument. It is now thought better to describe the *qualitative* aspects of processing. Features extracted by 'shallow' processing may lead to better memory. For example, words with unusual spellings (e.g. *phlegm*) may be remembered better than words with regular spellings. Recognizing that an item does not fit a particular category does not lead to such good memory as recognizing that it does so. The underlying process might be establishing the *distinctiveness* of an item.

Another mechanism is the *organization* of material. The effects of chunking on memory may be explained by the individual organizing material by relating items to the existing body of knowledge. A related memory strategy is to use **mnemonics**, in which deliberate efforts are made to organize material by the use of rhymes, or associating items with visual imagery (e.g. visualizing objects at particular places in the neighborhood. Visual imagery works by establishing links between new items and familiar ones. Recall of the learned item is facilitated by first recalling the familiar one (Topic F2).

F2 Retrieval from memory

Key Notes

Recognition and recall

Two main ways of retrieving material from memory are recall, when information is retrieved directly from memory, and recognition, when an item is identified as previously experienced. Recall can be free recall, or facilitated by giving retrieval cues (cued recall). Re-learning savings show that some memory was retained that could not be accessed by recall or recognition. Recall, but not recognition, tends to worsen with age. Unusual words are better recognized, but common words are better recalled (the word frequency effect). Recall, but not recognition, is dependent on the intention to learn. The process of memory search involves a conscious or unconscious search for retrieval cues. The tip-of-the-tongue phenomenon is a feeling that we can almost recall a word.

Cues in retrieval

The encoding specificity principle holds that retrieval is best when the cues and context during recall match the way in which information was encoded. Flashbulb memories are the clear recollection of the context in which a major event was experienced. Flashbacks are vivid re-experiences of a traumatic event. Memory is also dependent on the internal physiological or mood state: state-dependent learning. 'Deeper' encoding provides more retrieval cues. Autobiographical memory is memory for one's own personal experiences, which are frequently used to date world events. Childhood amnesia is the inability of most adults to recall events from their first 3 or 4 years. It may be partly due to lack of appropriate cueing, and partly to immaturity of the hippocampus.

Reconstructive processes in retrieval

Bartlett showed that recall is a reconstructive process: people often add information to make sense of what they recall. This might occur at encoding or at retrieval. Reconstruction affects eyewitness testimony, which is often distorted to fit with expectations or suggestions (e.g. from leading questions or the use of hypnosis). Many alleged instances of 'recovered memories' of childhood sexual abuse have been shown to be inaccurate or impossible, and have been described as false memory syndrome.

Related topics

Encoding in memory (F1)   Memory loss (F3)

Recognition and recall

Following Ebbinghaus (see Topic F1), two main ways of retrieving material from memory are generally distinguished. In **recall**, information is retrieved directly from memory. **Recognition** occurs when an item is identified as one we have learned or seen before. Recognition memory may be tested by presenting target and irrelevant items individually (**yes/no recognition**), or by presenting together a list of target and irrelevant items (**forced-choice recognition**). In either case, both the number correct and the time to respond or to locate target items may be used as measures of recognition memory. The recall–recognition distinction is not quite as clear as once thought. For example, recall can be facilitated by giving hints (**retrieval cues**) about the target item. This is known as **cued recall** (as opposed to **free recall** when no hints are given). Ebbinghaus demonstrated that when a list cannot be recalled or recognized, it may be learned again more quickly than originally. Such **re-learning savings** show that some memory was retained that could not be accessed by recall or recognition.

Recall and recognition have certain, different characteristics:

- Typically, recall memory, but not recognition memory, worsens with advanced age (see Topic H7).
- From a mixture of unusual and common words, unusual words are better recognized, but common words are better recalled (the **word frequency effect**).
- Recall is dependent on the **intention** to learn (or depth of processing; see Topic F1), while recognition is almost as good after one presentation with no instruction to learn as after a deliberate attempt to learn.

How information is actually retrieved by recall is unclear. Some argue that retrieval cues are always involved; that the process of **memory search** involves looking for cues associated with information to be recalled. These cues may be **contextual**, **feature-related**, or **semantic**. The search process may be rapid and effortless, or may involve a conscious attempt to find characteristics associated with the item. During such a memory search, we may experience a **feeling of knowing**: a sense that we know the answer, but cannot grasp it. The **tip-of-the-tongue phenomenon** is characterized by a feeling that we can *almost* recall a word. We can often provide some accurate information about the word, such as its first letter, the number of syllables, or what it rhymes with.

Cues in retrieval

The usefulness of a retrieval cue depends on how it relates to the original encoding of the material. The **encoding specificity** principle holds that retrieval is best when the cues, and more generally the context, during recall match the way in which information was encoded, or the context in which it was encoded. For example, if material was learned focusing on word sounds, then the best retrieval cues will be to suggest words the item might rhyme with. If encoding focused on the meaning of items, then cues suggesting synonyms or meanings will be more effective. The effect of learning context is shown by how a tune or a scent can evoke vivid memories of events associated with them. Context cues are shown in free recall of items of a particular class. For example, asked to write a list of all the birds you can think of, the list will include clusters of related birds, such as of birds of prey, as thinking of one such bird cues retrieval of another similar one. A classic study of context-dependent memory asked divers to learn word lists on land or underwater, and tested them either underwater or on land. Recall was markedly better when they were tested in the same place as they learned the material (*Fig. 1*). Context cueing is also revealed in the phenomena known as **flashbulb memories**. People recalling a notable event (classically the assassination of J.F. Kennedy; more recently the O.J. Simpson verdict) can frequently describe in great detail the circumstances in which they heard the news. Not to be confused with flashbulb memories are **flashbacks**: vivid re-experiences of a traumatic event, characteristic of post-traumatic stress disorder (see Topic L4). These may be cued by specific events or context, or may appear to be spontaneous (but possibly with hidden cues).

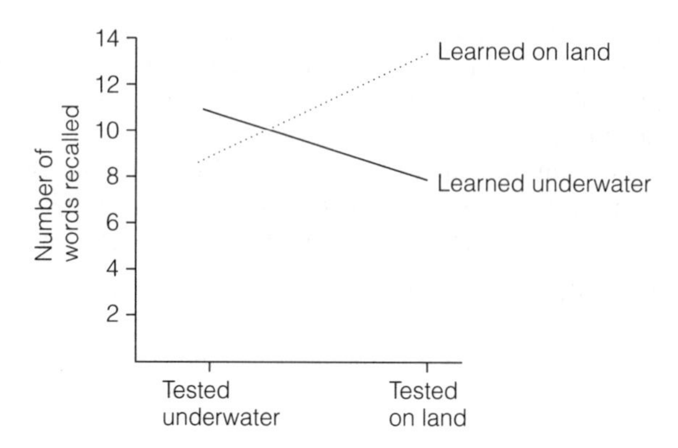

Fig. 1. Context-dependent memory. Recall was better when divers were tested on word lists in the same situation that they had learned them, than when they were tested in a different situation. Based on data from Godden, D.R. and Baddeley, A.D. (1975) Context-dependent memory in two natural environments: on land and under water. British Journal of Psychology, ***66:*** *325–331.*

Memory is also dependent on the internal state: **state-dependent learning**. For example, when sober, people often cannot remember what happened during an episode of heavy drinking. However, the next time they are drunk, they can recall the events. Recall of the memory depends on being in the same state of inebriation as when the memories were encoded. Similar effects have been demonstrated with mood: when happy we can recall more about happy periods in our life, and when sad more about sad periods. State-dependent effects can be demonstrated in laboratory conditions, with mood or other state manipulated by hypnotic suggestion or other means.

If recall is facilitated by contextual and other retrieval cues, the more cues are available, the easier an item should be to recall. Many of the effects of different ways of encoding (see Topic F1) can be interpreted as 'deeper' encoding providing more retrieval cues. **Elaborative rehearsal** describes the conscious effort during learning to make links between items to be recalled, and between those and other material already in memory. The more links that are made, the more likely it is that another item or event will cue recall.

Autobiographical memory is memory for one's own personal experiences. The ability to recall the dates on which events occurred seems not to be lost in the same way as other information (see Topic F3, *Fig. 1*). Instead of most information being lost early, forgetting is approximately linear (*Fig. 2*). We use our autobiographical memory to structure our recall of world events. For example, we can place an event in a particular period because we remember that it occurred while we were in high school, or more specifically because it was just after we went to college. One notable aspect of autobiographical memory is **childhood amnesia**: the inability of most adults to recall events from their first 3 or 4 years. It has been suggested that one reason for this is that the infant's world is so different from that of the adult (in terms of size and schemata) that context cues lose their meaning. A second reason is the immaturity of the hippocampus, which is involved in encoding autobiographical memories (see Topic F3).

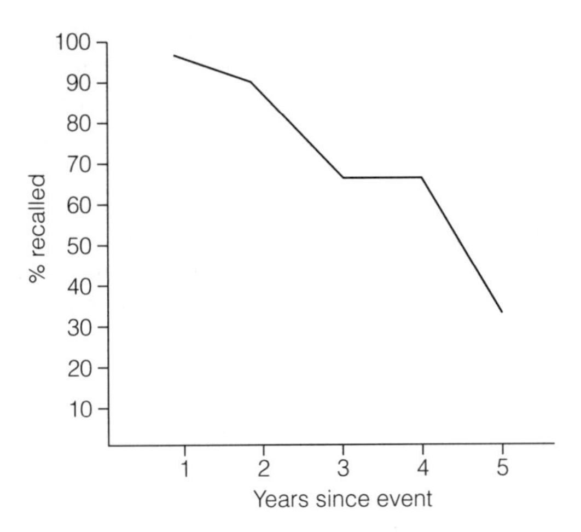

Fig. 2. Loss of information from autobiographical memory. Based on data from Linton, M. (1975) Memory for real-world events. In: Explorations in Cognition *(eds Norman, D.A. and Rumelhart, D.E.). Freeman, San Francisco.*

Reconstructive processes in retrieval

Bartlett (see Topic F1) argued that recall is a **reconstructive** process. He presented people with stories from other cultures, and asked them to retell them. People made two types of error in recall of information that did not fit with their own world view: they omitted it (e.g. supernatural events were ignored), or they added information that made sense in terms of their own culture. Either of these might be explained by a failure of encoding: material would be processed in relation to existing knowledge (Bartlett's schemata), so that some items would not be encoded at all, and others encoded in such a way as to cue recall of more clearly meaningful information (reconstruction). However, there is evidence that at least some reconstruction takes place at retrieval: information presented after a story has been read can become incorporated into its retelling.

Reconstruction also influences **eyewitness testimony**, as given for example in law courts, often many months after the witnessed events. Numerous laboratory studies have shown that accounts of events are modified in similar ways to Bartlett's stories. That is, they are made to fit with expectations and to make internal sense. Schank argued that we understand everyday events using an internal **script** that describes the course of such events. We use such a script to guide our recall of an actual event we witness, and the event becomes distorted to fit the script. Of even greater concern is the incorporation into eyewitness testimony of material that is suggested during questioning. The use of such **leading questions**, where the expected reply is included in the question, is strictly controlled in court proceedings. However, if leading questions or direct suggestions have been used during earlier interrogations by police or lawyers, these may become incorporated into the witness's evidence. The use of **hypnosis** in questioning is also of doubtful validity. It has been used in attempts to get better recall from a witness, but its effect is to increase suggestibility, enhancing reconstructive distortions.

An extreme case of the reconstruction (or construction) of memories is the **recovered memory–false memory syndrome** argument of the late 20th century. Following links being shown between early childhood sexual abuse and later psychological dysfunction, many people claimed to have recalled such incidents, leading to criminal charges being brought against their parents. These recollections were often made during psychotherapy, after a link was suggested, directly or indirectly, by a therapist. Close examination of these 'recovered memories' has shown in many cases that they cannot possibly be true. Perhaps the saddest aspect of this is that, in common with distorted eyewitness testimony, these memories are experienced as *true* memories by the person reporting them.

F3 MEMORY LOSS

Key Notes

Forgetting

Ebbinghaus argued that the approximately logarithmic forgetting curve shows that memory decay is the gradual weakening of the memory trace (or engram). Slow-wave sleep reduces forgetting, suggesting active retroactive interference. Previously learned material also interferes with memory (proactive interference). Interference probably takes place on recall. Most interference is probably due to response competition, and intrusions occur when a recalled competing response is wrongly recognized. According to psychoanalytic theory, all memories are permanent but some are kept out of awareness by repression.

Amnesia and memory storage

Surgical removal of parts of the medial temporal lobes of the brain produces anterograde amnesia (inability to form new memories) without retrograde amnesia (loss of previously formed memories), or loss of the ability to learn motor skills or form conditioned responses. This challenged the view that memory is diffusely distributed throughout the cortex, and suggests two types of learning: explicit (or declarative) memory (conscious memories that can be described in words) and implicit (or procedural) memory (which can be expressed behaviorally but without the necessity for awareness). Animal research and studies of other pathological conditions that produce different patterns of amnesia have clarified the role of brain structures in memory. The rhinal cortex is involved in the formation of new long-term memories; the hippocampus in long-term retention of spatial information; the amygdala in memory for the emotional content of experiences; the cortical association areas in specific sense-related memories; the prefrontal cortex in memory for temporal ordering; and the mediodorsal nuclei of the thalamus in object recognition and long-term memory formation.

Related topics

Encoding in memory (F1)
Retrieval from memory (F2)
Disorders of personality, identity, and memory (L6)

Forgetting

Ebbinghaus (see Topic F1) demonstrated that the amount of information remaining in memory decreases in a characteristic way depending on the time since learning (the **retention interval**). He interpreted this approximately logarithmic **forgetting curve** (*Fig. 1*) as showing that memory **decays** passively, because the **memory trace** (or **engram**) gradually weakens. One effect of rehearsal would be to strengthen the engram, so that rehearsed items would remain in memory longer. It is probably impossible to test this directly, as it would require preventing other possible processes from interfering with memory. Further, reconstructive processes could apparently reinstate memories. The effect of sleep on memory argues against simple decay. A person who goes to sleep shortly after learning material will forget less than one who stays awake, although this effect is smaller than was once believed.

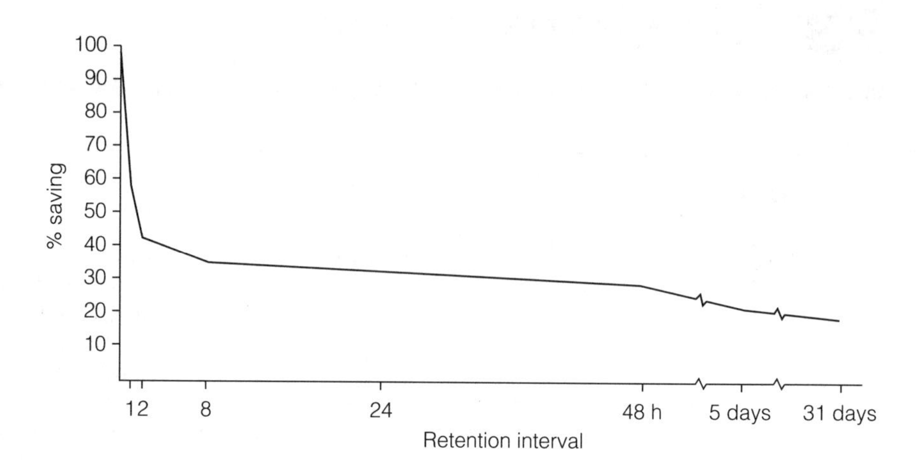

Fig. 1. The forgetting curve. The amount of memory retained is measured by the relearning savings method. Data from Ebbinghaus, H. (1885) Über das Gedächtnis. *Dunker, Leipzig; Trans: Ruyer H. and Bussenius, C.E. (1913)* Memory. *Teachers College, Columbia University, New York.*

There is much support for a competing explanation of forgetting: **interference theory**. According to this approach, information is forgotten because new information interferes with its retrieval. This explains the effect of sleep: those who go to sleep are exposed to much less new information than those who stay awake. However, given that sleep is not the passive state it was once believed to be (see Topic B6), this effect is limited. In fact, forgetting is greater when learning is followed by rapid eye-movement sleep than by slow-wave sleep. The effect of new information on memory is called **retroactive interference**. Previously learned material also interferes with memory: **proactive interference**. Both proactive and retroactive interference are greater when the interfering material is similar to the target material. This shows that the effects are not due to displacement of material within memory, and suggests that the interference takes place on recall. Furthermore, if the new material is consistent with the old material, memory for both may be enhanced.

Early explanations of interference suggested that it is due to **response competition**. Two similar items in memory will be cued by many of the same retrieval cues, and so target items will be less likely to be recalled. However, this suggests that as accuracy for target memories decreases the number of **intrusions** (items from the wrong set falsely recalled as members of the target set) from the interfering list will increase. While this is true for a few trials with an interfering word list, with larger numbers of trials the rate of intrusions drops to zero. Retroactive interference may cause the **unlearning** of the target material (rather like the extinction of conditioned responses; Topic E1). However, unlearning is unlikely as it should only apply to retroactive interference, but attempts to demonstrate it also affect proactive interference. Another view is that retrieval always consists of recall followed by recognition. Most interference is due to response competition, and intrusions occur when a recalled competing response is wrongly recognized. With more trials with the interfering list, its items become better recognized, and less likely to be falsely recognized as target items.

A completely different approach to forgetting comes from psychoanalytic theory (Topic K3). **Repression** is an active, motivated process, and one of the primary defense mechanisms that protects the individual from traumatic events or knowledge. According to this view, all memories are permanent, but some are kept out of awareness. Although they influence conscious thought, they can only be retrieved by the techniques of psychoanalysis. Although it is clear that memories of particular events may become difficult to access at times, particularly under emotional stress, there are no convincing experimental demonstrations of repressed memories (see also Topic F2). Dissociative disorders (see Topic L6) also often include unawareness of aspects of behavior or personality. On the other hand, traumatic events are often followed by flashbacks (Topic F2) and inability to forget, often for many years.

Amnesia and memory storage

The study of individuals with various types of memory loss, or **amnesia**, has told us a lot about memory, particularly about memory **storage**. Much understanding of the brain structures involved in memory comes from studies, mostly conducted in the 1960s, of a patient known as H.M. In 1953, H.M. had major brain surgery to relieve frequent and severe epileptic convulsions. This involved the bilateral removal of parts of his **medial temporal lobes**, including most of his **hippocampus** and **amygdala**, and parts of his cerebral cortex. Following the operation, H.M. had normal memory for events before the operation (i.e. no **retrograde amnesia**), although he did have some difficulty with the months immediately leading up to it. However, he did not appear to be able to form new memories at all (**anterograde amnesia**). Specifically, while he had a normal short-term memory (STM) span, shortly after he stopped thinking about something he would forget it. This amnesia appeared to be **global amnesia** (it included all sensory modalities), and left him unable to recognize people he had met since the operation, or to remember where he was or how long he had been there. It was subsequently discovered that H.M.'s amnesia did not extend to perceptual-motor learning (e.g. he readily learned to trace a pattern viewed in a mirror) or to the formation of conditioned responses (Topic E1), and he retained these behaviors over many months. However, he could not remember learning them. These results have largely been replicated with patients with who have processes damaging these brain areas.

The importance of the studies of H.M. are that they focused research on the neural basis of memory. Before then, the predominant view derived from Karl Lashley's proposal that memory is recorded diffusely throughout the cerebral cortex. According to another adherent of this view, Donald Hebb, the transition from STM to long-term memory (LTM) (Topic F1) is a process of formation of reverberating **neural circuits** that represent the memory. H.M.'s case showed that memory was not diffusely distributed throughout the brain, and that structures such as the hippocampus and amygdala played a crucial role in the consolidation of memory into LTM. The distinction between, for example, the retention of learned motor skills and the memory of learning those skills suggests that two kinds of memory are involved. One is **explicit** (or **declarative**) memory: conscious memories that can be described in words (e.g. January has 31 days). The other is **implicit** (or **procedural**) memory, which can be expressed behaviorally but without the necessity for awareness (e.g. we may be able to ride a bike but cannot describe exactly how we do it).

Other pathological conditions can produce different patterns of amnesia. **Korsakoff's syndrome** (Topic L6), following prolonged consumption of large quantities of alcohol, includes both anterograde and extensive retrograde amnesia. The main damage in Korsakoff's syndrome is in the **thalamus** and/or **hypothalamus** (Topic A3), and other injuries to nuclei in these structures produce similar patterns of amnesia. Patients with injuries to the **prefrontal cortex** do not generally show general retrograde or anterograde amnesia, but show specific inability to recall the temporal order of events and to perform ordered tasks. Some Korsakoff patients show such temporal memory deficits, and they have diffuse damage in prefrontal regions. **Post-traumatic amnesia** is memory loss following a blow to the head, usually severe enough to have caused concussion (temporary loss of consciousness), or **electroconvulsive therapy** (Topic L7). There is permanent retrograde amnesia for events immediately preceding the injury, and anterograde amnesia for events shortly after, when the patient is experiencing confusion.

Further detailed work on brain structures and memory has involved experiments with animals. Anterograde amnesia has been modeled in monkeys and rats with object recognition tasks following experimental selective brain lesions. It was at first assumed that the crucial damage was to the hippocampus, but after much research it was established that the location of these effects is in the **rhinal cortex**, part of the cerebral cortex adjacent to both the hippocampus and the amygdala. *Table 1* shows the brain structures now believed to be involved in memory, and the particular functions they apparently perform. Research on the neural bases of memory continues, and the summary presented in *Table 1* will doubtless change.

Table 1. Some of the brain structures that are implicated in memory

Brain structure	Location	Function
Rhinal cortex	Medial surface of temporal lobe of cerebral cortex	Formation of new long-term memories
Hippocampus	Limbic system structure at base of forebrain	Long-term retention of spatial information
Amygdala	Limbic system structure at base of forebrain	Memory for emotional content of experiences
Association areas	Cerebral cortex (Topic A4, *Fig. 1*)	Sense-related memories
Prefrontal cortex	Anterior to primary motor cortex (Topic A4, *Fig. 1*)	Memory for temporal ordering
Mediodorsal nuclei	Thalamus (Topic A3, *Fig. 1*)	Object recognition; long-term memory formation

F4 MENTAL REPRESENTATION

Key Notes

The nature of representation

A representation stands for something else in its absence. Internal representations are used to store information in memory, and there is debate about what form they take. There is some (but not unanimous) agreement that images and propositions are necessary forms of representation. A third type, the mental model, has also been suggested.

Paivio's dual coding theory

The dual coding theory employs different memory codes for verbal and sensory information; verbal representations are called logogens and sensory representations are called imagens. Referential links enable the two types of code to operate in conjunction with each other.

The nature of images

The function of mental images is a matter of some debate, but when people are asked to make use of them they behave as though the image was like any other percept. For example, it takes the same time to scan a mental image as to scan an actual image, and the time it takes to rotate a mental image depends on how much rotation is required.

The propositional hypothesis

The propositional hypothesis holds that knowledge is stored in an abstract form, by propositions, which are independent of natural language or sensory modality. Propositions express the relationships between concepts and are manipulated by logical rules. According to the propositional hypothesis, mental images are by-products of cognitive activity, not the basis for it.

Kosslyn's model

Kosslyn suggested a way in which both propositions and images might play a part in representation. Visual images are said to be stored separately from propositional information, and are generated in a special medium by processes that depend upon that propositional information. The visual medium places limitations on image size and resolution, which is consistent with experimental evidence.

Mental models

The mental model is a third type of representation that differs from an image by being more abstract and containing less specific detail. Models and images are described as superordinate to propositions and may be generated from them, but there is evidence that people can operate at whichever level is best suited to the task.

Related topics Form perception (D4) Cognitive development in infancy (H2)

The nature of representation

In order to accumulate and make use of our knowledge of the world we must be able to **represent** it in some way. The term representation is used to refer to anything that stands for something else in its absence. Drawings, photographs, maps, and words are all examples of things which, although entities in their own right, principally serve as representations of other things. It is characteristic of representations that they only represent *some* aspects of that which they stand for. Thus, a verbal description and a picture will differently represent a bowl of strawberries and cream, and neither representation will convey what strawberries taste like.

The examples listed above are all instances of **external representations**, which, as we can see, may be **pictorial** or **linguistic**. Because pictures have the same spatial structure as the things being pictured, they are said to be **analogical representations**. Linguistic representations, on the other hand, have an arbitrary relationship with the thing they represent. External representations are stored on media such as books, film, tape, and computer disk.

Internal representations can similarly be divided into two classes: analogical representations (or **mental images**) and **propositional representations**. Analogical representations are particularly associated with information that is sensory in nature, while propositional representations are language-like, regardless of the nature of the original information. Internal representations are, of course, stored in our memory.

Paivio's dual coding theory

Paivio suggested that there are two systems of coding internal representations in memory: the **verbal** system and the **nonverbal** system. These systems are connected, but have specialized capabilities. The verbal system deals with information that occurs in linguistic form (basically spoken and written language) and is **serial** in nature. The representational units of the verbal system are called **logogens**. The nonverbal system deals with spatial information and events (i.e. changes over time) and inputs from other senses; its representational units are called **imagens**. There are **referential links** between the two systems that enable us, for example, to name things we see. Logogens have been described as functioning both as informational structures and as response generators, and in this respect they are rather like schemata (see Topic H2).

Evidence that supports the dual coding hypothesis comes from a number of sources. For example, if people are presented with a rapid sequence of words and a rapid sequence of pictures, and then asked to recall either words or pictures in any order, they will recall more pictures. However, they will recall the *order* of presentation of the words more readily than that of the pictures. Furthermore, recall of objects presented as pictures *and* words (and therefore coded in both systems) is better than that of objects presented by words alone.

The nature of images

It seems self-evident that we can manipulate images 'in our mind's eye' and that we can use such operations in very practical ways. Two well-known types of experiment have been used to explore ways in which we can use images. These make use of the tasks of **mental rotation** and **image scanning**.

A typical mental rotation task involves presenting an observer with two objects that are identical, except that one (the target) has been rotated into a different orientation and may also have been mirror reversed (*Fig. 1*). The observer's task is to say whether the rotated object is mirror reversed, and most people appear to do this by mentally rotating the target object until it has the same orientation as the comparison drawing, and then making their judgment. This mental rotation takes time, and the more the target has to be rotated, the longer observers take to give their answer.

Fig. 1. Mental rotation task. The right-hand object of pair (a) must be rotated further than the right-hand object of pair (b) in order to bring the pairs into the same orientation. Thus, it takes longer to make a judgement about pair (a) than pair (b).

Image scanning tasks require observers to memorize a picture, such as a map, which has a number of distinctive features marked on it. After the picture is removed, they are asked to imagine a 'speck' traveling between the locations. The relative times for these imagined journeys turns out to be proportional to the distances between the locations, just as might happen if the observers were scanning the original picture.

In both these types of experiment, people behave as if they were working with a mental image that had some of the properties of the original visual stimulus. However, this does not in itself prove that we store mental representations in the form of images, and whether or not we do has been the source of much debate.

The propositional hypothesis

The propositional view is that knowledge is represented not by images but by much more abstract forms called **propositions**, which are explicit entities that express the relationships between concepts. Propositions are not dependent upon language or sensory modality, and are sometimes referred to as **mentalese**; that is, the basic code in which all mental activities are carried out. Operations on propositions are carried out by way of the logical system known as **predicate calculus**. This provides a convenient notation in which the relationship between objects is represented as a **predicate** and the objects themselves are represented as **arguments**. The general form of this notation is:

[Relationships between elements] [Subject element] [Object element]

that is:

[predicate] [argument,argument]

So the proposition *the dog bit the postman* would be represented as:

[bite] [dog{subject element}, postman{object element}]

Propositions cannot be observed directly and as such they are hypothetical constructs. However, they can be used to describe complex relationships, and can be combined to represent images and sequences of words. The essential point is that propositions are *neither images nor words* but an abstract representation of knowledge. Propositional theorists suggest that mental images are simply epiphenomena; that is, secondary events that occur as the result of other cognitive processes but which are *not part of them*. Thus, when we perform tasks that seem to require the manipulation of images we are actually using our knowledge of how visual images ought to behave. It has also been suggested that the linkage between logogens and imagens in the dual coding theory requires a third code which could be propositional.

Kosslyn's model

Kosslyn's model of visual imagery uses a computational metaphor, and attempts a synthesis of the imagery and propositional ideas. Visual images are said to be generated in a visual medium, which is analogous to a computer screen. This medium has boundaries, and is capable of representing spatial relationships. Like the eye, the medium has its sharpest resolution in the center, and has limited resolution overall, restricting the amount of detail which can be reproduced. And like the computer screen, the image begins to fade as soon as it is generated and must be regenerated (or 'refreshed') if it is to persist. Images are generated in this medium by a set of processes that use information stored in **image** files and **propositional** files that reside in long-term memory. Image files contain information about what objects look like. Some image files contain information about overall shape (called a **skeletal image**) while other files contain details of *parts* of the image. Propositional files contain information about the properties of objects, and those files which contain information which is central to the representation (the **foundation part**) are linked to the image files containing the skeletal image. To construct an image, the system first locates the propositional file and checks for a link to the skeletal image. If that link is intact, the system then proceeds to place the image in the central (high resolution) region of the medium, unless other coordinates are specified, and to locate and place other details of the image as required. Operations such as rotation and zooming in and out are also possible.

A range of experiments have been carried out to investigate the properties of the visual medium. One example uses a task called visual scaling. Individuals are asked to imagine pairs of animals of different relative size, such as an elephant and a rabbit, and are then asked specific questions about each animal. The model predicts that images that take up a larger part of the visual screen will contain more detail, and therefore questions about the larger image will be answered more quickly than those about the smaller image. This is in fact what happens. If the questions were being answered by referring to two abstract propositional representations there is no reason why this difference should occur.

Mental models

The approach suggested by Johnson-Laird also uses a computational metaphor and makes use of three forms of representation, propositions, images, and mental models, each of which is distinctively defined. Propositions are taken to be abstract representations of meaning, as before, but are said to be fully expressible in natural language rather than in some hypothesized mentalese. Images are representations of specific objects and include many of their perceptual properties, while **mental models** are more abstract and generalized analogical representations of objects, and are rather like prototypes (see Topic D4).

The model uses an analogy between these three forms of representation and the hierarchical way in which computer programing languages are organized. The fundamental programing language is called machine code, which operates in a very detailed fashion. Writing programs in machine code is complex and time-consuming technical work, and so higher-level languages have been developed which make programing more straightforward. The computer translates the higher-level languages into machine code for itself, in order to execute the program. According to Johnson-Laird's view, mental models and images perform this same, higher-level function and remove the need for us to operate at the basic propositional level. Thus, instead of debating whether representation is achieved by propositions *or* images, the focus of attention can move to the question of *how* people use different forms of representation for different purposes.

The distinction between propositions and mental models has been investigated using determinate or indeterminate descriptions of spatial relationships, which correspond to propositions. Determinate descriptions contain precise verbal information about the location of every object in a test array, and subjects can therefore form a mental model of the array from the propositions. Indeterminate descriptions, however, contain ambiguous information and may be consistent with several sets of relationships, so that subjects cannot form a unique mental model from them. People who were given determinate descriptions were, when questioned, able to make inferences about spatial relationships in the array that were not contained in the original descriptions, indicating that they had formed a mental model. Furthermore, they did poorly when asked to recall the original descriptions, which suggests that use of the model supersedes the original propositions. In contrast, people given indeterminate descriptions were unable to make additional inferences because they were unable to construct a model. However, they were much better at recalling the original descriptions because of having to rely on the propositional information throughout.

F5 PROBLEM SOLVING

Key Notes

Problem-solving cycles

Problem-solving cycles are descriptions of the problem-solving process as a sequence of stages. They are useful both as tools for guiding effective, practical problem solving and as a way of identifying areas for research.

Well-structured problems

Well-structured problems provide the basis for important computational theories of problem solving. Their structure can be defined by a problem space that contains all possible pathways from an initial state to a goal state. Problem solving is characterized as a progression along pathways by way of intermediate knowledge states generated by mental operators. This way of solving problems can be simulated by computer programs, but people often devise shortcuts, or heuristics, to simplify the process.

Ill-structured problems

Most problems we encounter are ill-structured and have no clearly defined paths to their goal state. In many cases, the solutions to these problems suddenly appear in consciousness having apparently been generated by insight. The three-process view of insight suggests that it arises from selective encoding, selective comparison, or selective combination of information.

Obstacles to problem solving

Problem solving may be obstructed by a mental set that leads someone to apply an inappropriate solution technique because it has worked for other problems that appear similar, but are not. It can also be hampered by functional fixedness, which is the tendency to see an object as having only one use rather than several different ones.

Aids to problem solving

Analogical thinking allows us to solve new problems by referring to old problems with the same structure, although people often make the mistake of applying it to problems which are similar only in content. Incubation may also assist problem solving by encouraging those processes thought to be responsible for insight.

Related topics

Form perception (D4)

Encoding in memory (F1)

Problem-solving cycles

If we have no ready means of achieving a goal or answering a question then we have a problem to solve. There are two aspects to the study of problem-solving behavior. One of them is to examine the *logical* ways in which solutions to problems may be discovered, and to specify procedures that will help people become more effective problem solvers. The other is to discover how people *actually* solve problems and to understand the thought processes involved. Problem-solving cycles break down the problem-solving process into a logical sequence of stages, and thereby help both to describe the kinds of activities that lead to effective problem-solving behavior and to identify useful areas for research.

Sternberg's seven-stage cycle is representative of several which have been published.

1. **Identifying the problem.** It is not always obvious that a situation is (or is becoming) problematic, and therefore recognizing that there is a problem is not always a simple matter.
2. **Defining and representing the problem.** The way in which a problem is represented can have an important bearing on how difficult it will be to find a solution.
3. **Formulating a strategy.** Different types of problem naturally require different strategies, and the distinction between well-defined and ill-defined problems (see below) is particularly important for the selection of a solution strategy.
4. **Organization of information.** Some problems are stated in such a way that little additional data is needed, while others may need a good deal of additional information in order to implement the strategy.
5. **Resource allocation.** Effective problem solvers pay attention to how their resources (particularly time) are allocated to the problem-solving process. Experienced problem solvers, for example, devote more time to advance planning and to maintaining a view of the entire process than do novices.
6. **Monitoring.** Effective problem solvers monitor their progress towards a solution so that mistakes are detected at an early stage.
7. **Evaluation.** The purpose of an evaluation stage is to determine the strengths and weaknesses of the way in which the other stages were implemented. This can lead to new insights and the formulation of new strategies that can be used in the future.

The cycle is not intended to be prescriptive and, in practical terms, successful problem solving usually involves a degree of flexibility. For example, if the monitoring process reveals that a chosen strategy is not yielding progress towards a solution it may be necessary to return to stage three, and this in turn may have implications for resource allocation. Stages two, three, and four describe areas that have attracted the most psychological research.

Well-structured problems

Well-structured problems are those which have clear paths from their **initial state** to the **goal state**, with explicit rules about what can and cannot be done along the way. Newell and Simon described a computational model for solving this type of problem. Within their framework, all possible pathways from the initial state to the goal state are contained within the **problem space**. Individuals are said to solve problems by employing **mental operators** to generate **knowledge states** that enable them to progress through the problem space from the initial state towards the goal state. Mental operators embody the legal moves and any restrictions that might apply under particular conditions. The Tower of Hanoi problem (*Fig. 1*) is a good example of a well-structured problem.

Using the computational metaphor, Newell and Simon suggested that the fundamental strategy for solving problems was to break them down into a sequence of **operations** that could be organized into a **program.** A program may contain operations that are organized hierarchically into **sub-routines**, and which are repeated over and over again until a specified condition is reached; such sequences of operations are called **algorithms**. A program thus solves a problem by calculating all possible operations between the initial state and the goal state and choosing the optimum (i.e. shortest) combination.

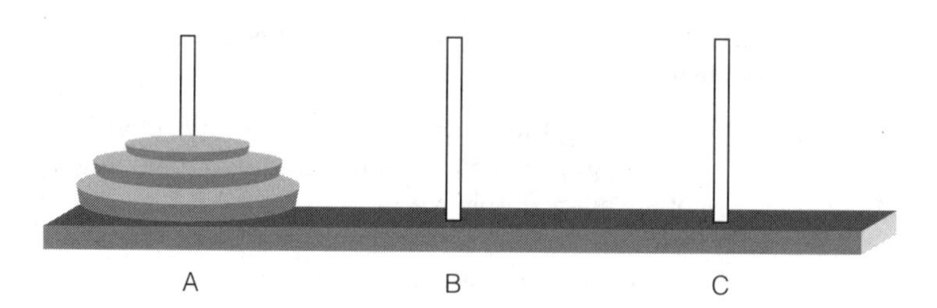

Fig. 1. The Tower of Hanoi: initial state. The objective is to move the disks one at a time from peg A to peg C in the fewest moves. Only the top disk may be moved at any time and a larger disk may not be placed on top of a smaller one.

It is important to realize that this model does *not* describe how people actually solve problems; it describes how certain types of problems might, ideally, be solved. The limited capacity of our working memory (Topic F1), amongst other things, means that we can only consider a few options at a time. What people actually do is devise shortcuts, known as **heuristics**, which sometimes produce an effective solution but (in contrast to an algorithm) sometimes do not. One type of heuristic, known as **means–end** analysis, is particularly common. It involves the problem solver considering how far away the solution is, and selecting an operation that will reduce the distance. In effect, this creates a series of **sub-goals** by breaking down the problem into smaller steps. Such strategies need effective monitoring; if attention is shifted from the entire problem to the sub-goals there is a risk that the final solution may not be the most effective one.

People can learn to apply heuristics and to become more effective at solving well-structured problems. For example, if subjects learn to structure the Tower of Hanoi problem into sub-goals, they can solve more complex versions involving five and six disks more readily than controls without such experience. However, heuristics do not always transfer from one problem to another even if the two problems are **isomorphic** (i.e. have exactly the same structure) because, as we shall see, the way in which a problem is presented can have a key effect on whether or not the isomorphism is evident.

Ill-structured problems

The computational approach deals with a class of problems which are really only puzzles, and all the knowledge required to solve them is stated as part of the problem. Everyday problems, on the other hand, are generally much more complex and often require additional knowledge. Such problems do not have defined problem spaces since the complete set of permissible operators is often unknown, and there may be no clearly defined path to their solution. Nevertheless, people solve such problems regularly, often showing considerable ingenuity, without being able to articulate exactly how they thought of a route to the solution. This phenomenon, known as **insight**, was first described by the Gestalt psychologists (Topic D4). Insight often involves some kind of re-conceptualization or novel view of the problem which goes beyond existing associations, and although an insightful solution may seem to appear from nowhere, it is often preceded by a good deal of careful thought. Solving problems that require insight appears to differ from solving well-structured problems in two ways. Firstly, people are more accurate at predicting their success at solving well-structured problems compared with problems requiring insight. Secondly, when people are working on well-structured problems they can indicate how close they are to a solution before they reach it, but when working on insight problems they have no feeling of *progression* towards a solution, simply a sudden realization that it has been reached.

The Gestalt psychologists believed that insight was 'something special' but did not suggest an explanation for it. An alternative view is that there is nothing special about insight, and that it is a manifestation of one of three processes. **Selective encoding** involves distinguishing between those aspects of the problem that are relevant and those that are not. Once the crucial pieces of information have been identified the solution to the problem may become evident. **Selective comparison** involves novel perceptions of how new information relates to things we already know. The use of analogies is one example of selective comparison. **Selective combination** refers to the combination of information in a novel and productive manner. Research indicates that people tend to use selective combination most often, and that individuals may also have a tendency to employ one strategy in preference to the others.

Obstacles to problem solving

As we have already mentioned, people may have difficulty solving problems even when they have worked out the solution to identical (isomorphous) problems. External factors that contribute to this include the introduction of new objects, and changes to the number and complexity of rules. Problems may also be difficult to solve if the individual approaches them with a particular **mental set**, which is the tendency to apply a problem-solving technique that has proved successful for similar problems but is inappropriate for the one in hand. The classic demonstration of this is provided by a series of problems devised by Luchins. The task is to measure out a quantity of water using three containers of different sizes (*Table 1*). The first five problems in the sequence may be solved using the formula: Target amount = B – A – 2C. Problems 6–10 *can* be solved using the same formula, but there are more simple ways of arriving at the correct answer. If people are asked to solve all the problems in the sequence then around 75% will use the more complex strategy for the later problems. However, more than 90% of people who are asked to solve only the last five problems will use the simpler strategy; they do not have the established mental set that would prevent them from seeing the easier solution.

Table 1. The Luchins three-container problem. To solve the first example, imagine container B filled with 25 cm³ of water. Fill container A from container B; B then contains 22 cm³. Now fill container C from B, empty it, and then fill it again. Container B now contains 12 cm³. All the problems can be solved in this way, but problems 6–10 can also be solved without filling container B at all

Problem number	Capacity of containers (cm³)			Target amount (cm³)
	A	B	C	
1	3	25	5	12
2	19	60	6	29
3	17	75	8	42
4	8	50	3	36
5	11	48	14	9
6	15	45	5	20
7	12	28	4	8
8	40	104	8	48
9	26	70	6	32
10	10	44	8	18

Another type of obstacle, **functional fixedness**, arises when people have a fixed idea of the function of an object, and are unable to see that it may have other uses which would enable them to solve the problem. An early task from the Gestalt school illustrates this well. Participants are given a large box of matches, a candle, a length of string, and some thumb tacks. Their task is to fix the candle to the wall so that it can burn without dripping wax on to the floor. The solution is to empty the box of matches, remove the inner tray, fix it to the wall with the tacks and then stand the candle in it. People find this problem difficult because, it is supposed, they do not perceive the novel use for the inner part of the matchbox. According to the Gestalt view, the solution to this problem would emerge as a result of insight. The three-process theory would suggest that a successful solution requires both selective encoding (the string and the matches are irrelevant) and selective combination (the novel re-combination of part of the matchbox with the thumb tacks). Neither explanation, of course, can predict whether any individual will actually solve the problem nor, if they do, how they will do so. However, it has been found that if the materials for the problem are laid out with the matchbox empty and separated into two parts (thus facilitating selective combination), the problem is more likely to be solved, and solved more quickly.

Aids to problem solving

The positive transfer that occurs when knowledge or skills acquired in one setting are applied to another one can facilitate problem solving. Functional fixedness and mental sets provide examples of *negative* transfer, which illustrates the fact that transfer is only beneficial (other than by chance) if we can recognize in some way that a new problem is **analogous** to one that we have solved before. Unfortunately, many people have trouble recognizing analogical relationships between problems unless they are specifically told to look for them, and the difficulty is greater if the content of the two problems is very different. Nevertheless, the essence of an analogical relationship between problems is that they have similar *structural relationships*. Similarity of *content* is irrelevant, as the Luchins problems illustrate. Some theorists suggest that effective use of analogical thinking for solving new problems requires a mechanism (called **analogical mapping**) by which key structural features of the old (or base) problem are compared with those of the new one. If the match is good, then knowledge from the base problem can be transferred to the new domain. However, it is by no means certain that mapping is a necessary component of analogical thought, since people can produce analogical solutions in a new domain even if they are only given the solution part of a base problem and thus have little, if any, structural information about it.

Another way to facilitate problem solving is to gather information about the problem, explore its structure, and then set the matter aside for a period of time. This strategy is known as **incubation**. Much of the evidence concerning incubation is anecdotal, but there have been some convincing descriptions, for example from mathematicians, of its benefits. The three-process account of insight suggests two ways in which incubation could help problem solving. Firstly, removing material from working memory may assist selective encoding since unimportant details in the problem description are likely to be lost, making the structure of the problem more evident. Secondly, new material becomes assimilated into existing schemata with the passage of time, and this will assist selective comparison. However, there is some uncertainty about the benefits of incubation and, particularly, about whether it is a necessary component of problem-solving behavior. When a person is attempting a problem they find difficult incubation may be helpful, but it is unlikely to improve the quality or speed of solving problems that a person finds easy.

F6 THINKING AND REASONING

Key Notes

Language and thought

The question of whether thought precedes language or *vice versa* is still open to debate, but there can be no doubt about the importance of early interaction with the environment and with other people in the development of both. The linguistic relativity hypothesis, which asserted the influence of language on thought, is now thought to be largely incorrect.

Categories and concepts

Concepts are fundamental cognitive structures that represent the ways in which we categorize our experiences. There are two classes of model that describe the ways in which concepts are related to each other. Network models utilize serial processes and fail some important experimental tests. Parallel distributed processing (PDP) models show much more of the flexibility of human thinking, but do not readily account for the speed at which concepts can be learned and modified.

Deduction

Deductive reasoning is a branch of logic that enables us to draw valid conclusions from a set of propositions plus evidence. Although humans can learn to follow these rules, under natural circumstances they often make logical errors because of their tendency to modify logically valid conclusions to fit realistic circumstances.

Induction

We use inductive reasoning to make causal inferences and to form hypotheses about the world around us. The probabilistic nature of induction makes it vulnerable to biases which arise from the heuristics we use to infer likelihood. Analogical reasoning enables us to use past experience to generate new relationships between concepts.

Related topics

Vision (C2)
Problem solving (F5)
Cognitive development in infancy (H2)
Self-perception (I4)

Language and thought

The traditional view of the relationship between language and thought, which can be traced back to Aristotle, is that thought is more fundamental than language and that languages develop in order to express the kinds of thought that people have. The work of Piaget (Topic H2) supports the view that as their intellectual development proceeds, children acquire an increasingly sophisticated use of language so that they may express new ways of thinking. Language is thus a method for representing thought; it exerts no formative influence on it. What can be talked *about* derives from action, and is therefore constrained by the child's stage of development. For example, children in the pre-operational stage talk as they play, but they do not *converse* (i.e. their speech is egocentric) because they cannot think of the world from another person's point of view.

Vygotsky, however, took the view that language and thought develop separately, and that while language has its origins in social interaction, thought develops from action. During the first two years of life, infants develop schemas as a result of their actions on the environment, and this forms the content of pre-linguistic thought. At the same time, pre-intellectual language develops through social interaction, as the child practices the production of speech sounds. When speech starts to develop, the child begins to think in words, and pre-linguistic thought and pre-intellectual speech merge. The content of this early, egocentric speech does not distinguish between thinking and talking, but Vygotsky suggested that, as development proceeds, egocentric speech becomes internalized speech about actions and plans and forms the basis for verbal thinking, while spoken language becomes a social tool for communicating with others.

The **Sapir-Whorf hypothesis** maintains that the vocabulary and grammar of the language we speak determines how we think. As a child learns the language of its culture it is, according to this hypothesis, also learning that culture's particular view of the world. The hypothesis is often stated in two forms. The strong form (called **linguistic determinism**) is that language determines how we think. The weak form (called **linguistic relativity**) is that language influences thought.

One of the best-known attempts to provide evidence for linguistic determinism centers on the number of words for snow in the Eskimo language. These words name the different sorts of snow which are important to the traditional Eskimo way of life, and it was argued that their existence enabled Eskimos to think about snow in a different way from Europeans. Furthermore, it was said, someone who did not have the words to describe different types of snow would not be able to recognize the differences between them. In fact this is not true. People who go skiing, for example, are perfectly capable of discriminating between different kinds of snow even though it is not a feature of their normal environment, and they have to construct descriptive phrases rather than using the single words that Eskimos have.

There have been experimental tests of linguistic determinism using color naming, which is a convenient task because colors can be identified independently of verbal labels by using physical measurements (Topic C2), and the number of basic words for color varies between languages. Such tests also fail to provide evidence for linguistic determinism since they show that people can discriminate between colors even though they do not have words to talk about them. While there is some evidence for the weaker hypothesis of linguistic relativity, the current assessment is that the influence of language on thought is probably small.

Categories and concepts

One of our most fundamental cognitive abilities is to be able to categorize our experiences since this enables us to recognize the recurrence of things, people and events, and to make use of our experience to plan and act consistently. The link between categorization and language is of the utmost importance because language enables us to give names to categories, to express the essential attributes of categories to form concepts, and to describe relationships between concepts.

A **concept** may be defined as the set of necessary and sufficient conditions for membership of a category. While it has been convenient to use this strict definition for the construction of formal models, the fact is that many everyday concepts have a somewhat looser specification that results in fuzzy boundaries and some consequent ambiguity over whether something belongs in one category or another. Examples of this might be the classification of a tomato as a vegetable or a fruit, or the question of whether a garment is a jacket or a coat. Furthermore, some concepts (e.g. game) do not seem to have defining attributes at all. We should also note that there is some debate about whether categories exist independently of our ability to perceive them (the **realist** view), or whether they are entirely constructed from our experience (the **anti-realist** view). The existence of natural categories (such as animals) provides support for the former view, but some thinking (in perception, for example) tends to favor the latter (Topic D1).

The ways in which concepts are related to each other have been described in a variety of formal descriptions that can be classified as either serial or parallel processing models. **Semantic network models** use serial processing mechanisms to operate on concepts that occupy **nodes** in a **network**. The links between nodes represent properties or other kinds of relationships. These models utilize clearly defined natural categories, and the example in *Fig. 1* illustrates this using two types of link: the 'is a' link and the 'has the property of' link, known as ISA and HASPROP, respectively. One of the useful characteristics of this hierarchical organization is that properties of concepts may be inherited by concepts below them in the tree. Thus, if 'cats' have the property of furriness there is no need to describe each type of cat as furry since they inherit the property.

The example in *Fig. 1* is a serial network, because when in use information is extracted from it in a step-by-step fashion. Two predictions from the serial type of model have proved particularly interesting. The first is that questions about nodes that are equally far apart should be answered equally quickly. This turns out not to be the case, but the suggestion is that this is because the links between nodes are not equally important; that is, they have unequal **activation strengths**. If processes within a network involve the spreading of activation then one of the

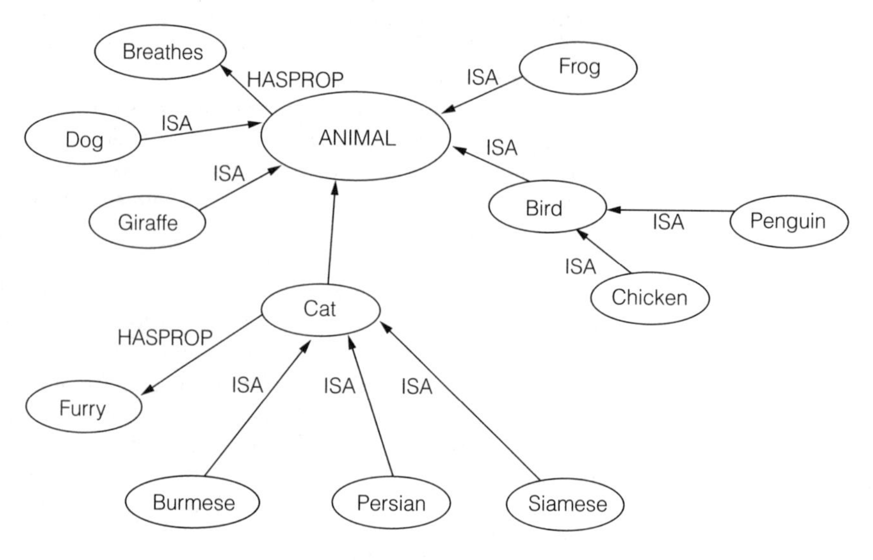

Fig. 1. Part of a semantic network. ISA links represent sub-cagegory relationships; HASPROP links are property relationships.

consequences may be the change in activation strengths to reflect use, or the modification of relationships as a result of learning. The second prediction is that questions about nodes that are close together (e.g. is a penguin a bird?) should be answered more quickly than questions about nodes which are further apart (e.g. is a penguin an animal?). The fact that this does not always happen casts doubt on the serial network as a useful model.

Parallel processing models (often called parallel distributed processing (PDP) models, or connectionist models) use a network structure that is modeled more on the nervous system than on a computer program. The nodes in these networks (sometimes called units) have the properties of activation and inhibition in a fashion analogous to neurons, and the fundamental difference is that concepts are represented by **patterns of activation** spread (or distributed) across a portion of the network and not by a single node. The same portion of a network can store several concepts without interference, using different activation patterns running in parallel. Computer realizations of PDP networks can learn (or, at least, be taught) by associating inputs with outputs, and by using a process of **backwards propagation of error** that corrects activation patterns that have led to incorrect responses. They can, therefore, operate without explicit sets of propositional rules and can store anti-realist categories, making them much more flexible than serial models. For instance, if we came across a small, furless animal which had three legs, purred and went miaow we would have little difficulty in categorizing it as a hairless cat which had lost a leg. Network models that accommodated such possibilities would be very cumbersome, but PDP models can readily cope with categories that have fuzzy boundaries.

Although PDP networks can provide useful accounts of cognitive structures and processes they are not without their difficulties. When they are being taught, PDP networks may learn very slowly (that is, they may need many trials) and this is at variance with the human ability to construct a detailed representation of a complex event (such as a football match) which has only been experienced once. Furthermore, when people are given information which contradicts previously learned patterns they can change their response very rapidly, while PDP networks may take rather longer. However, PDP networks offer the attractive prospect of computer-based models with an architecture which could map on to the structure of the brain, and their use in research continues to expand our understanding of a broad range of cognitive functioning.

Deduction

Deductive reasoning is the process of drawing logical conclusions from stated principles (called **logical propositions**) and evidence. People often make mistakes in deductive reasoning and this is demonstrated by studies of **conditional reasoning**, which is a common form of deduction based on the 'if–then' proposition: **if** an antecedent condition is met **then** a consequence follows. Two deductively valid and two fallacious conclusions can be drawn from an if–then proposition (*Table 1*).

In the **Wason selection task**, participants are given an if–then proposition and four two-sided cards. The task is to deduce whether the proposition is true or false by turning over exactly those cards which bear the necessary information (*Fig. 2*).

Most people recognize the need to apply the *Modus Ponens* argument and turn over the first card to ensure that the word 'blue' is on the reverse. However, few also recognize the need to apply the *Modus Tollens* argument and turn over

Table 1. Valid and fallacious inferences from a conditional proposition

Type of argument		Conditional proposition	Evidence	Inference
Affirmation of the antecedent *(Modus Ponens)*	Valid	If it is raining then Fran will get wet	It is raining	Fran will get wet
Denial of the consequent *(Modus Tollens)*	Valid	If it is raining then Fran will get wet	Fran is not wet	Therefore it is not raining
Denial of the antecedent	Fallacious	If it is raining then Fran will get wet	It is not raining	Therefore Fran is not wet
Affirmation of the consequent	Fallacious	If it is raining then Fran will get wet	Fran is wet	Therefore it is raining

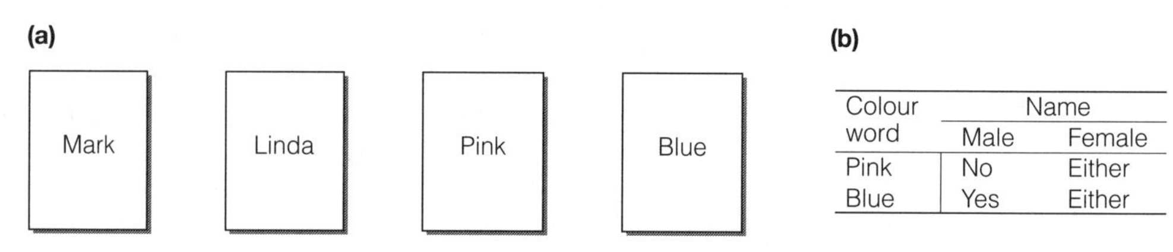

Colour word	Name: Male	Name: Female
Pink	No	Either
Blue	Yes	Either

Fig. 2. (a) An example of the Wason selection task. Which two cards must be turned over to test the rule 'If there is a man's name on the front of the card, then the word "blue" is written on the other side'? (b) Possible combinations of words for the four-card problem. The rule can only be tested by looking at cards for which there is a definite statement. Cards with a female name are of no interest because they can be combined with either color, and cards with the word 'blue' on them are of no interest because they can be combined with both male and female names.

the third card to ensure that there is *not* a male name on the other side, and opt instead for one of the other two fallacious alternatives.

Two types of explanation have been put forward for the mistakes that people make in logical reasoning: abstract rule theories and concrete rule theories. Abstract rule theories suggest that people represent deductive arguments by internal generalized rules that do not take into account the context of the problem. Mistakes may be due to **comprehension** errors, where the original premises of the problem were misunderstood; **heuristic** errors, where the co-ordination of stages in the reasoning process failed in some way, or **processing** errors due to failure of memory or attention processes. Concrete rule theories, on the other hand, suggest that the internalization of rules is sensitive to the context in which the rules are to be applied and that logically fallacious conclusions may be very reasonable in particular contexts. Take for example the proposition 'If you buy a new car from Dealer X then they will give you back £500 in cash'. You may reasonably believe that if you don't buy your car from this dealer they will not give you £500, yet this denial of the antecedent is technically fallacious. Similarly, if the dealer gives you £500 I could reasonably conclude that you have just bought a new car from them, even though by affirming the consequent I am committing a logical error. There is a lot of evidence which demonstrates that people construct their own **pragmatic rules** to help them deduce what might *reasonably* be true in the kinds of situations in which they find themselves.

Induction

Reasoning from specific observations to general conclusions is called **inductive reasoning**, and while deductively based conclusions can be regarded as logically certain, inductively based conclusions can never be better than highly probable. Inductive reasoning enables us to make causal inferences and to formulate hypotheses about the world around us based on our experience. On the whole, people make good use of the information in **causal inference** problems and generally follow the inductive principles first described by John Stuart Mill as the method of agreement and the method of difference (*Table 2*).

Table 2. Estate agents' observations on movement of property prices. How would you interpret the causal relationships by looking for agreement and differences?

Location 1	A new motorway is to be built one mile away A rehabilitation home for young offenders is to be built in the next street	Residential property prices drop by 25%
Location 2	A new motorway is to be built one mile away A new hospital is to be built in the next street	Residential property prices drop by 25%
Location 3	No new roads are planned An office block is to be built in the next street	Residential property prices are stable

Inductive reasoning always involves judgements of likelihood and we use **heuristics** to help us make probability estimates. The misapplication of heuristics can give rise to two types of error or bias. The **representativeness** bias occurs when making inferences about category membership. For example, participants in an experiment were presented with this description:

> 'Linda is 31 years old, single, outspoken and very bright. In college she majored in philosophy … and was deeply concerned with issues of discrimination'.

They were then asked to estimate the probability of each of these statements being true:

1. Linda is a bank clerk.
2. Linda is a bank clerk and an active member of the feminist movement.

Most subjects rated statement 2 as the most probably true, because the description of the woman is more similar to (i.e. representative of) the people described in statement 2 than those in statement 1. The reasoning is fallacious, however, because the probability of being a member of one category (bank clerks) cannot be less than the probability of being in that category joined with another.

The **availability** bias arises when we base decisions on the availability of information in memory. When something is easily recalled there is a tendency for people to think it is more likely than it is. Thus, if someone whose house value had dropped as a result of nearby road building was asked to judge the statements in *Table 1* they may well come to a different conclusion than a more neutral observer. In general, the incidence of dramatic, distinctive, or even simply recent events tends to be over-estimated as a result of this bias, a fact which has been observed amongst (for example) physicians and clinical psychologists.

At the conclusion of an inductive reasoning process we will often test our conclusions by looking for additional evidence. A rational procedure would be to seek evidence which might test (i.e. refute) the conclusion (see *Fig. 1*), but in fact people tend to look for evidence that confirms the conclusion and selectively discount evidence which may run counter to it. This **confirmation** bias is very pervasive and plays an important part in self-perception (Topic I4).

Another form of induction is reasoning by **analogy**. Verbal analogies are often presented as puzzles of the form 'lion is to pride as sheep is to (a) gluttony (b) sloth (c) flock (d) vanity?' where the task is to infer the relationship between the first pair of items and then use it to choose the correct alternative and complete the second pair. Much of the time taken to solve these problems is taken up by encoding them (that is, working out the form of the analogy) rather than by the reasoning operations themselves. In a wider context, analogical reasoning can facilitate problem solving when people bring their experience of similar problems to bear on a new one (Topic F5).

G1 THE STRUCTURE OF LANGUAGE

Key Notes

Human language

Only human beings make natural use of language. We speak many thousands of different languages, but they all serve the same communicative purpose and are constructed on the same general modular pattern.

Phonemes

We are capable of making hundreds of different speech sounds, but any particular language is built up from a small sub-set of these sounds. These are the phonemes of the language; those speech sounds which make a difference to the meaning of words.

Words and morphemes

Words are the smallest linguistic unit which can stand alone. They are composed of one or more meaningful sub-units, known as morphemes. Free morphemes, such as *dog* or *pick*, are independent words. Bound morphemes combine with other morphemes to change the meaning of a word, as in *pick* and *pick-ing*.

Syntax and sentences

Words are combined by the grammatical rules of a language (syntax) to create sentences. The number of words in a language is limited; the number of sentences is not. Sentences can thus express the meanings of an indefinite number of ideas or propositions.

Phrase structure

Sentences are composed of sub-units known as phrases. A phrase structure diagram is used to show the way in which a sentence is put together. Different sentences with different phrase structure diagrams may have the same meaning; the diagram shows the surface structure of the sentence. This surface structure is related by syntactic rules to the deep structure of the sentence; its propositional meaning. To understand a sentence, it is necessary to appreciate both the surface and the deep structure.

Related topics

Speech recognition and production (G2)
Reading (G3)
Language comprehension (G4)
Learning to talk (G5)
Communication (J1)

Human language

More than anything else, the ability to use language distinguishes human beings from other animals. The study of all aspects of the psychology of language and its use is known as **psycholinguistics**. There are many thousands of languages currently spoken and they differ radically from one another in many ways. However, they all have certain crucial characteristics in common.

The object of spoken language is communication. This may involve conveying information, giving commands, asking questions, or a variety of other functions. In order to achieve the goal of effective communication in a flexible and efficient way, all languages have a multilevel structure in which small numbers of components are combined together according to rules in order to express a potentially infinite number of meanings.

Phonemes

The most basic components of a language are **speech sounds**. These are produced as a result of air passing from the lungs through the vocal tract, and being modified by the vocal cords (the larynx) and by rapid movements of the tongue and lips. Human beings are capable of making many hundreds of distinguishable speech sounds. Any one language, however, only employs a small number of these differences. The basic sound units of a language are known as **phonemes** and they differ from language to language. There are about 40 phonemes in English; other languages use more or less. Phonemes are the building blocks of a language; they are defined not in physical terms but in relation to meaning in that language. In English, /r/ and /w/ are distinct phonemes (symbols for phonemes are conventionally enclosed between slashes), not because they sound different, but because there are pairs of words (such as *wig* and *rig*, or *twice* and *trice*) which differ in meaning *only* because of the difference between the two sounds. In other languages, such as Chinese, these two sounds are not distinct phonemes. Similarly, there are differences in speech sounds which are phonemic in, for instance, Urdu but not in English. All languages are based on a discrete set of phonemes rather than on the far larger, amorphous set of speech sounds.

Words and morphemes

In all languages, this small stock of phonemes is used to build up many thousands of words. A **word** is the smallest linguistic unit that can stand on its own. Words, unlike phonemes, convey meanings. The relationship between the sound of a word and what it means is arbitrary. There is nothing cat-like about the word *cat*; the language would work just as well if feline pets were called *brugs* or *glingwacks*.

Words are constructed from smaller units known as **morphemes**; these are the smallest linguistic units that carry meaning. **Free morphemes** such as *table, also,* or *fry* are words themselves and so can stand alone. **Bound morphemes** cannot stand alone but are combined in systematic ways to modify the meaning of a word. Examples in English are: *un-, -ing, -ly* which change the meanings of words in predictable ways, as can be seen in the word *un-surpris-ing-ly*. The commonest bound morpheme in English is the **suffix** *-s* which is attached to the end of a noun to indicate the plural.

The majority of words are **content words**. These are words which convey meaning even out of the context of a sentence; examples are: *hawk, drink, clever, quickly*. There are a much smaller number of **function words** which play an organizational (grammatical) role in a sentence and depend for their meaning on the sentence as a whole; examples are *and, into, the*.

Syntax and sentences

Just as a limited number of phonemes can be combined to produce a very large number of words, so words themselves can be combined to produce a theoretically infinite number of sentences expressing an endless variety of ideas. All languages use **syntax** (grammatical rules) to achieve this. English syntax depends largely on word order. Thus, *The dog bit the man* and *The man bit the dog* mean different things; the two sentences are composed of exactly the same words but English syntax dictates that the subject is different in the two sentences. Other languages use other devices to achieve the same object.

Different languages have very different sorts of syntax, just as they use different phonemes and have a different vocabulary. Nevertheless, linguists believe that there are basic similarities underlying the syntax of all languages. Only English syntax will be described here.

The **sentence** is the primary unit of analysis in linguistics and psycholinguistics. Syntactic rules apply to sentences but not to any larger linguistic units. A sentence can be grammatical (syntactically correct) or ungrammatical, but it makes no sense to talk about a paragraph, composed of grammatical sentences, being ungrammatical.

The job of sentences is to express the meaning of **propositions**. Propositions are abstract entities which correspond to one of the ways in which we store knowledge (see Topic H3). Grammar is employed to connect the words in the sentence in a particular way so as to capture a specific propositional meaning.

A simple sentence will be about some concept, the **subject** of the sentence, and will then state something about the subject; this part of the sentence is known as the **predicate**. In the sentence *The baby was sick* the subject is *The baby* and the sentence predicates that it *was sick*. Here, the verb requires only one item (the subject) to yield a complete sentence. Other propositions have a more elaborate structure and the sentences corresponding to them are consequently more complex. For example, the verb *eat* needs two components, as in *The baby eats the slug*; while *give* requires three: *The baby gives the slug to the cat*. The verb is the core of sentence meaning, and so determines how many players there should be in the little drama that the sentence describes.

Phrase structure

Just as words are built up of morphemes, so sentences are constructed from **phrases**; syntactically organized groups of words which act as units in a sentence. Examples are: *behind the cupboard, the man in the blue T-shirt, coming to tea*.

In order to understand how we produce and understand sentences it is essential to work out how the sentence is made up of phrases. This is normally shown in the form of a tree diagram known as a **phrase structure diagram**. *Fig. 1* shows a phrase structure diagram for a simple sentence. The string of words which constitutes this particular sentence is shown at the bottom of the diagram. The

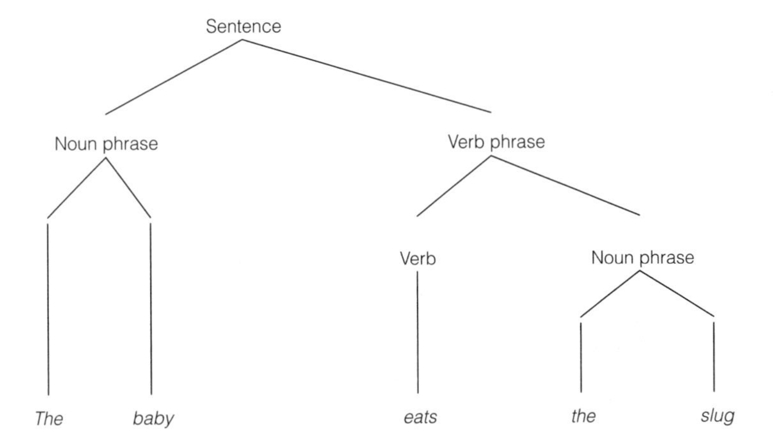

Fig. 1. Phrase structure diagram of a simple sentence.

topmost level states that the entire diagram describes a sentence rather than, for example, a phrase. In the intermediate layers of the diagram, the syntax of the sentence is analyzed to show how it is composed of a noun phrase relating to the subject, and a verb phrase dealing with the predicate. The verb phrase, in turn, is composed of a verb and a second noun phrase. Sentence structure is modular, with a small number of components which can be arranged in a great variety of ways to express an indefinite number of propositions.

Exactly the same diagram would describe many other sentences of the same syntactic form, such as *The man drives the car*, *The sun warms the earth*, etc. The meaning of a sentence arises *both* from its syntax *and* from the meaning of the particular words of which it is composed.

Figure 2 presents the phrase structure diagram for the passive form of the sentence in *Fig. 1*. These two sentences have essentially the *same* meaning (express the same proposition), but they have a very different structure. Thus, two syntactically different sentences can express the same proposition. In order to understand a language, listeners must be able to work out the phrase structure of the sentence; this is sometimes called the **surface structure**. They must also appreciate the underlying propositional meaning; often called the **deep structure**.

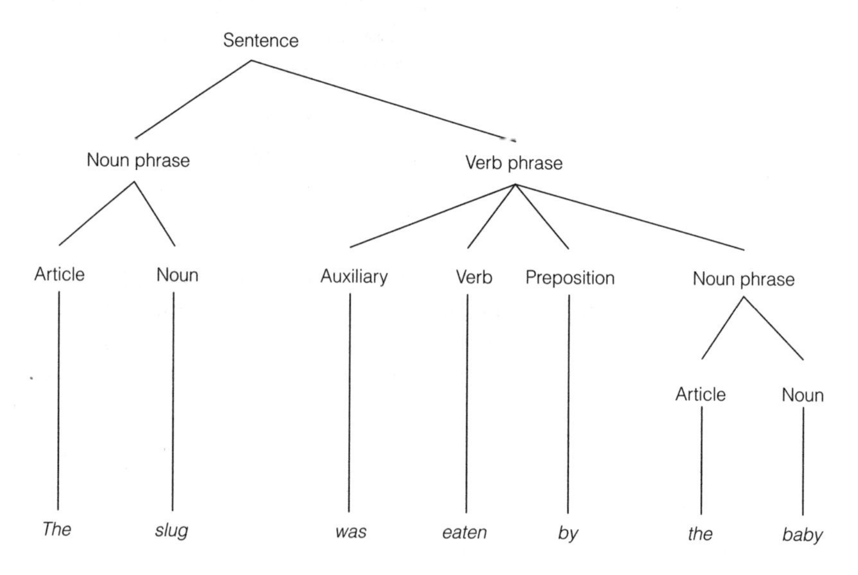

Fig. 2. Phrase structure diagram of the equivalent passive sentence.

Sentences can, of course, be far more complex than any mentioned so far. We can readily understand the following sentence: *The shark, which had attacked the diver, had been hooked by the fisherman, who believed that his line had become snared.* The basic proposition here is: *The fisherman hooked the shark*, but the sentence is elaborated with embedded clauses describing the shark and the fisherman's beliefs. The phrase structure diagram would be complicated but would be of essentially the same kind as those illustrated above.

The difference between the two sentences in *Fig. 1* and *Fig. 2* is that the first says something about the baby while the second focuses on the slug. In everyday use, sentences with different surface structures may relate to the same proposition but have different functions.

The baby ate the slug.
The slug was eaten by the baby.
Did the baby eat the slug?
The baby did not eat the slug.
Eat the slug! (spoken to the baby)

All these sentences (and many more) refer to the same, rather unpleasant, state of affairs, but approach it from a different point of view or **propositional attitude**. The surface structure of a sentence will therefore reflect the underlying proposition and also the speaker's purpose in using the sentence. How the surface structure and the underlying structure are related is controversial, but it is clear that, to understand sentences in use, we need to appreciate both aspects.

G2 SPEECH RECOGNITION AND PRODUCTION

Key Notes

Communication and language	The function of speech is to convey our ideas to other people through the medium of sound. Speaker and listener need to cooperate with one another in order to achieve effective communication.
Speech recognition	Identifying the phonemes in the speech stream is difficult because, when we speak, we articulate phonemes in an overlapping fashion and modify their pronunciation according to the surrounding phonemes. To help with these difficulties, the auditory system uses special mechanisms for the perception of speech.
Word identification	Listeners use the phonemic information derived from the speech stream to identify words. Theories of word identification suggest that, in addition to this bottom-up processing, listeners also use their knowledge of words to help them in the identification process (top-down processing). Both types of information are used flexibly to identify spoken words.
The mental lexicon	Our knowledge of the sound, meaning, syntactic role, and pronunciation of words is stored in the brain. This mental lexicon is a richly interconnected database of linguistic information. Access to the mental lexicon is typically extremely rapid.
Speech production	Less is known about the processes of speech production, much of the information being derived from speech errors. Planning speech occurs clause by clause and proceeds from the semantic to the phonological level in a series of stages which overlap in time.

Related topics The structure of language (G1) Language comprehension (G4) Learning to talk (G5)

Communication and language

We speak to communicate with others and they in turn interpret our words. Speech production begins with an idea in the speaker's mind; this is converted into a linguistic form which is translated into speech sounds. In essence, speech recognition is the reverse process, beginning with the perception of speech sounds and ending up with an idea in the listener's mind.

Language is typically used in a social context and, like most social activities, can only proceed effectively if there is cooperation between speaker and listener. In particular, speakers are expected to provide information that is relevant to the situation, that provides enough new information, but not too much, and that is accurate and clear given the context of the conversation. The way in which this is achieved depends, among other things, on the **common ground** between the speaker and listener; that is, how much knowledge they share. Where there is little common ground, as between total strangers, the speaker may need to provide a good deal of background information. Close friends can communicate effectively with utterances such as: *Bill's done it again* or *Let's have the same as last time*.

Although speech production logically precedes speech recognition, far more is known about the latter and it will be covered first.

Speech recognition

Words are composed of phonemes, so the first stage of speech recognition is to identify the phonemes in the input. In normal speech, phonemes may be produced at a rate as high as 15 per second. In order to achieve these speeds, phonemes are **co-articulated**. This means that they are pronounced in an overlapping fashion so that the sound pattern for one phoneme is altered by the phonemes that precede and follow it. Also, although we perceive speech as a discrete series of sounds and words, the speech stream is in fact a continually changing pattern of sound with very few gaps in it. We get some appreciation of this from listening to an unfamiliar foreign language.

Because of these difficulties, the auditory system deals with speech differently from other forms of auditory input. It has been argued that there is an entirely independent modular system for dealing with speech. Whether this is correct or not, there is a strong left-hemisphere advantage for perception of speech corresponding to the lateralization of language function (see Topic A4).

Identification of phonemes is essential for word recognition, and perception of phonemes is **categorical**. This means that we find it much easier to distinguish between two sounds representing *different* phonemes than between two sounds which are equally different physically but belong to the *same* phoneme (the same category). Categorical perception of phonemes has been demonstrated in pre-verbal infants, which suggests that it is a fundamental feature of speech recognition.

Word identification

For the reasons given above, the mechanisms for identifying phonemes cannot be error free. In addition, under normal conditions, there will be masking noises which obscure many phonemes, and the speaker will sometimes make speech errors. In order to recognize words, this degraded signal has to be enhanced. This can only be done by using knowledge of the words in the language together with contextual information. This is known as **top-down processing**, as opposed to the data-driven **bottom-up** processing arising directly from the speech stream.

Evidence for top-down processing comes from experiments in which subjects listen to sentences from which a phoneme has been deleted and a neutral sound substituted. They do not usually notice that the phoneme is missing; it is restored on the basis of word knowledge. In similar experiments, subjects are presented with sentences of the following sort (the asterisk shows where a phoneme has been deleted):

It was found that the *eel was on the axle.
It was found that the *eel was on the shoe.
It was found that the *eel was on the table.

Context determines what the subjects hear; in the first sentence they tend to hear *wheel*, in the second *heel*, and in the third *meal*. These and similar studies show the importance of top-down processes.

Cohort theory is a widely accepted theory of word recognition which incorporates both bottom-up and top-down processes. It proposes that there are **word recognition units** which are activated as soon as the first sounds of a word are heard. If the first phoneme of a word is /f/ the recognition units for all words beginning with that sound will be activated: the **word-initial cohort**. As additional phonemes are identified, the cohort of activated recognition units is reduced until only one unit, corresponding to the word to be recognized, remains active. The model predicts that words will be recognized at the point where their sequence of phonemes differs from that of all other English words; this is called the **recognition point**. Thus, in hearing the word *flimsy* the recognition point is reached at /*m*/, since no other words (other than derivatives) begin *flim-*, and recognition will be complete before the end of the word. Cohort theory also proposes that top-down contextual effects play an important role in word recognition, reducing the cohort of possible words still further.

This theory can be tested in word monitoring tasks where a subject has to respond to specific words in a spoken sentence. It is found that the target word is identified very fast, well before it is complete; furthermore, the better the context provided by the sentence the faster the recognition. In cases where the context provides strong cues, the word is identified well before the recognition point, which means that context operates early in word recognition.

Cohort theory lays much emphasis on the initial sounds in a word. In its original form, it would predict that a word could not be recognized if the initial phoneme was missing. A revised version of the theory remedies this problem and makes the theory similar to other theories based on network models. All current theories assume that top-down and bottom-up processes interact, that a number of candidate word recognition units are activated early in the identification process, and that processing is parallel, in order to account for the speed of recognition.

The mental lexicon

When we identify a familiar word we immediately recover its meaning. Similarly, if we are given a meaning, such as *the pet that meows and catches mice*, we can readily produce the word *cat*. Furthermore, we know that *cat* is a noun and that *catch* is a verb which requires two nouns to make a sentence. This shows that we must have an internalized knowledge base in which the meanings of words, their syntactic role, their sounds, and the way to pronounce them are all associated. This word store is known as the **mental lexicon**; a typical undergraduate may have around 50 000 **lexical entries**. For the sake of simplicity, it is assumed that the mental lexicon is the same for both speech recognition and production.

The process by which a word recognition unit activates the appropriate location in the mental lexicon is known as **lexical access.** Lexical access is very fast, typically about 200 ms. Furthermore, we can say very rapidly that a sequence of sounds such as *combrate* is *not* a word; this cannot possibly be done by carrying out an exhaustive search of all the items in the lexicon. These findings show that the mental lexicon is highly organized. Part of this organization must be semantic (dependent on meaning) since presentation of a word such as *doctor* will speed recognition of words with related meaning such as *nurse* and *hospital.* There is also evidence of other forms of organization in the lexicon including phonological associations such that words which sound similar are interconnected.

Understanding language does not end with lexical access. The further stages of comprehension are dealt with in Topic G4.

Speech production

Speech production proceeds from an intended meaning to sound, essentially the reverse of speech comprehension. It is hard to devise experiments on speech production and thus most of the evidence comes from the analysis of spontaneous speech. Speakers typically pause at clause and phrase boundaries; this suggests that speech is planned in chunks of this size.

Speech errors are a valuable source of information, although there are problems with bias in the collection of examples of errors. One common form of error is a selection error in which the wrong word is selected from the lexicon. Such errors always involve the same part of speech (a noun is not replaced with a verb) and the erroneous word is usually similar in meaning or sound to the target. Thus: *He came tomorrow* (yesterday) or *The emperor had several porcupines* (concubines). Such errors are good evidence for the organization of the mental lexicon along semantic, syntactic, and phonological lines.

Exchange errors are also common. These may be at the level of words (*We have a laboratory in our computer*), or of morphemes (*She wash upped the dishes*), or of phonemes (*the queer old dean* – the dear old queen). Such exchanges almost always occur within a clause and they involve similar items (noun for noun or consonant for consonant).

The form of these errors reveals something about the processes that generate speech. Theories of speech production agree that processing occurs clause by clause and that at least the following stages are involved:

- **semantic**: the intended meaning, which is not in linguistic form;
- **syntactic**: an appropriate grammatical form is chosen;
- **morphological**: words are selected from the mental lexicon and appropriate bound morphemes (such as *-ing* or *-s*, see Topic G1) are attached to them;
- **phonological**: the sound structure is determined, including stress and rhythm.

These stages are not sequential; processing goes on at all levels in planning an utterance, though at any one point the 'higher' stages will be more advanced than the 'lower' ones. While there is much still to be learnt about speech production, it is clear that it is based on the standard linguistic units of phoneme, morpheme, and word, and that it is not unlike the reverse of speech recognition.

G3 Reading

Key Notes

The nature of written language

Language may be written in a variety of ways. European languages, and English in particular, are alphabetic; in principle, each letter corresponds to a speech sound or phoneme. In English, however, this correspondence is poor. Reading any script can be defined as getting meaning from print.

Eye movements in reading

The eyes move over the page in a series of jumps (saccades). Skilled readers reading normal text make enough fixations to be able to get fairly full information about all the words on the page.

Word identification

Word identification in skilled reading is a bottom-up, data-driven process; this is different from the situation with speech, which is much influenced by top-down processing. Identification of letters does not precede word recognition; both processes go on simultaneously and mutually reinforce one another; this speeds word identification.

Routes to the lexicon

In reading there are different routes to the mental lexicon. One route involves a direct connection between an internally represented string of letters and a location in the mental lexicon. Alternatively, the reader may work out the sound of the word and then use the spoken word system to achieve lexical access. This indirect route does not work on a letter by letter basis, but makes use of analogies between words and correspondences between sounds and groups of letters.

Learning to read

Teaching reading is a central objective of education. Most children learn to read adequately, but a significant proportion do not, and a few suffer from specific reading problems or developmental dyslexia. The main problems facing a child learning to read an alphabetic script are to do with word identification.

Related topics

The structure of language (G1)
Speech recognition and production (G2)
Language comprehension (G4)

The nature of written language

Developmentally and historically spoken language comes first; however, the written word is of crucial importance in the modern world. The psychology of reading is therefore both important in itself and revealing about the psychology of language in general.

There are three major types of writing system, each of which works on a different set of principles:

- **logographic**, in which each written symbol represents a whole word and there is no connection between symbol and sound. Chinese is a logographic script;

- **syllabic**, in which each symbol represents the sound of a syllable and words are built up syllable by syllable. Japanese makes a partial use of a syllabary of this sort as do some Indian languages;
- **alphabetic**, in which the symbols (letters) represent phonemes. All European scripts are alphabetic. In some cases, such as Italian, the correspondence between letters and sound is very close. In English, there is a strong relationship but, in some cases, the correspondence between spelling and sound is poor, as in words like *knight* and *two*.

Reading may be defined as getting meaning from the written word, and someone is **literate** if they can do this as readily as they can derive meaning from the spoken word. This section will deal with word recognition in reading; the question of text comprehension is covered in Topic G4.

Eye movements in reading

We have the impression that our eyes move smoothly over the page when we read. In fact, if we are skilled readers, our eyes fixate a particular word for about 200 ms and then jump eight or nine characters in a rapid **saccade** and make another fixation. About 10% of these movements are **regressions** in which the eye moves back to an earlier point in the text. Detailed studies of eye movements suggest that a skilled reader can take in information from 1½ to 2 words on each fixation. Such a reader will make about 250 fixations in a minute and will read some 300 words. Since written words can typically be identified within 200 ms, there is time to extract fairly full information about each word in the text. The idea that skilled readers only identify a few words in a text is generally mistaken.

Word identification

Word identification in skilled reading is normally an automatic, bottom-up process. The first stage might be expected to be recognizing the letters of which the word is composed. However, experiments show that letters can be identified *more readily* when they occur in words than when they are presented *by themselves*. This **word superiority effect** shows that word identification cannot follow letter identification, but must go on at the same time, in such a way that partial identification of a word confirms the partial identification of its letters. There are a number of successful neural network models of this process, all of which assume that the letters in a word are processed in parallel, and that word and letter identification occur simultaneously. The two processes facilitate one another and this speeds up both word and letter recognition.

Word identification in reading is different from the equivalent processes for speech. Reading is normally a self-paced task, the letters are clearly presented (unlike the phonemes in speech), and the letters in written words can be processed in parallel, while phonemes in speech have to be dealt with sequentially. These differences account for the different types of processing involved.

Routes to the lexicon

The output of the word identification process is an internal representation of a string of letters. This is in an abstract format which discards information about typeface or case; experiments have shown that, in appropriate circumstances, subjects can identify words and yet be unaware of whether they were presented in upper or lower case.

There are at least two different ways in which this internal representation of a written word could access the mental lexicon (see Topic G2). The **direct route** consists of a connection going straight from the letter string to an entry in the lexicon so that, for example, the string of letters *d-o-g* directly activates the *canine pet* location in the mental lexicon. This route does not involve the sound of the word. The Chinese *must* read in this way since Chinese characters stand for whole words and do not indicate how they are pronounced; in the same way, in English the symbol *&* gives no indication of sound but still means *and*. The **indirect route** is mediated by the sound of the word. In alphabetic scripts, letters correspond to sounds so that the reader knowing these correspondences can, in principle, work out what the letter string *d-o-g* sounds like. This gives access to the lexicon using the same processes as for a spoken word.

Readers of English make use of both routes. Evidence for use of the direct route comes from words such as *colonel* or *quay* which are spelt in such an irregular way that they could not be read by the phonological route. **Homophones** are words such as *peak* and *peek* which are spelt differently but pronounced the same; we readily understand which string of letters has which meaning. If such words were read by the phonological route there would be no way of knowing which meaning was intended, since only the sound of the word would be available.

English readers must also be able to use the indirect route. We are able to sound out nonwords such as *tep* and *rass*, which can only be done by taking account of the sounds the letters represent. Occasionally, we will come across words which we know in speech but have never seen in print before; here, the direct route is not an option.

However, English spelling is sufficiently irregular that letter by letter sounding out of words would often fail to produce a recognizable pronunciation. When groups of letters, such as *-ight*, are taken into account, more regularities appear and English spelling no longer appears so irregular. **Analogies** can also be helpful. If I know the word *leak*, I should be able to find a pronunciation for *beak* or *weak* or even for a nonword like *deak*.

The indirect route does not therefore depend on letter by letter translation from letter to sound. Rather, it uses groups of letters and analogies to construct a pronunciation of a word. It remains different from the direct route because it operates on elements smaller than whole words, and because it makes specific reference to sound in accessing the lexicon.

There is neuropsychological evidence in support of the existence of these two routes to the lexicon. Some brain-injured patients appear to be impaired in reading because of problems with the direct route, others have specific difficulties with the indirect route. These deficits are known as **acquired dyslexias**.

Learning to read

Skilled readers use the direct route for identifying most words; the situation is different when we are learning to read. In this century, for the first time in history, universal literacy is a general objective; teaching children to read is seen as the single most important educational aim. With adequate schooling, the great majority of children learn to read. However, a substantial minority leave school **functionally illiterate**; that is, unable to cope with the normal reading demands of everyday life. There are a variety of reasons for this: inadequate teaching, emotional problems, family difficulties, low intelligence, etc. There is a smaller group of children who are of normal intelligence and do well in subjects other than reading but who have quite specific reading problems; they are known as **developmental dyslexics**.

In order to read, it is necessary to discriminate between the letters of the alphabet. Virtually all children achieve this, although there may be difficulties where letters are mirror images of one another ('mind your p's and q's'). Failure to identify letters is *not* a major cause of reading problems, although children who are familiar with the letters of the alphabet when they arrive at school typically learn to read easily. This is because they tend to come from families with an interest in reading, and also because ready familiarity with letters makes the process of word identification faster and more accurate.

Failure to identify words is a more important source of reading difficulties. When they first begin to read, most children start by recognizing a small number of common words as single items, without any decomposition into letters. This may be a good way to get started on reading but it has a number of disadvantages: each word has to be individually recognized, which imposes a strain on memory; words may be recognized by inappropriate characteristics (e.g. *elephant*, 'the longest word I know'); the child has no way of tackling new words, either a word is known or it is not. To advance in reading, children need to 'crack the code' relating letters on the page to sounds. Once they have done this they can make attempts at words they have not seen before; teachers say that they have **word attack skills**.

From the very early stages of learning to read children make use of simple analogies (of the sort described above) to read new words. In addition, they learn to identify sequences of letters where there is a good correspondence between sound and writing. Thus, they may learn that the letters *-eat* correspond to a particular sound; this then allows them to read *beat*, *cheat*, *meat*, and a range of other words. In order to do this, however, they need to be able consciously to analyze *spoken* words into their component sounds. If the child does not recognize the spoken similarity between *beat*, *cheat*, and *meat*, she cannot take advantage of spelling similarities in order to read the words.

This ability to analyze spoken words is known as **phonemic awareness**. One test of phonemic awareness is the 'odd-man-out' test, in which sets of spoken words such as *pin*, *win*, *ran*, *tin* are presented and the child has to identify the 'odd-man-out'. Children who perform well on such tests *before* they can read, later learn to read more easily than those who perform badly; it seems that phonemic awareness is a vital prerequisite for learning to read. Developmental dyslexics are commonly deficient in phonemic awareness.

Children may be taught to read using either the **whole word** (or **look-and-say**) method, or by the use of **phonics**. The whole word method teaches them to recognize complete words without worrying about the letters of which they are composed. Phonics approaches, of which there are many, emphasize the relationship between letters, groups of letters, and sounds. In practice, most teachers make use of both methods according to how the child is progressing, but evidence strongly suggests that phonics, especially systematic phonics (in which the relationship between letters and sounds is explicitly emphasized), is the most effective approach.

Children who can read words with ease very rarely have further problems with reading; comprehension usually follows readily once lexical access is automatic. Learning to read is essentially learning to read words.

G4 LANGUAGE COMPREHENSION

Key Notes

Listening, reading and comprehension

Lexical identification of written or spoken words is only the starting point of understanding language. Post-lexical comprehension processes operate in a similar way for spoken and written material, despite the differences in mode of presentation and the different cues that they provide. Theories of comprehension largely ignore these differences.

Parsing

In order to understand a sentence it is necessary to parse it; that is, to work out its grammatical structure. Parsing occurs as the words of a sentence are being recognized. In order to do this effectively the parser must guess an appropriate structure early in the sentence. This is probably done by using rules of thumb which generally predict the simplest structure for a sentence unless evidence arrives which contradicts this assumption. Models of the parser are tested by using sentences in which the initial words are misleading about the true structure of the sentence. Parsing generally proceeds without taking the meaning of the sentence into account.

Inferences and comprehension

Sentences express propositions and, once they have been parsed, the sentences tend to be forgotten and only the propositions that they express are remembered. These propositions have to be linked together and to a more general knowledge of the world by a process of inference. Simple inferences are used to decide who or what particular pronouns refer to; these link together sentences in continuous text. Other inferences depend on general knowledge, and an indefinite number of such inferences could be made for any sentence. It is likely that the only inferences made at the time that the sentence is processed are those directly needed to understand it.

Related topics

The structure of language (G1)
Speech recognition and production (G2)
Reading (G3)
Learning to talk (G5)

Listening, reading and comprehension

The previous two topics describe the process of understanding written and spoken language up to the point of lexical access. Comprehension is a **post-lexical** process, the starting point of which is the output of the mental lexicon; this can be thought of as a sequence of words. In comparison with readers, listeners have the advantage of information from intonation, gesture, and contextual support. On the other hand, readers can go at their own speed, can retrace their steps, and have a less defective input. Despite these differences, comprehension processes appear to be very much the same whether the input to the lexicon is spoken or written; most theories therefore ignore the input modality. (Because most of the research has been done with comprehension in reading, the rest of this section will refer to 'readers'. However, the findings generally apply to both listeners and readers.)

Parsing

In order to understand a sentence it is necessary to work out its syntax; essentially to construct a phrase structure diagram (Topic G1). This is known as **parsing** and, because of the speed with which it is done, it must occur as the words are heard or read. Therefore, the parsing mechanism must try to make sense of partial sentences before all the evidence is in.

For example, the words *The man* ... could be the beginning of an indefinite number of grammatical sentences. It is not possible to consider all of them and eliminate ungrammatical ones as more words come in. We need to work out a single structure and modify it only if it turns out to be inadequate. To do this we must make use of our knowledge of syntax; we must also have some effective rules of thumb for generating likely phrase structures. One way of testing how the parser works is to devise sentences which appear, up to a critical point, to have one syntactic form and then turn out to have another. These are known as **garden-path sentences** because they mislead the reader about syntax; this gives valuable clues about how the parser works.

The **garden-path model** proposes that parsing depends *only* on sentence structure and is unaffected by meaning. As we read a sentence, we choose the simplest possible model of sentence structure by adopting two related rules. **Minimal attachment** states that new words are attached to the phrase structure in a way that involves the minimum number of new units (phrases or clauses). Thus, in the sentence *The girl knew the answer by heart*, *'the answer'* is assumed to be the direct object of *'knew'* and therefore part of the same phrase. This guess turns out to be wrong in the sentence *The girl knew the answer was wrong* where *the answer* is the beginning of a new noun phrase; in such a sentence the parser would have to pause at *was* and recompute a phrase structure. The second rule, **late closure**, means that new words will always be attached to the phrase currently being analyzed, rather than assuming that they belong to another one. This ensures the correct interpretation of a sentence such as that in *Fig. 1a*, but fails in the sentence shown in *Fig. 1b*, where *a mile* belongs to another sentence.

Use of these strategies allows the parser to construct a single interpretation of a sentence, increasing speed of comprehension and greatly reducing demands on memory, since only one phrase structure has to be considered at any time. The theory has been tested by measuring eye movements as people read garden-path sentences. The results show that there are pauses for extra processing only at points where there is a conflict between the actual structure of the sentence and the structure produced by minimal attachment and late closure, suggesting that the model is appropriate.

The garden-path model supposes that meaning plays little part in working out syntactic structure. The evidence here is conflicting. Where the syntactic clues are strong, it seems likely that parsing proceeds independently of meaning, but where the cues from meaning are powerful they do help to guide parsing. For example, it takes longer to decide who was the actor in the sentence *The brother was hated by the sister* than in *The sister hated the brother*; but there is no difference with the sentences *The swimmer was rescued by the lifeguard* and *The lifeguard rescued the swimmer*. In addition, readers can make use of punctuation, and listeners of intonation, to guide them in parsing a sentence. Parsing therefore cannot be an entirely independent syntactic process as garden-path theory suggests.

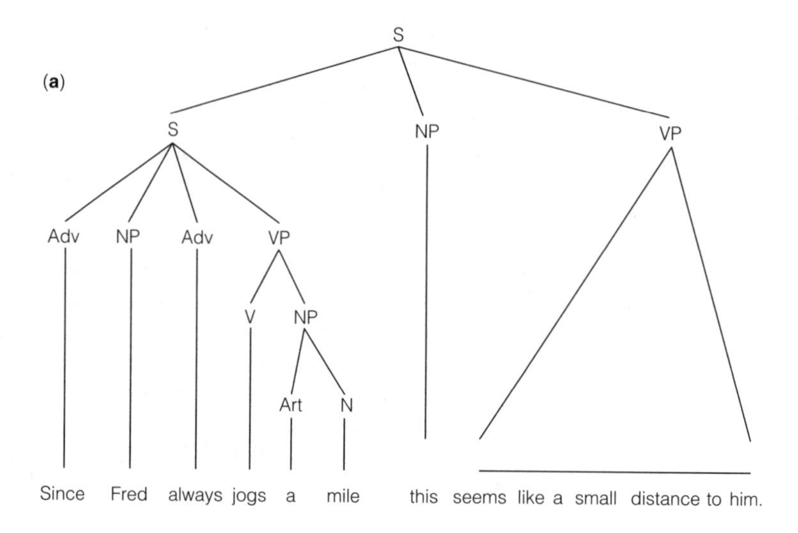

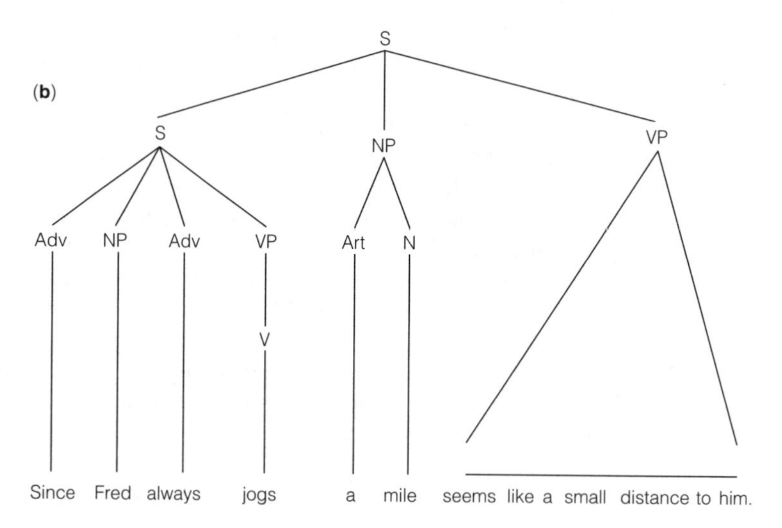

Fig. 1. Sentence in which late closure results in (a) a correct structure diagram; (b) an incorrect structure diagram. S = sentence; VP = verb phrase; Adv = adverb; V = verb; NP = noun phrase; Art = article. No analysis is shown of the phrase marked with triangle.

Once meanings have been attached to words and the syntax of the sentence has been worked out, the reader or listener is in a position to determine the meaning of the proposition which the sentence expresses. Little is known about this process; it must, in principle, be the reverse of that by which a speaker composes sentences from propositions. Until more is understood about the mental representation of knowledge it is unlikely that this aspect of comprehension will be properly understood.

Inferences and comprehension

Comprehension involves more than parsing individual sentences. Sentences generally have to be related to one another and/or to knowledge of the world if they are to be properly understood. Experiments on text recall show that subjects store the propositional content of a sentence but not its syntactic (grammatical) form, which appears to be lost once the sentence has been parsed and understood. The process of text comprehension therefore consists of connecting these propositional representations into a coherent mental model.

The following brief story illustrates some of the features of this process:

Jane was invited to Jack's birthday party. She wondered if he would like a kite.
She went to her room and shook her piggy-bank. It made no sound.

The individual sentences, or rather the propositions that they express, are not very obviously related; yet the story is comprehensible because of a series of inferences which go far beyond what is written.

Some of these inferences are internal to the text; we infer that the pronoun *she* refers to Jane, and that *it* in the last sentence means the piggy bank. Such inferences are called **anaphora** and are generally fairly straightforward. In the story above, gender alone establishes what the pronouns refer to. In other cases, however, we need to use our knowledge of the world in order to understand an anaphoric reference. In the sentence *Peter was being questioned by the detective; he had been accused of murder,* we assume that *he* refers to Peter, rather than the detective, because of what we know about the police.

More generally, we make constant use of **inference** to interpret texts. The apparently unconnected sentences in the story about Jane and Jack are held together by what we know about birthday parties, the need to buy and give presents, and the significance of the silent piggy-bank.

There is considerable debate about the role of inferences in text comprehension. In the sentence *He chopped down a tree*, there is no mention of an implement but, in a memory test for a series of such sentences, *axe* will often prove as effective a recall cue as *chopped*. This has been taken to mean that the inference that an axe was used to cut down the tree was made at the time the sentence was read. One problem with this idea is that there is no limit to the number of inferences that might be drawn from a sentence and most of these inferences would be unnecessary, generating nonessential processing. Furthermore, any inferences which are detected by means of recall tests may in fact have been made at the time of recall and not when the sentence was read.

It is more likely, therefore, that only a small number of inferences are made at the time that a sentence is originally processed. Some of these inferences are straightforward and automatic; anaphoric reference for example. Others must be made to understand the sentence properly; it is only by this sort of inference that we can appreciate the radically different states of affairs described by these two sentences:

The policeman raised his hand and stopped the car.
The goal-keeper raised his hand and stopped the ball.

Faced with a disjointed text, such as the piggy-bank story, the reader may be forced to make goal-directed inferences serving the specific purpose of understanding the text and constructing an appropriate mental model.

G5 LEARNING TO TALK

Key Notes

The elements of language learning

Children learn to talk their native language without specific tuition and with remarkable ease and rapidity. This achievement is so remarkable that it has been claimed that it must be due to innate factors. It is possible, however, that a fuller understanding of the social context in which language is acquired will provide explanations which do not depend on very specific innate components.

Phonology

Young babies produce an enormous range of speech sounds. Learning about phonology involves eliminating the sounds which are not phonemes in the child's native language and acquiring the phonological rules of that language.

Learning words

Children's first words occur around the age of 1 year and develop out of pre-verbal communicative exchanges between the baby and her caregiver. These words develop as part of familiar routines in the child's world.

The naming explosion

Between 18 months and the age of six, children learn an average of six new words a day. This extraordinary feat is at least partially explained by the social context in which word learning occurs, which provides considerable support to the process, and a variety of cognitive strategies employed by the child.

Learning syntax

From the moment that children begin to use two-word utterances they show a sensitivity to syntax, and by the age of six they are using most grammatical constructions with ease. This is not done by imitation since most of their utterances are novel. It is not clear whether any specifically linguistic innate structures are necessary in order to account for the naming explosion or the acquisition of syntax.

Related topics

Complex learning (E4)
The structure of language (G1)
Speech recognition and production (G2)
Language comprehension (G4)

The elements of language learning

Learning to understand and speak our native language is one of the most impressive of human accomplishments and is achieved at a remarkably early age. The newborn infant merely babbles, the 2 year old has adopted the phonemes of her native tongue and is beginning to establish a vocabulary of words, the 5 year old knows several thousand words and has also learnt much of the syntax of her language.

These are striking accomplishments which are far from fully understood; however, research has provided an outline of how they may be achieved. The critical issue is the degree to which language can be acquired through normal learning mechanisms (see Section E). It is often claimed that there is a substantial innate component, particularly in the acquisition of syntax; but the detailed nature of this component is rarely specified and therefore strong examples of this position are more statements of faith than explanations. In resolving this issue, it is essential to consider the context in which language is acquired. The purpose of language is communication and this may be the key to understanding its acquisition. In particular, we need to take into account the linguistic and nonlinguistic interactions between parent and child, and the distinctive ways in which adults talk to young children. It may be that these factors provide sufficient support to account for the learning of language without appealing to a specialized innate component.

Phonology

This is the first, and probably the simplest, element in language learning. A 6-month-old infant produces a wide variety of speech sounds, including many which are not part of the phonemic structure of her native language and which she will not be able to articulate or discriminate as an adult. By the age of 18 months, she will be producing mainly the speech sounds which are phonemic in her language; she has learnt to *ignore* other distinctions. This process begins very early; babies a few days old show a preference for listening to speech in their caretaker's language. Learning about phonology continues throughout childhood. Two or three year olds notoriously mispronounce words. Some errors are mere simplifications, e.g. *bubba* for *bubble*. Others are more revealing: a child may say *fick* for *thick*, but *thick* for *sick*. Such errors are clearly not due to an inability to produce the sound; rather, they are evidence that the child is learning the rules of phonology of her language but making some errors in the process. This process of learning phonology continues for a number of years with the child developing more and more complex rules.

Learning words

The earliest words arise out of pre-verbal communication. Two- or three-month-old infants engage in turn-taking communicative exchanges with their caregivers. They gaze at one another, and the caregiver talks to the infant, using a few carefully articulated words, and then pauses for the baby to 'reply'. The initiative, at this age, lies largely with the caregiver. By 7 months, interactions between infant and caregiver are more equally shared and begin to involve objects and people other than themselves; this is known as **secondary intersubjectivity** and appears most obviously where both are looking at or playing with an object of joint interest. In these situations, the caregiver will usually name the most salient objects or actions. Speech to children of this age is often called **motherese** (although it is used by almost everyone talking to young children). Sentences are short and grammatically simple, important words are repeated, and intonational patterns are exaggerated. Both the context and the form of such speech are likely to make it easy for the child to associate word and meaning, and at around 9 months infants appear to understand a few words or expressions such as *Wave bye-bye* or *Do you want a drink?* which are used in familiar, ritualized situations.

At the same age infants begin to produce their first 'words'. These are vocalizations which may not sound much like an adult word but which are used consistently by the infant and are understood by their caretakers. The majority of these very early words are likely to be nouns, but there are great individual differences in the frequencies of different categories of words. They relate to familiar and interesting objects and routines in the child's world. They are likely to include such things as *doggie*, *sock*, *all-gone*, and (especially) *No!*. The child may use a word to label something, but this function accounts for only about half the uses of first words. They are also used to note changes (*all-gone*), to refer to things which are not immediately present, and to get others to do what the child cannot do for herself (*milk*).

The naming explosion

Around the age of 18 months most children will know some 50 words. The number starts to increase very rapidly at about this age, and by the age of six they typically have a vocabulary of about 10 000 words; they therefore learn about six words a day between these ages.

Quite apart from the demands on memory, learning the meaning of words poses major problems. When a child hears a new word, how does she decide what it refers to? Her father says *There's a sheep*. How is the child to know what is being referred to? The woolly animal, the field it is in, its ear, the nice sunny day, or an indefinite number of other possibilities? It seems impossible; but a proper analysis shows that there are a variety of sources of information to help with the task.

First, the situation itself is usually strongly constrained by context. The father points at the sheep, itself a salient object. The critical word is commonly strongly emphasized and may be repeated. The remark may be part of a series of exchanges all of which are about animals, so attention is already focused on animal words. All this will help the child attend to the correct referent for the word.

Learning vocabulary is closely tied to categorization and, by the age of one, children are already using categories similar to adult ones. They divide the world up into discrete objects, some of which are animate and can act on other objects; others are inanimate and are incapable of independent action. These very basic categories have evolved to enable us to represent the world effectively and may be common to all advanced mammals. Such a nonverbal representation of their surroundings provides children with a structure on which to map not merely words but also simple sentences such as *The dog chases the ball*. Because they categorize objects in this way, children are unlikely to consider such logically possible items as *the sheep's left foreleg and tail* as likely meanings for a word.

Furthermore, children seem to assume that no two words can have the same meaning, and that one object can only have one name; so the new word *sheep* will not be attached to anything whose name the child knows already, the dog at the gate or the bird in the air. This situation is complicated by the fact that the same object may in fact be referred to in differing ways. Depending on circumstances the same object may be *Tibbles*, *the cat*, *an animal*, or *our pet*. Children show a strong tendency to label objects at intermediate levels of abstraction, except in the case of personal names. These naming biases must make it easier for the child to make a correct choice when trying to decide what a word means.

Of course children do not always work out the meaning of a word correctly. Children often use a word too generally; thus *dog* may be used for any animal which is not too large, or (more embarrassingly) all men may be called *Daddy*. Such **overextensions** are common during the first few months of learning words but then become comparatively rare. As the child learns more and more words, the occasions for overextension are reduced.

Finally, as children become familiar with their native language, they begin to use syntactic cues to work out the meaning of new words. Thus, in English, a word ending in *-ed* is likely to be part of a verb, and one preceded by *the* is probably a noun. This will further limit the possible referents for a new word.

No one of these cues alone can account for how children can work out what words refer to, but a combination of them in familiar and repeated conversational situations goes some way to explain how this seemingly impossible feat is achieved.

Learning syntax

Direct imitation of adult speech plays an important part in learning words. By contrast, children do not learn syntax by imitation. Language is fundamentally creative, and very young children produce combinations of words which they have probably never heard, such as *pussy all-gone* or *sheeps goed home*. Detailed records of children's early speech show that few children show much direct imitation of adults; most of their utterances are their own invention.

The sorts of mistakes that children make in learning syntax are often revealing. For example, in English, the past tense of a verb is indicated by the inflection *-ed*; however, there are many common verbs which are irregular, such as *go–went* or *tell–told*. Initially, children may use these forms correctly, but then they begin to produce **over-regularized** versions such as *goed* or *telled*. These may persist for some years until they are eventually replaced by the correct, irregular form. Such over-regularizations are not very common, but they are informative about the acquisition of syntax since they show how children learn rules rather than individual words. They also show how they (mis)apply the rules that they have learnt.

The first evidence for the learning of grammar comes as soon as children start putting two or three words together, generally around the age of two. These utterances are not random associations of words but show regularities. (Examples are taken from English because it is the most fully researched language. Studies of other languages suggest that the broad principles of learning syntax are the same as in English, although the details are very different.) There is considerable individual variation but a common pattern is to combine one frequently used word with one of a number of other words, as in *all-gone drink, all-gone Daddy, all-gone pussy*. Generally, they employ word order to indicate propositional roles; so *throw ball* and *ball throw* will not be used to express the same thing. There is even evidence that children at the one-word stage are sensitive to word order in the sentences that they hear. Since word order is a crucial syntactic device in English, and in many other languages, these regularities are important evidence for sensitivity to syntax at the earliest stages of language production.

After the age of about two, the **mean length of utterance (MLU)** (MLU counts the number of grammatical morphemes rather than words; thus *dog-s* or *jump-ed* are each two morphemes) begins to increase rapidly, although here again there are marked individual differences (see *Fig. 1*). To begin with, function words are omitted, probably because of the memory and planning problems involved in producing multi-word sentences. Following this '**telegraphic**' stage, more complicated forms appear such as *That's the book* or *I'm walking*, and finally such elaborate forms as tag questions are mastered: *I saw you, didn't I?* or *You're joking, aren't you?*. Even adults have difficulty in specifying the rules for such phrases, yet children acquire them with little apparent difficulty. They must somehow derive abstract grammatical rules from the often ungrammatical speech which they hear around them. Furthermore, it is very rare for parents to

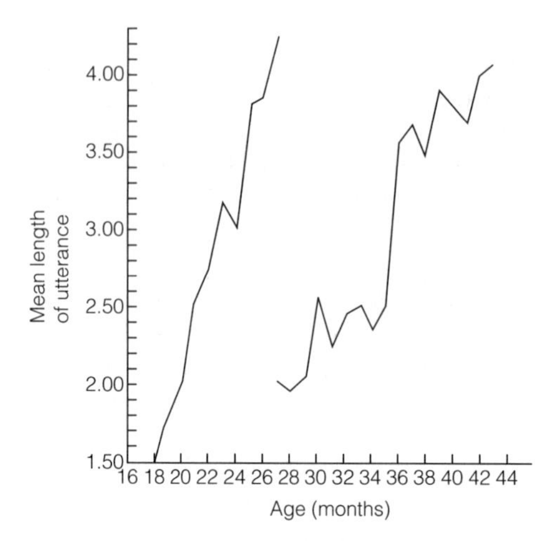

Fig. 1. Individual differences in the increase in mean length of utterance in the first years of life. Reproduced from Brown, A. (1973) A First Language. *Harvard University Press, Cambridge, MA.*

correct children's grammar, so children do not generally receive the feedback that would be useful in working out the rules.

These facts lead some researchers to think that there must be an innate component specifying possible types of grammar, and children use this information to work out the particular grammar of their own language. Others believe that, as in the case of learning words, there are various other factors which could enable children to learn grammar without this specific innate information. The arguments are complex and technical, and the issue remains to be resolved.

H1 THE NATURE OF DEVELOPMENT

Key Notes

Stages of development

Life-span developmental psychology studies cognitive, social and emotional changes during the lifetime of the individual. Four main stages are identified: prenatal, childhood (including infancy), adolescence and adulthood.

Developmental processes

Development involves differentiation, orderly progression, and cumulative change. Genetic factors and environmental influences interact to produce the phenotype (our final individual characteristics). Important processes are maturation and learning, especially social learning.

Related topics

Evolution and biological psychology (A1)

Heredity, the environment, and intelligence (K8)

Stages of development

Developmental psychology is the study of how the behavior and mental processes of organisms change over time. Development starts from the moment of conception and continues until death. It is usual to examine human development in four main stages of life:

- **Prenatal**. Before birth, cells differentiate to form specialized tissues and organs, and the basic structure of the body is formed. During this period we are most vulnerable to environmental factors that can adversely affect physical development.
- **Childhood**. This may be divided into *infancy*, the first 2 years or so, when rapid development of motor, perceptual, and cognitive functions takes place; *early childhood*, the remaining pre-school years, when much early social development occurs, often in the context of play; *middle childhood*, the period before adolescence, when further major cognitive development occurs.
- **Adolescence**. Around 12–18 years is the transition from childhood to adulthood. A key change is sexual maturation. Other features are increasing independence from care-givers, and changes in social relationships.
- **Adulthood**. During adulthood and into old age, we develop further as we occupy a variety of changing roles in life. Goals may change, and with advancing age, physical and cognitive performance often deteriorate.

The study of this **life-span development** encompasses physical, cognitive, social and emotional changes. We will concentrate on cognitive and social development during infancy and the later phases of childhood, and on some aspects of development during adolescence and adulthood. We look at language development in Topics G3 and G5.

It must be noted that developmental progress shows high individual variability. Particularly in adulthood, economic, social and cultural changes over the years have reduced the extent to which it is possible to describe a 'normal' pattern of development.

Developmental processes

Most types of development have features in common. Development involves:

- **Differentiation**. This is clearest in physical development in the prenatal period, as tissues become differentiated. It is also apparent in aspects of motor activity in infancy (e.g. the development of hand use from grasping objects with the whole hand to manipulating them using the opposed thumb), and in perceptual development.
- **Orderly progression**. Many aspects of development cannot take place unless necessary earlier changes have occurred. For example, the development of locomotion over a period of some 15 months in infancy follows an orderly progression from the infant lifting the chin, through crawling, standing, and walking. The notion of orderly progression is a crucial one in theories of development as we will see in later topics in this section.
- **Cumulative change**. Early experiences and development affect responses to later developmental influences. For example, the development of perceptual abilities depends on exposure to a rich environment in early infancy; the family environment can produce changes that predispose the individual to develop depressive illness.

The mechanisms of development include **genetic factors**. The underlying course of development is determined by the genes that we inherit from our parents (see Topic A1). The **genotype** that these provide is modified by a variety of **environmental influences** to produce the **phenotype** (our final individual characteristics). Examples of this interaction are seen throughout this volume, and we discuss the **nature–nurture** debate that occupied much of the last century in Topic K8. Environmental factors include family structure and interactions, education, and peer group and other sociocultural pressures. **Maturation** is a genetically programed sequence of changes. It is most obvious in physical development, but is necessary for all development. For example, a behavior cannot develop until the underlying neural pathways have matured. Maturation can be affected, advanced or delayed by environmental factors. All the types of **learning** that we looked at in Section E have a role to play in development. Of particular importance is **social learning** (including imitation and modeling), taking place within the family or with other children. Maturation is greatly affected by **nutrition**. Malnutrition in infancy restricts physical growth, and can have far-reaching effects on brain development, and consequently on cognitive and social development. These factors all interact to produce individual variation in the rate and extent of development.

H2 Cognitive development in infancy

Key Notes

Piaget's theory of cognitive development

Piaget emphasized orderly progression and cumulative change. During infancy, the child is in the sensorimotor stage, which has six sub-stages: exercising reflexes (birth–1 month); preliminary circular reactions (1–4 months); secondary circular reactions (4–8 months); coordination of secondary reactions (8–12 months); tertiary circular reactions (12–18 months); and invention of new means (18–24 months).

Object permanence

Object permanence is the understanding that objects continue to exist when they are out of sight. This develops through sub-stages 3–5. At sub-stage 4 the infant will search for completely hidden objects, but will show the 'AB error': searching for a hidden object in a usual hiding place despite having watched it being hidden elsewhere.

Schemata

For Piaget, development consists of the formation and coordination of schemata: mental ways of understanding the world. Schemata develop from simple behaviors to coordinated sequences of action. The development of schemata (adaptation) occurs through assimilation, the process of acquiring new information and incorporating it into existing schemata, and accommodation, when existing schemata are modified to encompass new information.

Criticisms of Piaget's theory

A child's failure in a Piagetian task may be due to lack of motivation, or physical or perceptual immaturity (a discrepancy between competence and performance). Making tasks easier to manipulate or more interesting may permit younger children to perform them.

Vygotsky's theory

Vygotsky proposed that the origin of cognitive development lies not in maturation, but in social and cultural influences, acting through language and social interaction. The zone of proximal development (ZPD) is a range of tasks that are too difficult for the child to manage alone, but which can be performed during interaction with another person.

Information processing approach

Cognitive development depends on the existence and maturation of basic information processing mechanisms. Stimulation and active exploration are crucial to development. Habituation (ceasing to respond to repeated stimulation) is a useful phenomenon for testing infant cognition. Selective attention and memory are essential for development, and must be considered when testing development.

Related topics

Reading (G3)
Learning to talk (G5)
Cognitive development in childhood (H3)

Piaget's theory of cognitive development

In the middle years of the 20th century Jean Piaget developed the most influential theory of child development. He emphasized the properties of orderly progression and cumulative change. The child's development proceeds through a series of stages (see *Table 1*). The stages always occur in the same order, no stage may be missed, and the child never reverts to an earlier stage. Children differ in the age at which they reach each stage. Development is not a passive process: children are active agents in their own development. Development is characterized by *qualitative* changes, rather than simply growth.

Table 1. The main stages of Piaget's theory of cognitive development

Developmental stage (approximate age)	Main features
Sensorimotor (birth–2 years)	A sequence leading to the emergence of symbolic representation: • Reflex actions. • Reactions based on infant's own body. • Reactions focused on effects on objects. • Goal-directed coordination of actions. • Greater variety of actions; focus on results. • Ability to think of an object in its absence.
Preoperational (2–7 years)	• Can use symbolic representations. • Can perform deferred imitation. • Cannot place objects in order of size. • Cannot at first classify things by color or shape; subsequently by only one feature. • Cannot understand conservation of quantity; only later of number. • Cannot reverse operations. • Egocentrism.
Concrete operations (7–11 years)	• Show logical reasoning. • Limited to real-world objects or events of which the child has personal experience. • Can perform hierarchical classifications, and **class inclusion** relationships (classifying by more than one feature). • Operations are reversible.
Formal operations (11 years on)	• Can form and test hypotheses. • Can reason about abstract ideas, and objects and events beyond their experience. • Can think about thought (introspection).

During infancy, the child is in the **sensorimotor stage** (sometimes written *sensory-motor*). Piaget viewed this as consisting of six sub-stages:

1. **Exercising reflexes** (birth–1 month). The newborn infant reacts to internal and external stimuli, while quickly learning to differentiate stimuli.
2. **Preliminary circular reactions** (1–4 months). The infant performs repetitive actions focused on the infant's own body (e.g. thumb sucking).
3. **Secondary circular reactions** (4–8 months). Actions that produce changes in the outside world (e.g. moving a mobile, or a rattle) are repeated, with the infant closely observing the results.

4. **Coordination of secondary reactions** (8–12 months). The infant intentionally coordinates sensory inputs and motor actions, anticipating outcomes, and repeating strategies that have previously been successful.
5. **Tertiary circular reactions** (12–18 months). Repeated actions are more varied, and the infant 'experiments' with new actions.
6. **Invention of new means** (18–24 months). The infant no longer relies on trial and error, being able to think through problems and mentally find solutions.

Object permanence

The properties of these sub-stages may be illustrated by the development of **object permanence**: the understanding that objects continue to exist when they are out of sight. In the first two sub-stages, infants will look at an object, but will not attempt to look for it if it is hidden from view. Only during sub-stage 3 does the infant start to search for objects, but only to a limited extent, for example continuing eye movements in the direction that an object was moving. At sub-stage 4 the infant will search for completely hidden objects, but will show what has been called the **AB error** (or A-not-B effect). A child at this age will play a hide-and-seek game, repeatedly retrieving an object hidden in one place (A). If the object is now concealed in a new place (B), while the infant watches, the child will look for it in the original place, A. Piaget's interpretation of this was that the child had not yet developed an understanding that the object has a continued existence when out of sight. At the next stage, some sense of object permanence is achieved. The infant will search for an object, but only where it has been seen to be hidden. If an object is moved from the hiding place without the child seeing, the child will not look elsewhere for it. Only in sub-stage 5 is real object permanence shown. The infant now will search in other places for the 'lost' object.

Schemata

In Piaget's theory, development consists of the formation and coordination of **schemata** (or *schemas*), which are mental structures: systems of knowledge, actions and thoughts used to understand the world, and which guide behavior. The earliest schemata are behaviors: infants learn that to touch an object they must reach towards it. As development progresses new schemata are formed, and schemata become coordinated so that sequences of action are possible, such as reaching for an object, grasping it, and bringing it to the mouth. During the sensorimotor stage schemata are limited to such behaviors. The development of schemata is described as **adaptation**, and takes place through two basic processes. **Assimilation** is the process of acquiring new information and incorporating it into existing schemata. **Accommodation** occurs when existing schemata are modified to encompass new, discrepant information.

Criticisms of Piaget's theory

Piaget has been attacked for relying heavily on observations of just a few children (particularly his own). His methods have also been criticized for lacking scientific rigor. His method of testing might have influenced the child's performance (see Topic H3). A fundamental problem is that observing that a child does not perform some action does not necessarily mean the child is cognitively incapable of doing it. There may be other reasons (for example *motivational*, the child might not want to perform the task; or *physical*, the underlying physical mechanisms may not have matured sufficiently). There is some support for the view that there is such a discrepancy between *competence* and *performance*. For example, in one study, some infants of 5–6 months failed an object permanence test when a toy was hidden by a heavy cloth, but passed it when it was covered by a light piece of paper. The ease with which the covers could be manipulated appeared to affect cognitive performance, but actually reflected a physical difficulty with the task.

Similar modifications in testing procedures have suggested even greater discrepancies from the ages at which Piaget claimed some cognitive abilities develop. For example, Piaget held that counting (or the extraction of the property of number) is impossible before the end of the preoperational stage (see Topic H3). However, 6 month olds have been shown to attend to a display with two objects when they hear two drum beats, and one with three objects when they hear three drum beats. This and other similar results suggest that even young infants have some understanding of number.

Vygotsky's theory

One alternative view to Piaget's is based on contradicting his emphasis on the child as an individual who matures through a series of stages. Lev Vygotsky proposed that the origin of cognitive development lies not in maturation, but in social and cultural influences. Acting through language and social interaction, culture permits development to occur (providing that an appropriate level of underlying biological maturity has been attained). The central concept in Vygotsky's theory is the **zone of proximal development** (ZPD). This refers to a range of tasks that are too difficult for the child to manage alone, but which can be performed during interaction with another person. The child develops by learning in collaboration with an adult or more capable peer. In practical terms, this suggests that teaching will be successful only if it addresses tasks within the ZPD. The importance of social interaction in development is perhaps easiest to observe in language development (see Topic G5).

Information processing approach

The **information processing approach** focuses on the processes by which children obtain, remember, retrieve, and use information to solve problems. **Stimulation** is crucial to development. For example, the development of neural pathways underlying visual perception depends on appropriate stimulation, and also on **exploration** of the environment. It is also well established that varied stimulation during early infancy is associated with better language and cognitive performance later in life. Infants do not attend equally to all aspects of the environment (**selective attention**: see Topic D4). For example, they attend to pictures of faces rather than to the same elements assembled in a meaningless array.

If the same stimulation is repeated, all organisms show **habituation**, gradually ceasing to respond to it. Habituation is a useful phenomenon for testing infant cognition. An infant who has habituated to a stimulus will stop attending to it. If the stimulus is changed in a way that is recognizably different the infant will again attend to it: **dishabituation**. This has been used, for example, to show that infants have a concept of object permanence earlier than Piaget supposed. As adults, we have no difficulty in understanding that an object partly obscured from view continues in the part that is occluded (the **occlusion** effect). Piaget's view was that this does not develop until sub-stage 6, around 18 months. However, infants as young as 4 months show signs of the occlusion effect. In one study (see *Fig. 1)* they were shown a display in which a rod moved from side to side behind a block. When they had habituated, they were shown two new displays, one in which a complete rod moved from side to side, and one in which a broken rod moved from side to side, both of them unoccluded. Infants

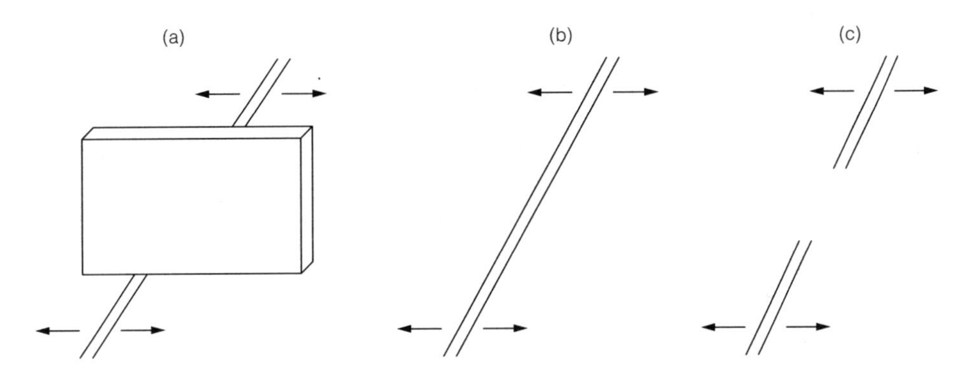

Fig. 1. Demonstration of the occlusion effect in infants. Infants are habituated to a rod moving from side to side behind a block (a). They are then tested with a complete rod (b) and a broken rod (c), both moving from side to side. Four-month-old infants looked longer at (c) than at (b). Reprinted from Kellman, P.J. and Spelke, E.S. (1983) Perception of partly occluded objects in infancy. Cognitive Psychology, ***15:*** *483–524. With permission from Academic Press, Orlando.*

looked longer at the broken rod, indicating that this was perceptually novel to them. This suggests that they had habituated to a percept of a complete rod, which is what would be expected if they showed the occlusion effect.

Memory (see Section F) is essential for development. At about 1 month, an infant can distinguish the mother's face from the faces of other people. One further difficulty with object permanence tests might be that young infants have only a short memory, especially for events occurring infrequently. The more quickly testing is performed after an object is hidden, the more likely the young infant is to find it in its new location.

H3 COGNITIVE DEVELOPMENT IN CHILDHOOD

Key Notes

Preschool years: Piaget's approach

In the preoperational stage, 2–7 years, the child cannot perform mental operations: higher order schemata that permit the fuller manipulation of internal representations. The child shows centration: an inability to attend to more than one feature of an object at a time, and egocentrism: an inability to adopt other viewpoints than their own.

Preschool years: information processing

The inability of younger children to perform Piagetian tasks is due to more general limitations of information processing. Attention, memory and language improve with age. Using tasks more suited to the development of these skills shows earlier apparent cognitive development.

Middle childhood: Piaget's approach

In the stage of concrete operations children show an increased ability to reason logically. However, they cannot handle abstract material. They start to perform tasks showing a weakening of centration, and lose their egocentrism. From about 11 to 15 years, a person passes into the stage of formal operations and can handle abstract relationships and hypothetical situations.

Criticisms of Piaget's approach

A child's abilities appear reduced by tasks not appropriate to the child's level of information processing. The transition from stage to stage is much more gradual than Piaget implied, and can be modified. Not all people reach the formal operations stage, and such thinking might be culture-bound.

Related topics

Cognitive development in infancy (H2)

Social and emotional development in childhood (H5)

Preschool years: Piaget's approach

The child leaves infancy with the ability to use **mental representations**; that is, to think about actions and objects in their absence. The child can also imitate actions that were observed earlier (**deferred imitation**). However, Piaget apparently demonstrated that there is a wide variety of cognitive tasks that the child of preschool age cannot perform (see Topic H2, *Table 1*). These **operations** were viewed as higher order schemata that permit the fuller manipulation of internal representations. Piaget believed that these operations were not available until around 7 years of age, and he called the period from 2 to 7 years the **preoperational stage**.

Piaget catalogued the types of task that the preoperational child apparently cannot perform. A number of these are examples of a general cognitive approach of **centration** (sometimes called *centering*). This is the apparent inability of the child to attend to more than one feature of an object at a time.

One example is a failure of **conservation**. That is, the child is unable to appreciate that the quantity of something remains the same when its shape is changed (see *Fig. 1*). For example, a child is shown two glasses of the same shape with the same amount of liquid in each, and will agree that they both have the same amount. The child then watches while the liquid from one glass is poured into a taller, thinner glass. The child will now indicate that the taller vessel contains more liquid. According to Piaget, preschool children also cannot place objects in increasing order of size (**seriation**); they cannot make **transitive inferences** (i.e. told that A is bigger than B, and that B is bigger than C, cannot see that as a consequence, A is bigger than C); and they cannot **reverse operations** (e.g. watching three different colored balls being dropped into a tube, they expect them to come out in the same order if the tube is inverted).

Another general feature of the preschooler's thinking is **egocentrism**, which is an inability to adopt other viewpoints than their own. A simple example is that, even if they can tell their own left hand from their right, they may reverse left and right on a person standing opposite them. Piaget introduced the 'three mountains task' to demonstrate this. A preschool child seated at one side of a model with three mountains is unable to identify the view that would be seen by a doll seated at a different side of the model.

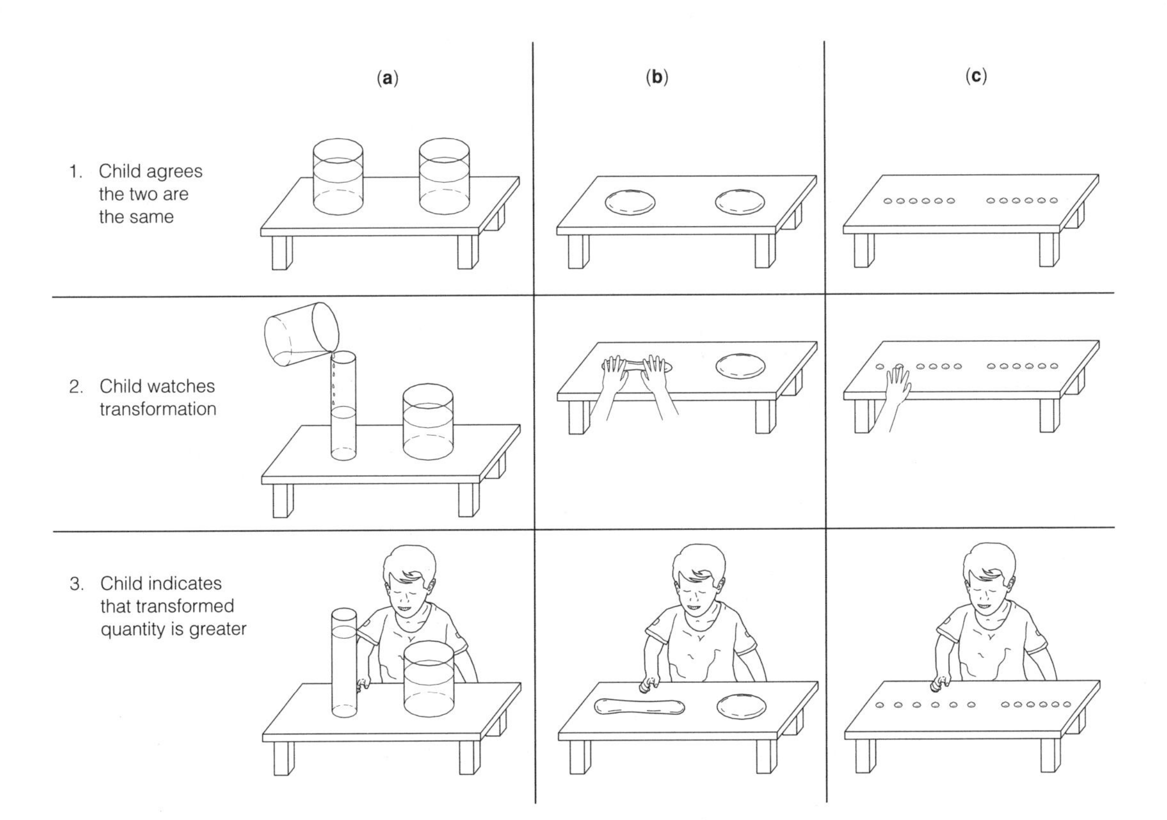

Fig. 1. Failure of conservation in preschool (preoperational) children. The child indicates in (a) that the taller vessel contains more liquid; in (b) that the rolled-out ball of clay contains more clay; in (c) that the spaced-out line of beads contains more beads.

Preschool years: information processing

While all these cognitive failings can be reliably demonstrated using Piaget's methods, it is now clear that, given more suitable methods, preschool children can perform many of these types of cognitive processes. They can place things in order of size, classify objects, and understand transitive inferences *so long as the task is a familiar one, and is presented in a way suitable to the child's abilities and interests*. For example, most 3- and 4-year-olds can perform seriation tasks as long as they are given only a small number of items to place in order. In variants of the three mountains task, children as young as 3 years are able to give appropriate descriptions of the other's view, especially if the views are made more distinctive.

It now seems likely that the inability of children to perform Piagetian tasks is due to more general limitations of **information processing**. Infants and preschool children show development of **attention**. The youngest children are highly distractible: it is difficult for them to maintain attention on aspects of a task that are essential for the required performance. Preschoolers are better able to maintain and control their attention, but cannot easily attend to features that are not highly distinctive. If a task is complex, or its features are not sufficiently distinctive, the child might fail to perform it as adults expect. **Memory** also shows a clear developmental course. The transition from infancy to childhood is accompanied by rapid development of language skills (see Topic G5), and this permits the child to remember better using words. Preschoolers do not spontaneously make use of **mnemonic** aids such as grouping similar items together, nor do they **rehearse** lists of items by repeating them to themselves, unless instructed to do so. They also tend to treat an instruction to remember something as the same as an instruction to look at it. In neither case do they make an active attempt to memorize information. More recent research, using materials and tasks that are more familiar to preschool children, and observing them in everyday activities instead of artificial laboratory tasks, has revealed better cognitive abilities.

Middle childhood: Piaget's approach

According to Piaget, during middle childhood children pass from the preoperational stage through stages in which they are able to use two types of operation (see Topic H2, *Table 1*). In the stage of **concrete operations**, from about 7 to 11 years of age, children show an increased ability to reason logically. However, their thinking is still primarily concerned with real events and objects: they cannot handle abstract material. They are, for example, limited in the extent to which they can generalize from one result to another. For example, even knowing that adding 1 to 2 or to 4 gives an odd number, they do not seem to understand that adding one to *any* even number gives an odd number. They have difficulty understanding metaphor. However, these older children come to perform many of the types of cognitive task that are seemingly impossible for preoperational children: conservation, classification, seriation and reversibility. During this stage they lose the egocentrism of the preoperational child.

At the end of middle childhood, between about 11 and 15 years of age, a person passes into the stage of **formal operations** (see Topic H2, *Table 1*), and this continues through adolescence into adulthood. Now, the child is fully able to understand and use abstract relationships, to deal with hypothetical situations, and to systematically explore the properties of the world in the manner of a scientist. This last involves both the use of **combinatorial logic** (thinking about all possible explanations for something, and of the effect of combinations of them), and **hypothetico-deductive reasoning** (forming hypotheses about the consequences of an action, and then testing those hypotheses).

Criticisms of Piaget's approach

Some of the same criticisms apply to these Piagetian middle childhood stages as to earlier stages. In general, a child's abilities appear to be reduced by the use of tasks that are not appropriate to the child's level of attention, memory, and motivation. Further, the transition from stage to stage is much more gradual than Piaget implied, and can be accelerated or retarded by environmental factors. A particular concern about the concept of the stage of formal operations has been that not all people reach this stage. Piaget recognized that in very adverse circumstances formal operations are never acquired. More recently it has been estimated that only half of all adults attain the stage of formal operations. Furthermore, many will use formal operations in some situations, but in others will not get beyond concrete operations. This may be for motivational reasons, or because the individual lacks the intelligence required. Formal operations might also be culture-bound. There is evidence that people in non-Western cultures perform poorly on ability tests requiring formal operations. It is possible that this type of thinking is dependent on reaching a particular level of education within a system that has formal operations as a goal. Piaget himself came to recognize the importance of education, interests, social influences, and culture on the ability to perform in tests of formal operations.

H4 Social and emotional development in infancy

Key Notes

Attachment

An attachment is an emotional tie that binds one person to another. Its main basis in infancy seems to be the provision of emotional comfort. Lack of strong attachment in infancy ('maternal deprivation') may lead to failure to thrive, emotional disturbance, and restricted intellectual development. These effects may be partly reversed if the infant is moved before around 2 years to a caring environment.

Types of attachment

Attachment has been investigated using the 'strange situation'. The infant's response to separation from the mother indicates secure, anxious/avoidant, or anxious/ambivalent attachment. Securely attached infants become more popular and outgoing preschoolers. It is unclear whether different attachment patterns result from maternal behavior or innate temperament differences.

Other caregivers

Attachment can be formed to any caregiver, and to more than one person. There are differences in an infant's attachment to mother and father. Children are more likely to turn to their mother for comfort, and their fathers for play. Infant–mother attachment may be affected by day care, depending on the quality of care provided, and how much time the infant is away from the mother. High quality day care can have beneficial effects.

Gender roles

Both biological and environmental factors contribute to the development of gender roles. Male and female infants are treated differently from earliest infancy. By about 18 months infants start to behave in gender-typed ways, and these behaviors are often differentially reinforced by adults.

Related topics

Cognitive development in infancy (H2)

Social and emotional development in childhood (H5)

Attachment

An attachment is an emotional tie that binds one person to another. According to Sigmund Freud, the basis of attachment is that it binds the infant to the person who provides for the basic needs, the primary caregiver. This has been interpreted as meaning, primarily, nourishment, and implies that the object of attachment will be the mother. Harry Harlow in the 1950s separated baby rhesus monkeys from their mothers, and raised them in isolation in cages containing two crude dummy monkeys, a hard, wire one and a soft, cloth-covered one. If an infant monkey was frightened, it ran to the soft dummy and clung to it, even if the other dummy was fitted with a bottle and nipple and provided the infant's nourishment. This suggests that the basis of attachment is to do with comfort rather than nourishment. The effects of rearing these monkeys alone were far-reaching. When later introduced to other monkeys they could not play normally, appeared withdrawn, and later could not function sexually.

These infant monkeys were reared with no contact with any other monkey. It might be argued that the effects are due to solitary confinement rather to the absence of attachment to a mother. Human infants also cling to a parent when frightened, and appear to become 'attached' to cuddly toys or even scraps of cloth. Studies of human infants raised in institutions where they received minimal personal attention showed similar consequences. The children failed to thrive, became emotionally disturbed, and showed restricted intellectual development. In contrast to Harlow's monkeys, these infants typically have many other infants to interact with, so the absence of a mother seems to be the crucial factor: **maternal deprivation**. According to John Bowlby, the mother–infant attachment is crucial for emotional, social and, in consequence, cognitive development. This attachment results from innate tendencies in the newborn infant to seek and maintain contact with an adult, which is usually the mother. Again, there is an animal model of this. Ethologists such as Konrad Lorenz demonstrated that newly hatched birds such as geese follow moving objects. If the following continues for several minutes, **imprinting** occurs: the young bird follows the adopted 'mother' as if attached. Further studies of infants starting life in institutions providing minimal adult human contact have shown that the effects of maternal deprivation may be partly reversed so long as the infant is moved before the age of around 2 years to a caring environment.

Types of attachment

The nature of attachment has been investigated by the use of a standard procedure known as the **strange situation**. The infant is observed in a strange room while a series of episodes of separation and reunion with the mother are staged, some involving the presence of an adult not known to the infant. Although infants can visually discriminate their primary caregiver from others by 2–3 months of age, they do not show fear of strangers until about 6 or 7 months. This stranger fear increases up to about 2 years, and then declines. In the strange situation, the attachment between mother and infant is classified as **secure**, **anxious/avoidant**, or **anxious/ambivalent** (see *Table 1*). The nature of attachment in infancy predicts behavior later in childhood. Securely attached infants

Table 1. Some of the characteristics of different types of attachment, as revealed by the strange situation

Type of attachment	Typical behaviors in presence of stranger
Secure	• Continues to play and explore in presence of stranger. • Shows some distress when mother leaves. • Greets mother enthusiastically on return.
Anxious/avoidant	• Appears detached from mother. • Shows little distress when mother leaves. • Ignores mother when she returns.
Anxious/ambivalent	• Stays close to mother. • Shows high level of distress when mother leaves. • Shows approach–avoidance conflict towards mother when she returns.

become more popular and outgoing preschoolers. At that age they also explore their environment more, show more interest in reading, are more cooperative, and perhaps less aggressive. Attachment theorists like Mary Ainsworth, who devised the strange situation, argue that this shows the importance of the quality of the mother–infant relationship in the development of the child. Mothers of secure infants have been shown to be more sensitive to the infant's needs, and to pick them up and hold them more than do mothers of anxious children. Attachment theorists argue that differences in mothers' behavior cause differences in the children's attachment behavior. However, it is possible that innate differences in children's **temperaments** produce different reactions in the strange situation and differences in the behavior of the mothers. Another possibility is that infant behavior predicts later childhood behavior because the mother interacts with the child in the same way in later childhood as in infancy.

Other caregivers

It is now clear that attachment can be formed to any person who cares for the child, male or female. Attachments can be formed simultaneously to more than one person, for example to both mother and father, and this is increasingly true as fathers play a greater role in child care. However, there are differences in an infant's attachment to mother and father. There is some evidence that infants are more distressed in the strange situation when tested with the mother than with the father. Fathers tend to behave differently towards children, focusing on physical activities and play, while mothers tend to concentrate on talking and quiet interactions. Children are more likely to turn to their mothers for comfort, and their fathers for play.

Recently, with more mothers returning to work soon after bearing a child, attention has turned to the effects of day care and other substitute caregivers. From what we have already seen, it may be deduced that placing infants under 6 months with strangers should not cause immediate disturbance. However, after that age the likelihood of stranger fear indicates that an infant needs to be familiarized with the caregiver. The effects of substitute caregivers is a controversial issue. In the short term, many studies have found that the infant–mother attachment is not affected by day care, while some other research has found that infants are less likely to have secure attachment to their mothers if they receive more than 20 hours of day care each week. The crucial factor is probably the quality of the care provided. It is clearly important that the caregiver provides the infant with the same sort of nurturance as the mother would. Thus, individual care, or care in adequately staffed centers is likely not to be harmful, while placing an infant in a crowded, impersonal center could produce some of the effects of maternal deprivation. Some recent evidence suggests that high quality day care can have beneficial effects. For example, infants who received such care in infancy performed better in school tests and in tests of social competence at 8 and 13 years of age than did those cared for solely at home.

Gender roles

We saw in Topic B4 how biological factors produce structural differences in males and females, and probably predispose males and females to develop different **gender roles**. However, environmental factors are extremely important in gender role development. In general, gender is assigned at birth on the basis of the appearance of the external genitalia. This gender assignment is the source of environmental differences for boys and girls. Male and female infants are treated differently from earliest infancy. This differential treatment reflects the stereotypes (see Topic J3) that parents and others share about gender differences, and the different hopes and expectations that parents have for boys and girls. In one experimental study, mothers were asked to play with a 6-month-old child who was called either 'Joey' or 'Janie'. Although the same infant was involved, the mothers interacted differently depending on the sex they thought the infant to be. If they thought it was a boy they played more vigorous and physical games than if they believed it to be a girl. By about 18 months of age infants start to behave in gender-typed ways, and these behaviors are often differentially reinforced by adults. Parents are likely to provide gender-role specific toys for children, and also to show disapproval if an infant plays with an 'inappropriate' toy.

Few psychologists now take an either/or position on the 'nature–nurture' issue (see Topic K8). The findings above strongly suggest that environmental factors are very important. On the other hand, gender differences are observable very soon after birth. Male infants tend to be more active and irritable than girls, and there are reports that males are more active even *before* birth.

H5 Social and emotional development in childhood

Key Notes

Socialization

Socialization is the process by which the child acquires the behavior and thinking of the society in which it is growing up. Three socialization mechanisms have been suggested: reinforcement; social learning (based on imitation and modeling); and the cognitive approach in which the child is an agent in its own socialization. Three types of parental style have been distinguished: autocratic, permissive and authoritative. Socialization appears to be best in children raised with the authoritative parenting style.

Moral development

For Piaget, children move from moral realism as preschoolers (basing judgments on consequences) to moral relativism (understanding of the importance of intention) at 11–12 years. But, if the tasks are simplified, many 5-year-olds will show that they can base moral judgments on intentions. Kohlberg's theory describes six stages of moral development. First, morality is defined in terms of physical consequences (preconventional), then by conforming to the expectations of others (conventional), then by internalized moral principles (postconventional).

Play

Play is an important mechanism in development. Infant play is solitary and exploratory; early childhood play is parallel; preschoolers' play is associative; middle childhood play is cooperative. Children prefer to play with others of the same sex. Through play, children learn about the nature of the physical and social world, and develop physical, cognitive, and social skills to manipulate it.

Theory of mind

A theory of mind is the understanding that other people have thoughts, feelings, and desires. It permits us to interact with others and to feel empathy. Development can be viewed as the formation of a theory of mind. For example, the loss of egocentrism is the development of the understanding that others have a different viewpoint.

Related topics

Cognitive development in childhood (H3)

Social and emotional development in infancy (H4)

Socialization

Socialization is the process by which the child acquires the behavior and thinking (the culture) that characterize the society in which the child is growing up. Societies differ in the values that predominate, and so socialization means different things in different cultures. There are also differences between subgroups and between families within cultures. So, for example, competitive industrialized societies have different goals and values from those based on subsistence farming. Within an industrialized society people play different roles (e.g. managerial or laboring), which each have their own culture. Families differ in their specific cultures: some value education and scholarship, and others do not; some value self-reliance, and others obedience. Each of these differences implies that the end result of socialization will be different.

However, the mechanisms of socialization are mostly common to all groups. For most of us it takes place largely within the family, with the parents as the main agents of socialization. Three types of mechanism have been suggested as the basis of socialization. **Reinforcement** approaches hold that rewards for appropriate behavior and punishments for inappropriate behavior produce socialization as a result of instrumental learning (see Topic E2). Most believe that such an approach is inadequate to explain the complexities of socialization. **Social learning theory** argues that **imitation** and **modeling** are important. The child imitates specific behaviors of the parents, and models itself on their approach, in the process developing the culture of the parents. The **cognitive approach** makes the child an agent in its own socialization, rather than being molded by reinforcement or imitation. The child develops an understanding of the culture, together with a developing sense of right and wrong.

For much of the 20th century it was assumed, following **psychoanalytic theory**, that the most important factors in the development of socialization, as of personality, were events in infancy. However, it became clear that differences in, for example, toilet training and breast-feeding have little or no effect on later behavior. The most important stage for socialization is childhood. A number of attempts have been made to classify different parental styles and to examine the effects of these on socialization. A general distinction has been made:

- **Autocratic** parents are strict disciplinarians who expect the child to obey rules unquestioningly.
- **Permissive** parents exert little control over children, impose few rules, and rarely use punishment.
- **Authoritative** parents exert control when it is necessary, explain the reasons for rules, and respond to the child's point of view.

Socialization appears to be best in children raised with the authoritative parenting style. Children subjected to strict discipline become dependent and defiant. If they have been subject to violent punishment they are likely to become violent themselves. Children of permissive parents tend to lack social responsibility, are also quick to anger, and are not independent.

Moral development

Closely allied to socialization is **moral development**. Piaget (see Topic H2) viewed moral development as the development of moral *reasoning*: that is, it is a process of cognitive development. Preschool children base their judgments of right and wrong on *consequences*. Piaget demonstrated this by asking children questions such as:

> 'Janet broke one dish trying to sneak into the refrigerator to take some jam. Jennifer broke five dishes trying to help her mother. Who was naughtier?'

Until the age of about 7 years, children will answer 'Jennifer', apparently basing this judgment on the consequences of the act not on the intentions of the actor. Up to this age, children treat rules as fixed and inviolable, imposed by some infallible authority. Piaget called this a stage of **moral realism**. After this age, children pass into an intermediate stage when they will form some understanding, through interaction with other children, that rules can be negotiated. They are also likely to answer 'Janet' to Piaget's question, showing some understanding that intention is important. At about 11 or 12 years they pass into a stage of **moral relativism**. They now have an adult level of understanding of the importance of intention and how right and wrong, and fairness, depend on circumstances.

Just as with cognitive development (see Topic H3) Piaget's approach has been criticized. If the language and cognitive demands on them are reduced, and they are asked, for example, about *deliberate* damage, rather than accidental damage, many 5-year-olds will show that they can base moral judgments on intentions. Further, Piaget's tasks did not reveal anything about the child's understanding of what behavior is appropriate in a particular situation. Lawrence Kohlberg based his theory of moral development on a series of moral dilemmas that he put to subjects of different ages. He produced a six-stage theory of how moral reasoning develops (see *Table 1*). As in Piaget's theory of cognitive development, these stages are universal, sequential, and irreversible. The individual moves from a position in which morality is defined in terms of the physical consequences of an act (the **preconventional level**), through stages in which morality is determined by conforming to the expectations of others (the **conventional level**), to a level at which morality is defined by internalized moral principles (the **postconventional level**).

Table 1. Kohlberg's theory of moral development

Stage of moral reasoning	Basis of moral behavior
Preconventional level	
Stage 1: punishment and obedience	It avoids punishment: physical consequences of actions determine its goodness/badness.
Stage 2: instrumental relativist	It produces reward: right actions satisfy needs or produce rewards; reciprocal action based on consequences.
Conventional level	
Stage 3: interpersonal concordance	It gains approval of others: intentionality is important.
Stage 4: law and order	It is doing one's duty: respect for authority; maintaining the social order for its own sake.
Postconventional (autonomous) level	
Stage 5: social contract (legalistic)	It is generally agreed by society; recognizes individual rights; rules are relative.
Stage 6: universal ethical principles	It follows internalized principles based on equality of rights and the basic justice of reciprocity.

Research using Kohlberg's dilemmas has shown that children move between these stages with age. However, people may apply, for example, postconventional principles in some circumstances, but conventional ones in others. Furthermore, few adults seem to operate consistently at stage 6. There is evidence that, using Kohlberg's criteria, men on average reach a 'higher' stage of moral development than do women. It has been argued that Kohlberg imposed his own values on to his theory, and different stages should be considered different bases for morality, rather than as better or worse.

Play

Children (like the young of many other species) spend a lot of time in play. It is considered to be an important mechanism in social (and cognitive) development. Infant play consists mostly of actions that serve to explore the environment and objects and people in it. Such play is **solitary**; if another person is involved (e.g. in peek-a-boo games), there is no reciprocation of play actions. In later infancy the child may spend a lot of time watching older children at play, but not joining in. Solitary play continues into early childhood, and is followed by **parallel play**. Now, a child will play alongside other children, but will not cooperate or otherwise interact. They may, however, observe and mimic what another child is doing. Preschoolers will play together with others, sharing or arguing; cooperating or disputing. This **associative play** typically lasts for short periods. During middle childhood children will engage in **cooperative play**. Now, they organize themselves, take turns or adopt roles, and generate rules for the games they play. The games often simulate adult life (e.g. playing house), or are physical (rough-and-tumble), especially amongst boys. Children prefer to play with others of the same sex. We have already seen that boys and girls have gender-related preferences for toys and games (see Topic H4).

Although play is by definition an activity performed for its own sake, or for pleasure, each of these forms of play has an important role in development. Through play children learn about the nature of the world, and develop physical skills to manipulate it. They learn problem-solving skills, which are an essential part of cognitive development. The emergence of social play teaches children how to behave in social situations. They become less egocentric, learn to share and cooperate, and develop basic interpersonal skills. Play has an essential role in the development of friendship and other interpersonal relationships.

Theory of mind

Since the 1990s some theorists have argued that underlying both cognitive and social development is the emergence of a **theory of mind**. This is the understanding that other people, as well as ourselves, have thoughts, feelings, and desires. Our adult theory of mind recognizes that others have these inner states, and will in many instances feel as we do (e.g. in response to painful stimuli). However, we recognize also that the inner states of others may be different from our own. We also understand that other people have a similar theory of mind, and that they know that we have one. The theory of mind permits us to interact with others; to understand how they will respond to what we do or say to them. It also allows us to feel **empathy** for others, that is, to respond with distress if we see them in distress.

The developmental processes that we have described in this and the last three topics can be viewed as the formation of a theory of mind. For example, the loss of egocentrism (see Topic H3) is the development of the understanding that others have a different viewpoint. Similarly, the development of social play shows the emergence of the realization that others share similar feelings and desires to our own.

H6 ADOLESCENCE

Key Notes

Cognitive development

Adolescents show a high level of general cognitive development. The ability to take the other person's viewpoint makes adolescence a time of idealism. This leads to adolescent protest. The adolescent belief that their own generation is the first one to recognize social and environmental ills is an example of adolescent egocentrism, which frequently becomes the 'personal fable': a belief in the uniqueness of one's experience, and of invulnerability.

Identity

Erikson argued that the main 'task' of adolescence is the development of a sense of identity. Adolescent protest can be seen as part of this process, and it frequently involves emotional turmoil. The main social influence on adolescents is their peers, with whom they show self-consciousness. Adolescence is characterized by conformity to peer group norms.

Friendship and relationships

Adolescents turn for support to friends, mostly of the same sex. Adolescence is marked by sexual maturity, with sexual and romantic attraction to others. Early-maturing boys have social advantages which may extend into adulthood. However, they may become more rigid, more conformist, and less happy with their personal relationships. Early-maturing girls, by contrast, do not have immediate social advantages. Dating in early adolescence is recreational, helps maintain status, and explores the nature of relationships. In later adolescence it may become a mate-selection process.

Related topics

Social and emotional development in childhood (H5)
Adulthood (H7)

Cognitive development

In Piaget's scheme of cognitive development many people enter the final, formal operations stage in late childhood (see Topic H3). Whether or not we accept the Piagetian account, the adolescent shows a high level of cognitive development in terms of attention, memory, flexibility of thought, and the ability to reason and experiment. To the extent that they have these qualities, adolescents are able to make rapid headway with formal education. Conversely, it is likely that formal education at high school level helps the development of these abilities. Not only do all adolescents not reach a stage of formal operations, but those who do so do not consistently apply formal reasoning to solve problems. This is particularly true in the case of personal problems, where the adolescent often falls back into modes of thinking that characterize concrete or even preoperational stages.

The ability to take the other person's viewpoint (the development of a *theory of mind*; see Topic H5) makes adolescence a time of **idealism**. They realize, perhaps for the first time, that the world is not ideal: that inequality and injustice are widespread, and that environmental and social problems are pressing.

They perceive that previous generations have not resolved these issues and that it is for them, and their generation, to do so. This involves them in discussion with others and often, particularly as they become older, in active participation in protest movements. This is one aspect of **adolescent protest**, which is frequently targeted at parents as the immediate representatives of failed previous generations. Adolescent thinking on social issues is frequently polarized; things are either right or wrong. This is an example of reversion to concrete thinking.

The adolescent belief that their own generation is the first one to recognize social and environmental ills is an example of **adolescent egocentrism**. Unlike childhood egocentrism, adolescents are aware of the views and values of others, but give their own views and values priority. When this is personalized it has been called the **personal fable**. This is the belief (or apparent belief) that what he or she is experiencing or feeling has not been experienced or felt by anybody previously (especially parents: 'you don't understand'). Another aspect of this personal fable is the belief that 'it can't happen to me'. One reason why adolescents sometimes engage in risky activities (e.g. dangerous driving; unprotected sex) is because they feel that they are invulnerable.

Identity

The psychoanalytic writer Erik Erikson, who was largely instrumental in turning psychologists to the study of lifespan development, argued that the main 'task' of adolescence is the development of a sense of **identity**. To the extent that children are dependent on their caregivers, they do not need a separate sense of who they are and what they are here for. Adolescents need to break this dependence on parental attitudes and values before they can base their identity on their own values and attitudes. Adolescent protest can be seen as part of this process, and it frequently involves emotional turmoil. It has been argued that adolescent turmoil is a feature of relatively modern, industrialized societies. In less complex cultural systems the child's future is clearly mapped. A son might proceed through helping his father on the farm to taking over the farm; a daughter through helping her mother about the home to becoming a wife and mother herself. Neither has much choice in what they will become, and little reason to be concerned about their future role. In complex cultures a seemingly infinite range of possibilities is available, frequently leading to an identity crisis that the adolescent must resolve.

The main social influence on adolescents is their peers. Faced with them, adolescents show another aspect of their egocentrism: **self-consciousness**. Each adolescent feels that he or she is the center of attention: that others are preoccupied with his or her appearance and behavior. It is the opinions of peers that are important. An important component of this is the desire to be like their peers. Adolescence is a time of deep **conformity**, although it is conformity to peer group values, behavior, modes of dress, musical tastes, etc. rather than to those of the parents or wider society. Thus, adolescent protest is actually in part a sign of conformity.

Friendship and relationships

During middle childhood children have friends, mostly of the same sex, with whom they play and otherwise pass time. For help, advice, and emotional support they turn mostly to their parents. Adolescents turn more and more for emotional support and for many types of advice to friends, again mostly of the same sex. This produces a psychological separation between adolescent and parent, and is part of the process of identity formation.

Physically, adolescence is marked by the development of sexual maturity. The adolescent becomes unable to ignore other-sex peers and begins to feel sexual and romantic attraction to them. At first many do not know how to respond to this, and teasing often characterizes interactions between the sexes in young adolescents. A complication arises because people do not all mature at the same rate. Early-maturing boys have considerable social advantages. They are more popular with their fellows, girls, and adults than are late maturers, who are rated as less attractive, childish, bossy and dependent. Some of the early maturers' advantage extends into adulthood. At 30 years, late maturers are still less confident, less settled, and have a poorer self-concept. However, they are also more assertive and show greater insight. Later still, the early maturers, while more successful in many ways, have become more rigid, more conformist, and less happy with their personal relationships.

Early-maturing girls, by contrast, do not have immediate social advantages. Perhaps because early maturity for girls occurs at an age when most peers are still in middle childhood they are not necessarily considered more attractive, and they are not ready to deal with the sexual interest that they may arouse in men. In consequence, they may become awkward and self-conscious. However, later in adolescence there appears to be little difference between early- and late-maturing girls.

The typical adolescent in Western cultures will date a number of members of the opposite sex. Dating in early adolescence is not considered to be primarily for the purpose of mate selection. It has several functions; including being a recreational activity (both generally and sexually), maintaining status amongst peers, and exploring the nature of relationships. Later in adolescence (although decreasingly so in recent decades), it does become a mate-selection process.

H7 ADULTHOOD

Key Notes

Transition to adulthood

In many ways the transition to adulthood is gradual. However, it is frequently marked with formal or informal ceremonies or initiation rites, or abrupt changes in status.

Stages of adulthood

While many 20th century writers described adulthood as a series of stages, clear stages are less obvious today, as more people choose to avoid conventional goals, and exercise greater freedom of choice in employment and retirement.

Establishing autonomy

One of the first tasks of adulthood is to achieve autonomy. We can identify different aspects of autonomy: independent living; identity; occupation; and emotional independence.

Relationships

The early adult years are characterized by mate selection. The factors that affect mate selection are: attraction; propinquity (physical closeness); similarity; and compatibility. In the US and the UK still, most people marry. Marital satisfaction requires continual adjustment to the different stages of marriage.

Aging

Aging is the process of decline in physical and cognitive capabilities. Some aspects of intelligence (e.g. spatial ability) and memory (recognition memory and short-term memory) are not appreciably worse in old age than in youth. Adjustment needs to be made to, amongst others, retirement and bereavement. Reactions to bereavement include: emotional reactions (grief); physical effects (susceptibility to disease); and intellectual responses (idealization of the dead partner).

Related topics

Adolescence (H6)

Disorders of personality, identity, and memory (L6)

Transition to adulthood

Transitions between most of the stages of development are gradual, and in many ways that is true of the transition from adolescence into adulthood. The characteristics of adolescence we looked at in Topic H6 do not suddenly give way to some more advanced way of thinking or living. In other ways, however, the move to adulthood is marked as different. In many societies special ceremonies or initiation rites, often involving tests of skill, strength or endurance, mark the transition. These do not, however, always mark a point in late adolescence. While many take place in the late teens others, for example the *bar mitzvah*, occur earlier. There are also less formal transition rites. In many Western cultures we celebrate *coming-of-age*, often related to when the young person gets the right to vote. Graduation ceremonies and acceptance into professional organizations are other examples. While these ceremonies do not mean that the person suddenly becomes developmentally an adult, they often do mark sudden transitions to different roles in society; for example, leaving education and seeking employment.

Stages of adulthood

Erik Erikson, in the 1950s, described three stages of adulthood:

- **Early adulthood**, the main task of which is the establishment of intimate bonds of love and friendship;
- **Middle age**, concerned with fulfilling life goals that involve career, family, and society; and
- **Late adulthood**, when we look back over life and accept its meaning.

Other researchers in the second half of the last century similarly described adulthood as a series of developmental stages. Today, such stage-like change is less obvious. More people choose to avoid goals concerned with families, more voluntarily leave work early, or change careers in mid or later adulthood, and many more do not accept the notion that late adulthood is devoted to looking back. Nevertheless, it is possible to describe features that are common to the adult life of many people. It must be remembered, however, that adults exercise choice in many of these areas, and what we describe is not universal.

Establishing autonomy

Adolescence is a period in which the individual struggles to establish an identity separate from that of the parents (see Topic H6). One of the first tasks of adulthood is to achieve **autonomy**. We can identify different aspects of autonomy:

- **Independent living**.
- **Identity**; the individual's sense of identity is further established, and is less dependent on peer pressure.
- **Occupation**; an occupation or career provides much more than income. It is a vehicle for further social and cognitive development; it often provides a new peer group; and it can help to shape the person's identity.
- **Emotional independence**; the young adult no longer relies heavily on parents for emotional support. This comes increasingly from within, but also from peers and from life partners.

Relationships

The early adult years are also characterized by **mate selection**. This is a shift away from the dating of adolescence and ideally represents a more careful choice of partner with whom one intends to spend the rest of one's life. A number of factors affect mate selection:

- **Attraction**; initially physical attraction, but giving way on longer acquaintance to personal characteristics.
- **Propinquity**; we are likely to select as mates people who are physically near to us, either geographically by residence, or socially through, for example, employment.
- **Similarity**; people are more likely to select as partners those of similar race, religion, age, education, physical stature and so on. The less similar partners are, the less stable a relationship is in the long term.
- **Compatibility**; while similarity goes a long way to ensuring compatibility, relationships involve accommodation to the partner's lifestyle and routine to enhance the comfort with which two people can share their lives. However, there is no good evidence that living together before commitment to a long-term partnership improves relationship satisfaction.

For most people who form long-term relationships, this is still, in the US and the UK, formalized in **marriage**. In the US in 1994, 50% of men were married by the age of 26 and 50% of women by 24. By 54 years of age only 8.2% of men and 5.9% of women had never been married. The trend is for later marriage and for a greater proportion to remain single. Once committed to a marriage, the state of the marriage, or **marital satisfaction**, is the single most important factor in determining a person's overall happiness. To maintain satisfaction requires adjustment to the different stages of marriage. Early in marriage most people find that the marriage is not quite as they expected, and marital satisfaction declines gradually in the early years. The birth of the first child is often associated with a dip in satisfaction, as it involves widespread changes in the parents' freedom, relationship, and identity. Many find it difficult to adjust to these changes. In some marriages the steady decline in satisfaction continues, while in others there is an increase either when the children start school or when they leave home after adolescence.

On average, a mother is 46 and a father 48 when the last child leaves home. This requires further adjustment, particularly for mothers who have devoted their lives up to that point to rearing children. By the age of 50, half of all women will have gone through the menopause, making this period one of extreme change. One way of coping with this is to engage in employment or other activities outside the home. For many other women, and for most men, preexisting employment outside the home provides continuity of interest. In either case, in general, adults in this post-parental period are happier than at any time since the early period of their marriage.

Aging

Aging is the process of decline in physical and cognitive capabilities. We think of aging as a process of later adulthood, which is certainly when aging can start to interfere with daily activities. The age at which such declines start, and the rate at which they proceed varies from person to person. They are also influenced by environmental factors such as diet, living conditions, and exercise (physical and mental). Just as the average life span continues to increase, so the age of onset of the aging process increases, and the rate of decline decreases. However, in many respects aging is a process that starts as soon as adolescence is completed. Physically, for example, it is seen in changes in the skin (the appearance of wrinkles), and in the decline in many aspects of physical performance from early adulthood onwards. The effects on fitness can be offset, at least in part, by exercise.

There is more uncertainty about age and cognitive abilities. Scores on some intelligence tests (see Topic K6) decline slowly from around 20–25 years of age, but scores on verbal subscales may be stable, or even increase, up to a quite advanced age. Memory (see Topic F3) shows clear deficits with age, but these are generally not marked until after the age of around 50. Even then, the decline is mostly in recall in long-term memory; recognition memory and short-term memory are not appreciably worse in old age than in youth. The recall difficulties of older people are greatest with material that is meaningless or unfamiliar. There are pathological causes of cognitive decline in older people which we look at in Topic L6.

Older people have other life changes to cope with. Involuntary **retirement** has been rated as one of the 10 most stressful events that adults face. We saw earlier the range of benefits that come from employment. These benefits may be lost on retirement, and alternative sources need to be found. Older adults are also more likely to face **bereavement**. The effects of this include:

- **Emotional reactions**; the wide-ranging and varied changes of **grief** include shock, depression, anger, guilt, anxiety, helplessness, and difficulties concentrating.
- **Physical effects**; most seriously a greatly increased risk of illness or death, probably resulting from decreased immune function.
- **Intellectual responses**; these may initially include denial, more commonly a search for explanations or meaning of the death of the partner, and often **idealization**, forgetting the dead partner's negative characteristics.

I1 ATTITUDES

Key Notes

The nature of attitudes

We hold attitudes about objects (people, groups, ideas) in our social world. An attitude has affective (evaluative feelings), cognitive (beliefs, knowledge, and opinions) and behavioral (tendencies to particular actions) components. The affective component distinguishes attitudes from knowledge or belief.

The functions of attitudes

An individual's attitudes tend to form a cohesive value system. They affect all aspects of social behavior, including group membership and perception of others. They provide a framework to make sense of the social environment, and they influence all stages of information processing.

Changing attitudes

Attitudes may be changed by persuasive communications, acting through central mechanisms (involving attention, comprehension, yielding, and behavior change) or peripheral routes (change occurring without attention). Attitudes may also change secondarily to changing behavior (counter-attitudinal advocacy). Explanations of this have centered on cognitive dissonance and self-perception theory.

Attitudes and behavior

Earlier findings of only a weak relationship between attitudes and actual behavior result from failing to make the attitude measures correspond with the behaviors. Measure and behavior must coincide in elements of action, target, context, and time.

Related topics

Perceiving others (I3)

Self-perception (I4)

The nature of attitudes

As social beings, our behavior is influenced by our understanding or interpretation of our social environment. That is, we do not respond blindly to others and their actions, but apply cognitive processes that in part determine how we respond. These processes are known as **social cognition**. One aspect of social cognition that has been extensively studied by social psychologists is the concept of **(social) attitudes**. We hold attitudes about, amongst other things, individuals, groups, social institutions, ideas, and actions. For example, we have particular attitudes about the President of the USA, lawyers, government, socialism, and abortion. Attitudes are important because they inform us about the type of person another (or oneself) is, and in particular might allow us to predict or explain behavior (see below). While various definitions of attitudes have been proposed, most consider that they are learned, and that they have three components (see *Fig. 1*):

- **affective** (evaluative feelings about the object);
- **cognitive** (beliefs, knowledge, and opinions about the object);
- **behavioral** (intentions or tendencies to behave in particular ways towards the object).

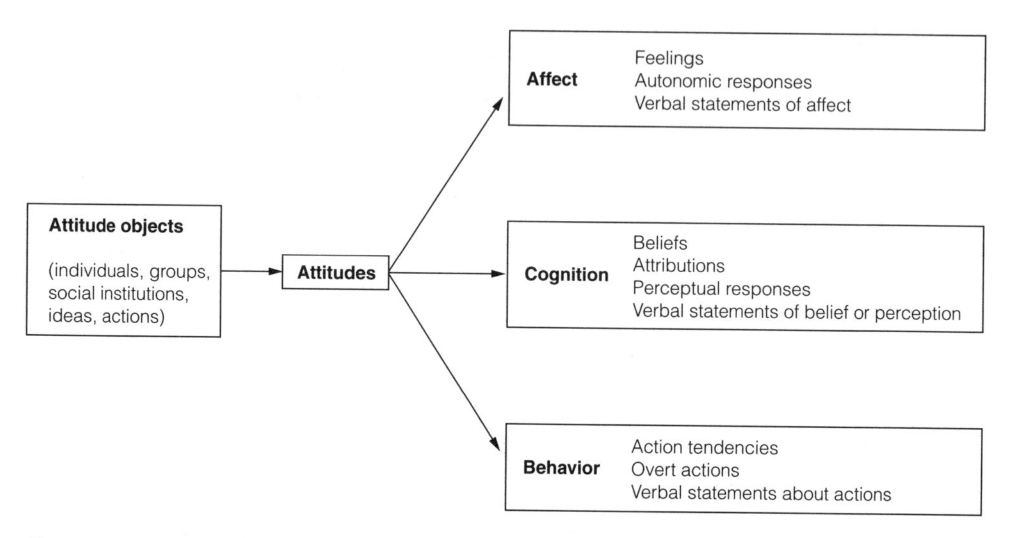

Fig. 1. Schematic view of the components of attitudes. Attitude objects are objective entities. Attitudes, and affective, cognitive, and behavioral components, are presumed intervening variables. Feelings, etc. are (in principle measurable) outcome variables.

The affective component is essential to distinguish attitudes from general knowledge and beliefs (for example that a rectangle has four square corners). Thus, while we might all agree (believe or know) that abortion puts an end to a potential independent human life, some people hold an attitude that this is murder (with associated negative feelings about the act), are likely to avoid having an abortion themselves, and may take part in anti-abortion protests (behavioral components).

The functions of attitudes

The set of attitudes that a person holds will generally form a more or less cohesive set. They may be said to reflect a person's underlying **values** or general approach to the world. Attitudes affect our social behavior. We may join groups (for example, political parties) that have aims that coincide with the behavioral components of one or more of our attitudes, and we select companions and friends who share our attitudes. Our perceptions of others are powerfully influenced by the attitudes they express (Topic I3). Expressing our attitudes can reinforce our self-concept (Topic I4). Attitudes provide a framework with which we structure our social environment, allowing us to build a consistent model of society. Attitudes guide our information processing at three stages.

- At **acquisition**, we actively search for information supportive of our attitudes, and avoid or devalue discrepant (inconsistent) information.
- During **encoding**, we may distort discrepant information to fit better with our attitudes, and will classify information according to our framework of attitudes.
- At **retrieval**, while there is some evidence that we are better able to recall attitude-congruent information than discrepant information; more generally, both strongly congruent and strongly discrepant information is better recalled than attitude-neutral information.

Changing attitudes

Changing attitudes is the business of advertisers, governments, politicians, and teachers. Others who may be involved in attempting to change attitudes are parents, peer groups, and law enforcement agencies. The most obvious way to try to change a person's attitudes is to use **persuasive communications**; for example, advertisements, political posters, and the like. Persuasive communications may work by a **central route** or a **peripheral route**. The processes involved in changing attitudes by the central route may be envisaged as a series of information processing steps, first outlined by McGuire.

- The recipient must **attend** to the message and
- **comprehend** it (processes together called **reception)**;
- next is **yielding**: the recipient has to accept the message;
- next comes **retention**: the changed attitude has to persist, otherwise it will not result in
- modified **behavior**.

Most research on attitude change has focused on the factors affecting the first three of these steps; the last two relate to maintenance and effect of attitude change rather than to change itself. Attention depends on motivation to attend, which is affected by how interesting the message is, and by the individual's preparedness to attend to information which may be counter-attitudinal. It is also adversely affected by the clarity of the message, and by interference or distraction. Comprehension depends on the complexity of the argument contained in the message, and on the recipient's ability to understand it (intelligence; see Topic K7). An important message characteristic affecting yielding is **source credibility**: the more expert, trustworthy, familiar, or liked is the person or body presenting the communication, the more likely is attitude change to occur. Intelligence reduces the likelihood of yielding, unless the supporting arguments are strong.

The peripheral route allows persuasion even without attention to the message. Here, the recipient is affected by message and situational factors such as source credibility, number (rather than quality) of arguments, and professionalism of the message. The recipient is said to apply these as **heuristics**, or simple decision rules that do not involve much cognitive processing.

Another way to affect attitudes is to change behavior directly. This could be attempted with conditioning methods, but more attention has been paid to inducing **counter-attitudinal behavior**. This may be achieved by providing incentives for people to behave in a particular way, for example giving tax breaks to married over unmarried couples; or by **forced compliance**: using legal or more direct means to get people to engage in a particular behavior, for example wearing seat belts in automobiles. Laboratory studies, mostly using **counter-attitudinal advocacy**, in which subjects are induced to make statements supporting a view they do not hold, have shown that attitude change depends on the reasons for compliance. In particular, those given large rewards show less attitude change than those given small rewards. The first explanation for this apparently paradoxical effect derived from Festinger's **cognitive dissonance theory**, which argues that, when a person recognizes discrepancies amongst their cognitions, they will experience tension or discomfort, and will be motivated to reduce the discrepancy. Here, the discrepancy between the attitude held and that expressed produces cognitive dissonance which, in the case of a large reward, is reduced by explaining the discrepancy as due to that external factor, with no change in attitude. If the reward is too small to justify this, however, the discrepancy may be reduced by a shift in attitude towards that advocated. An alternative view of attitude change under forced compliance comes from **self-perception theory**, which proposes that people often use their own behavior to interpret their own actions and infer their own attitudes (see Topic I4). In this instance, a large reward by itself justifies making discrepant statements. However, a small reward does not, so the subject infers that their attitude must really be closer to the one they advocated.

Attitudes and behavior

The main reason for changing attitudes is as a means of changing behavior. There is little point in an advertiser giving consumers a more favorable attitude towards a product if they fail to buy more of it. Similarly, we frequently use attitudes to predict a person's likely behavior. However, earlier research showed only very weak relationships between people's attitude statements and their actual behavior towards the attitude objects. More recently, it has become clear that these weak relationships result largely from failing to make the attitude measures correspond with the behaviors. For example, correlations between attitudes to birth control and actual use of oral contraception have usually been negligible. However, improving the correspondence between question and behavior increases the correlation, so that asking more specifically about attitudes to *their own use* of *oral contraceptives*, during the *next two years* results in quite a high correlation.

I2 CAUSAL ATTRIBUTION

Key Notes

The covariation principle

Causal attribution, the search for causes of social events, is a key process in social cognition. According to the covariation principle, an observer decides whether an event is caused by a person, an entity, or a circumstance. To do this, three types of information are used: consensus, distinctiveness, and consistency.

Correspondent inference theory

Correspondent inference theory holds that attributions vary along three dimensions: stability (in time and across situations), intentionality, and disposition (internal or situational cause).

The fundamental attribution error

There is a strong bias towards making dispositional rather than situational attributions: the fundamental attribution error. This might result from the greater salience of a person's behavior over the situation, or from a social norm for taking responsibility for our own actions. The tendency to make dispositional attributions is not fixed.

The actor–observer bias

We tend to attribute our own behavior to situational variables and that of others to dispositional factors: this is one aspect of the actor–observer bias. This might result from having more information about our own behavior, making us less likely to adopt the societal norm of a dispositional explanation. Alternatively, appearance and behavior are salient with regard to the other person, but much less obvious with regard to ourselves.

Related topics

Perceiving others (I3)   Self-perception (I4)

The covariation principle

One of the most important processes of social cognition is **causal attribution**. We look for causes of social events and people's behavior, and attribute the events and behavior to those causes. The process has been likened to a scientific or statistical one: we look for cause–effect relationships, deciding the factors most likely to have caused a particular behavior. According to Kelley's **covariation principle**, an observer watching a particular behavior, for example one person pushing another, makes a decision about which of three types of cause is involved. These are the:

- **person**: the behavior results from the personal characteristics of the individual performing the behavior;
- **entity**: the behavior is caused by the nature of the target;
- **circumstance**: the situation or circumstances in which the behavior occurs causes it.

To reach this attribution, the observer uses three types of information:

- **consensus**: the extent to which the behavior is shared by other people;

- **distinctiveness**: the extent to which other people are the targets of the same behavior;
- **consistency**: the extent to which the person's behavior is consistent across different targets and situations.

Particular combinations of information lead to particular causal attributions (see *Table 1*). For example, if consensus is high, especially if distinctiveness is also high (in the example, other people as well as the aggressor show aggression towards the target), then the observer will attribute the pushing to the entity: the target is unpopular, weak, or obnoxious. If consistency is high, especially if consensus is low (the aggressor shows such behavior to many people, but others do not), then the behavior will be attributed to the aggressor, who may be perceived as a bully. If distinctiveness, consensus, and consistency are all low (e.g., the aggressor shows general aggression, others do not do so, and it only takes place on this occasion), then the cause is the circumstance; the aggressor may be seen as drunk or upset.

Table 1. Illustration of the covariation principle

A pushes B: information available	Interpretation of information: Consensus	Distinctiveness	Consistency	Causal attribution
Nobody else pushes B	Low			Person
A also pushes other people		Low		
A has previously pushed B			High	
Others push B	High			Entity
A pushes only B		High		
A has previously pushed B			High	
Nobody else pushes B	Low			Circumstance
A pushes only B		High		
A has not previously pushed B			Low	

Particular combinations of available information are interpreted as levels of the three types of information, leading to the indicated causal attributions.

A number of criticisms have been leveled at the covariation approach. First, while people's attributions may appear to follow the covariation principle, this does not necessarily imply that they are *governed* by that principle. Second, even if we do use the three types of information, we do not apply them mechanically; our attributions are made within particular social and interpersonal contexts. Furthermore, we do not usually have access to all the information required to apply the covariation principle. Third, much of the evidence for the covariation principle is based on highly artificial situations in which strictly limited information is provided to subjects.

Correspondent inference theory

An alternative approach to attribution stems from the work of Fritz Heider in the 1940s. **Correspondent inference theory** holds that attributions vary along three dimensions:

- **stability**: people seek stable causes of events, since these give us the sense of being able to predict future events;
- **intentionality**: we try to distinguish between intentional and unintentional actions, since the former permits us to attribute responsibility for actions;

- **disposition**: we distinguish actions that result from characteristics of the individual (dispositional attribution) from those that are determined by circumstances (situational attribution).

The first step in the attribution process is to attribute intentionality, which is done by processing information back from the *effects* of the action. This processing involves making inferences about the actor's *knowledge* (did s/he know what effect the action would have?) and *ability* (did s/he have the power to control the action?). The observer then seeks to make an attribution of disposition, comparing the consequences of alternative actions with those of the observed action. If the observed action has very different consequences from the alternatives, a **correspondent inference** is made that the actor has a disposition to act in this way (e.g., s/he is by nature an aggressive person).

A number of criticisms have been made of correspondent inference theory. First, it holds that disposition attributions can only follow intentionality attributions. However, nonintentional dispositions can be identified (e.g., clumsiness, accident-proneness). Second, there is little evidence that people making attributions consider the alternative actions and their consequences; rather, they seek further examples of the chosen action.

The fundamental attribution error

The process of making causal attributions about behavior is subject to a number of biases. There is a strong bias towards making dispositional rather than situational attributions. This is known as the **fundamental attribution error**. The strength of this bias in the face of what seem to be clear indications of a situational cause for behavior is shown by laboratory studies. In one such study, American participants observed people making pro- or anti-Castro speeches. Even when they were informed that some pro-Castro speakers had been instructed to make that particular speech, the observers rated the speakers' attitudes as in line with the speech content. Thus, the speech content (behavior) was attributed to the speaker's attitude (disposition) rather than to the instructions (situation).

A number of explanations have been proposed for the fundamental attribution error. Heider argued that the person's behavior is more salient than is the situation, so is more readily available to the attribution process. Another view is that, in individualistic societies such as those in Europe and North America, we are expected to take responsibility for our own actions, providing a *social norm* for dispositional (or, more generally, internal) explanations. The tendency to make dispositional attributions is not fixed. Studying social sciences produces a shift in the direction of situational attributions, and situational attributions are more likely when a behavior is not consistent with an individual's usual behavior.

The actor–observer bias

An important exception to the fundamental attribution error occurs in making attributions about our own behavior. It has frequently been observed that we tend to attribute our own behavior to situational variables, while the same behavior in another would be attributed to dispositional factors. This phenomenon is one aspect of the **actor–observer bias**. For example, scientists have been shown to explain their belief in their own theories by referring to empirical evidence (situational), but opponents' theories in terms of personality or other dispositional factors. One explanation for the actor–observer bias is that we have more information about our own behavior, and so are less likely to adopt the societal norm of a dispositional explanation. However, there is no evidence that increasing familiarity with another person increases the likelihood of a situational explanation for that person's behavior. Alternatively, it has been suggested that the difference is one of *perceptual focus;* we have a different point of view of our own and others' behavior. In particular, what is most salient about the other person is their appearance and behavior, but these are much less obvious with regard to ourselves. Storms demonstrated the importance of this in a study in which pairs of strangers were videotaped having a discussion (*Fig. 1*). Each then viewed a videotape showing either only the other person or only the self, and was asked to describe his or her own behavior. It was found that, as usual, people made more situational attributions about their own behavior when they viewed the other person, but that they made more dispositional attributions when they viewed themselves from the other's perspective.

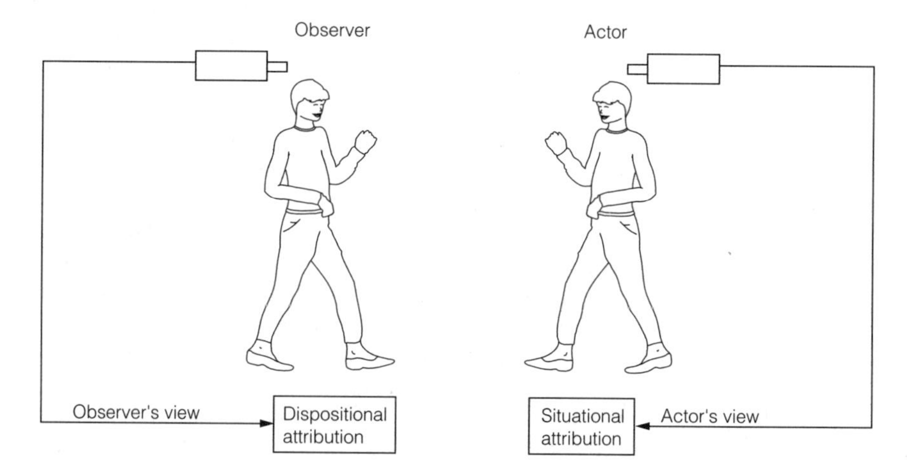

Fig. 1. The actor–observer bias. Viewed from his own perspective, the target person (actor) makes situational attributions about the interaction; viewed from the other's perspective he makes dispositional attributions.

I3 PERCEIVING OTHERS

Key Notes

Processes of person perception

While selective attention, motivation, expectancy, and experience affect the perception of others as social objects, person perception is not the same as object perception because we interact with other persons. Six processes have been described in person perception: attention, snap judgment, attribution, trait implication, impression formation, and prediction.

Impression formation

Consistencies between situations and between aspects of the same person allow us to form an impression of the whole person. Central traits imply a set of other traits and organize our impressions, while peripheral traits do not. We seem to have shared implicit personality theories which include beliefs about which traits go together. Impression formation is affected by primacy effects, halo effects, and expectancy, which may lead to a self-fulfilling prophecy.

Impression management

We actively try to present a particular image to the world. One consequence of failing to manage the impression we make is embarrassment. We may also experience empathic embarrassment when we witness another person failing to present the expected self, and we may collude with the other to assist in his or her impression management.

Related topics

Attitudes (I1)
Causal attribution (I2)
Intergroup relations (J3)

Processes of person perception

There is more to **person perception** (or social perception) than perceiving that another person *is* a person, and recognizing who that person is. Person perception refers to processes through which we identify others as *social* objects, rather than merely as objects. Even so, many of the general principles of perception have parallels here (see Topic D1). For example, we have to rely on surface properties of the person as a stimulus; we apply processes of selective attention (see Topic D7); and our perception of people is subject to factors, such as motivation, expectancy and experience, that affect the inferences we make from the surface properties. But person perception is not the same as object perception. Perceiving people can be thought of as attempting to *understand* them. Perceiving people is different because we change the people we meet, by interacting with them, in a way that we do not change objects.

Six processes have been described as taking place in person perception.

- **Attention**. We select which people, and which aspects of people, to attend to. We may describe a person in terms of physical characteristics and behaviors.
- **Snap judgment**. A more or less automatic inference is made from appearance or behavior, and may produce an affective (emotional) reaction, a

category judgment (such as 'intelligent' or 'extraverted'), or a stereotypical judgment (see Topic J3).
- **Attribution**. With further information and/or further consideration, attributional processes (see Topic I2) look beyond the surface characteristics to how much the person's present appearance or behavior reflects his or her characteristics, and how much it depends on situational factors.
- **Trait implication**. If we make a dispositional attribution, perceiving the person as having a particular internal characteristic, this can lead to the assumption that the person possesses other traits that we believe belong with the observed characteristic. This is another aspect of stereotyping.
- **Impression formation**. The characteristics and traits we perceive in another person are organized into an impression of that person (see below).
- **Prediction**. In the same way that social attitudes allow the prediction of behavior (see Topic I1), so impressions of other people lead us to make predictions about their future behavior.

Impression formation

In object perception, processes operate leading us to perceive objects as enduring and invariant (constancy; see Topic D5). Other processes lead us to construct perceptual objects from partial or ambiguous information, and to perceive *whole objects* rather than amalgamations of components (see Topic D1). Similar forces are at work in person perception. We look for consistency, not only between situations but between aspects of the same person. These consistencies allow us to form an impression of another person as a whole person. These processes were studied 50 years ago by Solomon Asch. He showed that people are prepared to write a description of another person simply on the basis of a few characteristics, or **traits**. Varying some traits can produce a large change in the description, while others influence the impression less. Traits which produce a large change, such as *warm* or *cold*, are known as **central traits**. Those which have little effect, such as *polite* or *blunt*, are known as **peripheral traits** (see *Fig. 1*). Central traits are seen as organizing our impressions and as implying a larger number of related traits than are peripheral ones. Bruner and Tagiuri later suggested that we have **implicit personality theories** which include beliefs about which traits go together. These implicit personality theories are largely shared, so that central traits will produce similar impression descriptions from a large number of respondents.

Bruner and Tagiuri concluded that our formation of impressions of others is not accurate, even if many of us will form the same impression given the same information. A number of related factors contribute to this inaccuracy. *First impressions* may influence how we deal with later information about a person. This can be demonstrated in **primacy effects**. Asch originally showed that the order in which traits are presented to respondents in an impression formation study can markedly affect what other attributes they will ascribe to the target individual. In particular, traits at the beginning of the list have a greater influence, and change the interpretation of later traits in the list. A second factor is the **halo effect**. This is a tendency to view a person in a generally good light, either because of a combination of identifying in that person a positive central trait with the primacy effect, or because we associate the person with a positive situation or experience. We can also identify the obverse of this, *negative* halo effects, which are an important component of stereotyping (see Topic J3). A third factor is the effect of **expectancy**. Believing that a person has a particular characteristic because of information we have received about them can not only

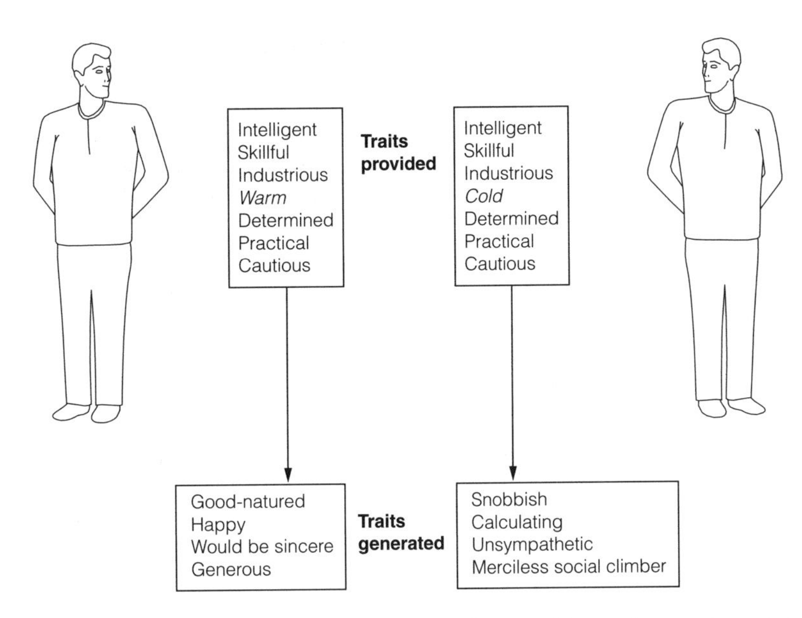

Fig. 1. Central vs. peripheral traits. Given lists of personal descriptions varying only in one 'central' trait, people are prepared to describe very different impressions of the person.

influence our impression of them, but can change our behavior towards them to such an extent that the expectation, even if based on erroneous information, becomes true: a **self-fulfilling prophecy**. In the classic demonstration of this phenomenon, Rosenthal and Jacobsen allowed teachers to overhear a conversation in which a randomly selected group of pupils were described as particularly intelligent. After 1 year, these children were found to have improved their actual IQ scores and to be performing better at school. It was concluded that the teachers were treating these children differently, leading to their expectations being confirmed. The social problem that can arise form this process, which has also been called **labeling**, is that when teachers and others have *low* expectations of others, for example in schools in areas of social deprivation, true potential may not be attained.

Impression management

As we said earlier, person perception is unlike object perception in that it is an interaction. One way in which this is so is that the person being perceived actively tries to present a particular image to the world; to use Goffman's term, they practice **impression management**. It is as if in a social context we put on a performance, as if on a stage. Since we might have a number of different roles in different social situations (e.g. worker, spouse, parent, 'one of the lads'), we strive to manage ourselves (e.g. dressing differently, using different vocabulary, behaving differently) so as to give an impression that is appropriate in each context.

One consequence of failing to manage the impression we make is **embarrassment**; we feel that we have lost the esteem of others by a deficiency in our presented self. In many types of social situation, we experience empathic embarrassment when we witness another person failing to present the expected self, or make the expected impression, particularly when that person is a friend or someone else we identify with. When we are interacting in such situations, we may collude with the other to assist in his or her impression management. We do this by failing to notice their errors, or by providing acceptable reasons for some action which would otherwise cause the other to 'lose face'. For example, when we are bored with somebody's conversation we will provide a plausible reason for leaving them ('my glass is empty', or 'my wife will wonder where I am'), rather than telling them the real reason.

I4 SELF-PERCEPTION

Key Notes

Self-perception theory

Social interaction is a major influence on the formation of the self-concept. The early 'looking-glass self' view that we develop through observing how other people behave towards us was replaced by Bem's self-perception theory. In order to know ourselves, we have to interpret ambiguous internal signals by applying attributional processes to our own behavior.

The self-serving bias

When we make attributions about our own successes, we are more likely to attribute them to dispositional than to situational factors. This self-serving bias may be a cognitive process, resulting from the individual's intention and expectation of success, or a motivational process, for example as an aspect of impression management or the maintenance of self-esteem.

Approaches to emotion

Older, noncognitive approaches to emotion are the peripheral physiological James–Lange theory and the central physiological Cannon–Bard theory. The James–Lange theory argued that emotions are our perception of bodily changes, while the Cannon–Bard theory held that bodily changes, behavior, and emotional experience all stem from thalamic centers. One, more recent, cognitive approach is Lazarus's view that emotions arise from particular patterns of appraisal of environmental stimuli.

Emotions as self-attributions

Schachter and Singer claimed to have demonstrated that sympathetic arousal could be experienced as either anger or euphoria, depending on whether those experiencing it attributed it to an angry or a euphoric situation. The resulting two-factor theory became highly influential in many fields, including physical attraction and the reduction of emotional responses by way of placebos.

Related topics

Causal attribution (I2)   Perceiving others (I3)

Self-perception theory

How do we form an understanding of ourselves as individuals? It has been argued that the **self-concept** is formed as part of a social process. Earlier in the history of psychology, it was thought that the self-concept develops through observing how other people behave towards us; we develop a '**looking-glass self**'. This view was challenged by Daryl Bem about 30 years ago. He argued that we come to understand ourselves in the same way that we come to understand other people: through attributional processes (see Topic I2). Bem's **self-perception theory** argued that we only have weak and ambiguous information about our internal states. In order to know how we feel or what we think about a social object, we have to interpret these ambiguous signals, and we do this by applying the cognitive processes of causal attribution, just as we do in understanding others. That is, our description of our own feelings or attitudes depends, at least in part, on the attributions we make about our own behavior. As we saw in Topic I1, self-perception theory provides an alternative explanation to cognitive dissonance theory for various instances of attitude change. Self-perception theory also explains instances of behavior change. In one study, Freedman and Fraser showed that, while it is hard to get people to agree to allow a large road-safety sign to be erected in front of their house, it was easy to get them to agree to display a small window sticker. People who had agreed to the sticker were later much more likely to agree to erect the large billboard. This has been interpreted as a change in self-perception: those who had agreed to the small sticker attributed this action to their own beliefs, and came to perceive themselves as active and involved in such road-safety issues.

The self-serving bias

One bias in attributional processes serves to maintain or promote the self-concept. As we saw in Topic I2, most of the evidence concerning the actor–observer bias concerns attributions about undesirable behaviors or outcomes. When we make attributions about our own *successes*, we are more likely to attribute them to dispositional than to situational factors. That is, while we blame our failures on external circumstances, such as task difficulty, we claim successes as results of our own characteristics, such as skill. It is unclear whether this **self-serving bias** is a cognitive process or a motivational one. Viewed as a cognitive process, the self-serving bias is interpreted as resulting from the individual's intention and expectation of success. It is perhaps easier to view this bias as an example of social motivation, for example that it is a facet of impression management, of the desire to present the best face to the world (see Topic I3). However, the self-serving bias has been observed even when the individual believes that nobody else knows about the attribution. In this case, another motivational factor, the maintenance of self-esteem, might be the explanation.

Approaches to emotion

The dominant theoretical approach to emotion changed dramatically in the course of the last century. From the late 19th century until about 1930, the most influential approach was an essentially self-perception view: the **James–Lange theory**. William James, in 1884, attacked the 'common sense' view that emotional feelings follow directly from the arousing stimulus, and emotional behavior follows emotional experience. He argued that emotions consist of our perception of bodily changes (originally peripheral physiological and behavioral) that follow directly from an arousing stimulus. (Lange simultaneously published a similar view, based on the perception of changes in the vasomotor system. This led James, in later accounts of his theory, to play down the significance of actions.) It follows from this that each emotion must be associated with a different pattern of bodily changes, and that a person who was completely insensitive to bodily changes would not experience emotion. James's evidence for his theory was largely introspective and open to alternative explanations. It was effectively demolished in 1927 by Walter Cannon, who argued amongst other things that fear, anger, pain, and hunger all share the same autonomic pattern of sympathetic arousal. Inducing these physiological changes in humans does not produce real emotions, and surgical procedures in cats and dogs that remove most of the afferent information from the organs does not prevent emotional expression. The first of these arguments is probably fallacious; we know now that the autonomic nervous system is capable of very finely differentiated responses, and there is evidence for some differentiation of autonomic and endocrine responses in emotions. The second point is not logical: since James did not predict that any particular emotion would be associated with sympathetic arousal, we would not expect inducing such arousal to produce a particular emotion. The third point assumes that emotional behavior in cats and dogs requires emotional experience; and in any case James expected emotional behavior to *precede* emotional experience.

Nevertheless, James's peripheral physiological theory was replaced by a central physiological theory first propounded by Cannon, and elaborated by Bard. In this view, centers in the thalamus (see Topic A3), responding to an arousing stimulus, control physiological, behavioral, and experiential components of emotions. For many years following, research centered on refining knowledge of the brain mechanisms involved in the processing of emotional stimuli and the production of emotional responses, research that continues to this day using modern research techniques such as, for example, functional magnetic resonance imaging (see Topic A4).

In the second half of the 20th century, however, in common with other areas of psychology, cognitive approaches started to appear in the study of emotion. There have been two main cognitive approaches. One, stemming from the work of Richard Lazarus in the 1970s, describes emotions as arising from particular patterns of **appraisal** of environmental stimuli. A stimulus is appraised as threatening or nonthreatening. Nonthreatening appraisals lead to the experience of positive emotions in the presence of the stimulus, while those appraised as threatening give rise to negative emotions. The particular emotion experienced depends on **secondary appraisal**, identifying more specific features of the situation. Lazarus applied this approach chiefly to the understanding of stress and anxiety, but more recently research has started to identify the patterns of appraisal underlying other emotions. The second main cognitive approach takes us back to causal attribution, and we turn to that next.

Emotions as self-attributions

In one of the most frequently cited psychological experiments, Schachter and Singer in 1962 claimed to have demonstrated that the same state of autonomic arousal could be experienced as either anger or euphoria (*Fig. 1*). The key condition that changed the experience was attributing the arousal to a particular apparently exciting cause. That is, following an injection of adrenaline (epinephrine), participants would experience unexplained arousal. They would engage in a search of their environment for a likely source of this arousal, and would use the apparent source to label their experience as one or other state. So, a participant paired with an experimental confederate behaving angrily would interpret his arousal as anger, and one paired with a euphoric confederate as euphoria. Schachter and Singer claimed that their results showed just this, but their data actually show that none of their experimental groups reported anger, and attempts to replicate the claimed findings have failed.

Despite this, the **two-factor theory** (arousal + attribution) of Schachter and Singer became very influential, and a variety of phenomena were ascribed to the attribution or misattribution of arousal. For example, Dutton and Aron used an ingenious method to demonstrate how physical attraction could be enhanced by arousal attribution. They had an attractive female interviewer approach men on either a solid, stable bridge, or on an exciting, rope-suspension bridge. At the end of the interview, she gave each man a contact number if they wanted to ask questions about the interview. More of the men approached on the suspension

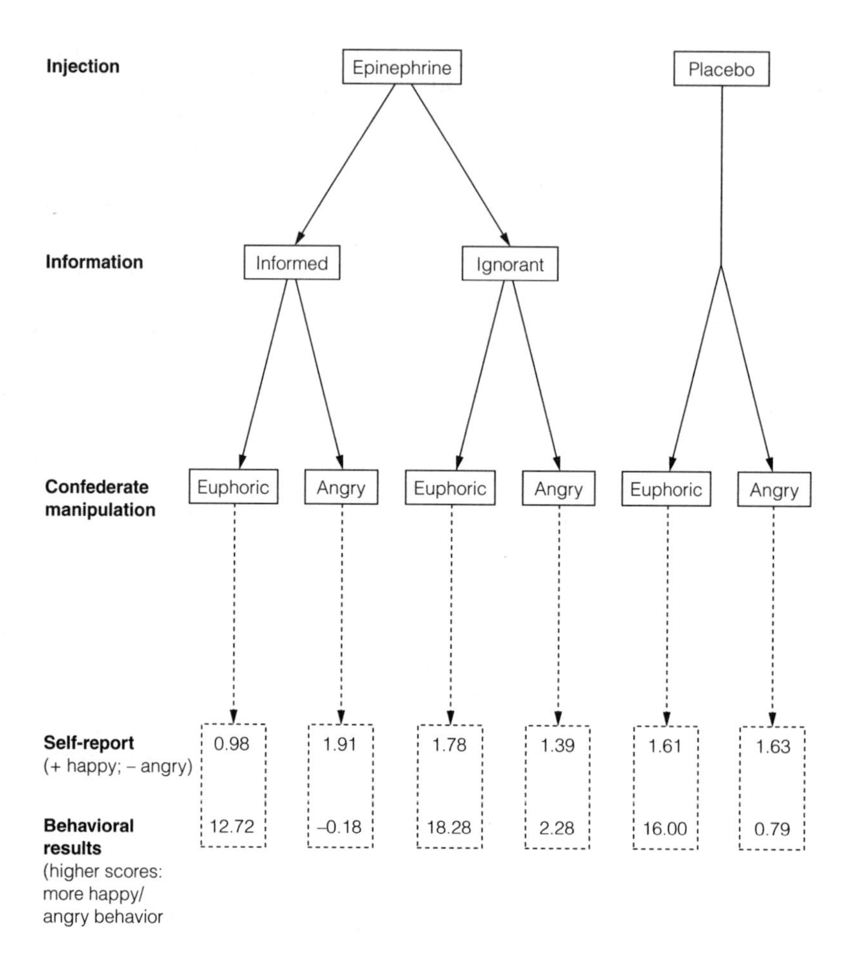

Fig. 1. The main conditions and results of Schachter and Singer's 1962 experiment. The prediction was that those kept ignorant of the arousing effects of the epinephrine injection would attribute their arousal to an emotion labeled by the confederate's behavior. The results show some support for this in relation to behavior, but self-report data show that no group reported anger. The conclusion that ignorant subjects 'had been readily manipulable into the disparate feeling states of euphoria and anger' is not supported.

bridge called, which was interpreted as showing that these men attributed their arousal (actually caused by the rickety bridge) to the attractiveness of the interviewer.

Other studies have attempted to show that adverse emotional reactions can be reduced by giving people a placebo (dummy) pill, and telling them that it causes arousal. It has been claimed that this can reduce, amongst other things, painfulness of electric shocks and the time taken by insomniacs to fall asleep. In each case, participants are said to have attributed their arousal to an external source, the placebo, so perceived their own response as intrinsically less aroused. However, many of these studies have been criticized, and others have failed to obtain such results. In general, there is actually very little evidence that attributional processes play a role in emotional arousal.

J1 Communication

Key Notes

The nature of communication

Communication takes place whenever information is transmitted from one place to another. Communication can take place through any sensory modality, and the specific modality or aspect of a modality that carries the information is called a communication channel.

Nonverbal communication

Speech is accompanied by nonverbal communication in auditory and visual channels. Nonverbal signals help in the conduct of a conversation, and speech cannot be properly understood unless we consider the nonverbal behavior that accompanies it.

Expression of emotions

Emotion theory has emphasized that some 'basic' emotions appear to have facial expressions that are the same in all cultures. We manage our facial behavior to communicate emotions, attitudes, motives, or action tendencies. The behavior we show in a particular situation is determined by display rules.

Deception

Deception involves an intention to deceive. Lies are hard to detect by unaided observers. People share common beliefs about the cues that reveal emotion, but most of these are not in fact associated with lying. Lie-detector machines (polygraphs) record physiological changes that are expected to accompany lying. They detect lying better than chance, but produce a lot of false positives.

Related topics

Language (Section G)
Perceiving others (I3)

The nature of communication

Communication takes place whenever information is transmitted from one place to another. For our purposes, we consider this to involve a **sender** organism and a **receiver** organism. Communication does not necessarily involve intentionality: the sender does not have to intend information to be transmitted, nor to be aware of its transmission. Neither does the recipient need the intention to receive information, nor awareness of its reception. Communication can be demonstrated when the behavior of one organism is influenced by the behavior of another. For example, the 'tail-wagging dance' of the honey bee described by von Frisch communicates to other workers the direction and distance of a source of nectar, and this is known because those workers fly accurately to this source. Rather than assume intentionality, we might argue that the sender responds to stimuli in the social or physical environment that elicit communicative behavior. Such **signals** have evolved to communicate between, particularly, members of the same species, and they have evolved because they confer an advantage on the individual organism's likelihood of passing its genes to the next generation (see Topic A1). These signals may be complex sequences of behavior, and are then known as **rituals**. Thus, we speak of courtship rituals used in mate selection, or ritualized aggression that maintains dominance hierarchies without harm to the animals. Of course, awareness and intentionality might themselves confer an evolutionary advantage.

Communication can take place through any sensory modality (see Section C), and the specific modality or aspect of a modality that carries the information is called a communication **channel**. Thus, language (see Section G) is a means of communication, but may be conveyed through auditory (spoken) or visual (written) channels.

Nonverbal communication

Speech is by definition the use of language in communication, but it is accompanied by **nonverbal communication**, which itself may involve auditory and visual channels. Nonverbal signals perform several functions. They help in the conduct of a conversation; for example, we may indicate that an utterance is a question by using a rising pattern of intonation (a **paralinguistic** channel), by tilting the head (**postural**), and by making eye contact with the other person (**gaze**). Gaze is particularly important in managing turn taking in conversation. When we start a long utterance, we tend to look away from the other person, making eye contact again when we near the end. This can be interpreted as signaling that we are about to finish speaking, and are 'offering the floor' to the other. Since selective visual attention depends on focal vision (see Topic C3 or D7), if we wish to gauge how the other is reacting to what we say, we need to maintain our gaze on them, particularly on their face, since the face is a rich source of nonverbal information.

Speech cannot be properly understood unless we consider the nonverbal behavior that accompanies it. For example, the tone of voice in which we say a simple sentence such as *It's cold in here* can communicate, amongst other things, general attitudes such as dominance or weakness, or more specific intentions such as blame, the intention to correct the situation, or a desire for physical contact. This is elaborated by signals in other nonverbal channels such as **gaze** (averted or eye contact), **facial expression**, **gesture**, **touching**, moving closer or further apart (**proxemics**), or **posture**. Nonverbal communication also takes place in humans in the absence of speech, and is capable of communicating an enormous range of attitudes, feelings, and behavioral intentions. Indeed, the verbal content of an utterance may be completely irrelevant; in our example it might not be cold at all.

Expression of emotions

Nonverbal behavior, especially facial expression, has been highly significant in the understanding of emotions. Many emotion theorists have proposed that there is a limited number of **basic** or **fundamental** emotions. The usual reason for this assumption is that the research of Paul Ekman and Carroll Izard has shown that some emotions appear to have facial expressions that are the same in all cultures, and are basic in the sense of being innately programed.

For some authors, the association of particular facial movements with particular emotions suggests that they literally express, or provide a 'read-out', of one's inner emotional state. However, it is clear that we manage our facial behavior in ways which might be called impression management (see Topic I3), or deception (see below). It fits better with the analysis of other signals to talk of facial behavior as *communicating* emotions (or attitudes, motives, or action tendencies), rather than as *expressing* them. It is known that the neural pathways controlling *spontaneous* facial expressions are different from those that control *voluntary* facial behavior. It has been proposed that the behavior we show in a particular situation, a combination of voluntary behavior and managed spontaneous behavior, is determined by **display rules** that determine *who* may show *what* emotions *to whom* and *when*. The concept of display rules has not been elaborated beyond rather global (even stereotyped, see Topic J3) expectations of different cultures and different status. It is certain that facial and other nonverbal displays are managed on a much more specific level than this, so that we in effect assess each interaction we take part in for multiple characteristics, such as status, gender, age, relationship, as well as known or perceived aims and expectations of the interactants and observers (see impression management, Topic I3).

Deception

It has been suggested that certain signals emitted to mislead another animal can be classed as **deception**. For example, individual birds of some species will behave as if they were unable to fly when their nest is approached by a predator, which sometimes has the effect of distracting the predator from the bird's young. It is tempting to say that the bird is *deceiving* the predator by behaving in this way. But that implies intentionality, and such behavior may merely be an automatic response to a particular stimulus. To be useful, the concept of deception needs to include a component of intention to deceive.

Lies are hard to detect by unaided observers. Most studies have shown that only 45–60% of lies are detected, and 'expert' detectors (e.g. police and customs officers) are no better than anybody else, although they usually *believe* that they are (an illusory correlation, see Topic J3). People share common beliefs about what are the verbal and nonverbal cues that reveal emotion. However, these are almost always cues that are *not* in fact associated with lying (see *Table 1*). For

Table 1. Believed and actual cues to deception

Behavior	What people believe	What happens during lying
Visual nonverbal cues		
Eye contact	Decrease	No difference?
Smiling	Increase	No difference
Touching self	Increase	No difference?
Blinking		Increase
Fidgeting	Increase	No difference
Dilation of pupils		Increase
Gestures	Increase?	No difference
Foot and leg movements	Increase	Decrease
Arm and hand movements	Increase	Decrease
Head movements	No difference?	No difference
Paralinguistic cues		
Hesitant speech	Increase	Increase
Speech errors	Increase	Increase
Pitch of voice	Higher	Higher
Rate of speaking	Slower	No difference/slower?
Duration of talking	No difference	Shorter
Delay before starting	Longer	Shorter?
Verbal cues		
Negative comments		Increase
Irrelevant content		Increase
Nonimmediacy		Increase
Self-references		No difference

In no case have all studies agreed; those where there is more doubt are marked with a ?. Empty cells indicate that there is little or no relevant research. Note that most studies of actual cues have been of relatively unimportant laboratory tasks, while those of believed cues mostly refer to important lies.

example, we believe that a person telling a lie will avoid eye contact, when in fact lies are associated with *more* eye contact.

If people are not good as lie detectors, can machines do better? Lie-detector machines, known as **polygraphs**, record physiological changes that are expected to accompany lying, usually skin conductance, heart rate, and blood pressure. The idea is that lying causes an increase in physiological arousal which the operator can then identify. Polygraphs do, indeed, detect lying at well above chance levels, but they do this at the expense of a lot of **false positives**; that is, a lot of truthful persons are classified as lying. The reason for this is that there is no specific pattern of arousal for lying, and anxiety caused by the procedure and questions relevant to a crime will cause physiological changes that are indistinguishable from those resulting from a lie.

J2 GROUPS

Key Notes

Social influence

The mere presence of other people may improve performance (social facilitation), or cause a decrement in performance. Explanations have been based on competition, evaluation apprehension, and arousal. Whether the audience effect is to enhance or disrupt performance depends on task complexity and amount of arousal.

Conformity and obedience

Conformity to a group might result from normative influence (wanting to be accepted by others), or from informational influence (others' judgments are used as evidence about reality). Milgram showed that social influence can induce obedience to instructions that are apparently injurious to others. Obedience is more likely when a person can increase the 'psychological distance' from the target, or when they shift from an autonomous state of responsibility to an agentic state in which they see themselves as the agent of others.

Crowd behavior

When crowds of people act in extreme ways (e.g. with panic or violence), the individuals seem to have lost a sense of personal identity (deindividuation). Deindividuation might depend on anonymity, which could lead to the individual feeling more likely to get away with the extreme behavior.

Group decision making

Group decisions were once thought to be riskier than individual decisions (risky shift), which was explained as the diffusion of responsibility. However, the general effect of group decision making is to make decisions more extreme (group polarization). This could result from informational influence or a process like impression management. If a group is highly cohesive, has a strong directive leader, is insular, and lacks systematic procedures, it may reach unrealistic decisions by failing to consider information that does not fit with its initial view (groupthink).

Leadership

It used to be believed that leaders had personal characteristics that would make them rise to leadership in any group (the 'great man' theory). The situational view suggested that the person who emerges as a leader depends on the nature of the group. The modern view is a transactional one, in which the emerging leader's behavior changes depending on the situation. Three leadership styles have been described. The authoritarian leader is a dictator, the democratic leader remains part of the group, and the laissez-faire leader leaves the group to its own devices.

Related topic

Intergroup relations (J3)

Social influence

The simplest form of social influence results from the mere presence of another person, or group of people. These effects have been described as **social facilitation**, since in early studies the effect was generally found to be an improvement in performance. For example, people will ride bicycles, learn mazes, or solve multiplication problems faster when in the presence of others. One early explanation of this was that this facilitation results from competition. However, it also occurs when the others are not *coacting* (performing the same task), but are simply *observing* the individual (**audience effects**). Furthermore, social presence effects may be detrimental to performance, particularly when tasks are more complex. Some findings (e.g. audience effects decrease when the 'observer' is blindfolded, but increase when the observer is described as an expert in the task) suggested that these results might be due to **evaluation apprehension**; that is, anxiety arising from being judged rather than simply observed. This suggests a cognitive basis for social facilitation. However, other species show social facilitation; even cockroaches will run faster when tested in pairs than when alone.

This led Robert Zajonc to propose in the 1960s that the common feature underlying all these effects is the generation of **arousal** by an audience. In humans, this arousal might result from various factors, such as evaluation apprehension, distraction, or competition, and in other species perhaps by simple competition. Arousal affects behavior in ways described by the **Yerkes–Dodson Law** (see Topic B1). In simple tasks, it facilitates the production of the dominant (correct) response, while in complex tasks its effect is sometimes to energize incorrect responses, leading to a deterioration in performance of more complex tasks.

Conformity and obedience

In other examples of social influence, it is not simply the quality of task performance that is changed, but the nature of the behavior itself. When we modify our behavior to fit in with that of others, we show **conformity**, or **majority influence**. Of course, it might be that we choose to associate with those who share our attitudes and other attributes (see Topic I1), but we can be strongly influenced by those we have no choice about associating with. In the 1930s, Muzafer Sherif showed that a small group of people quickly forms a consensus (**group norm**) about judgments of an ambiguous stimulus (the autokinetic phenomenon, see Topic D3). True conformity was demonstrated in classic studies in the 1950s by Solomon Asch. He put a naive subject with six experimental confederates, and gave them a very simple perceptual task: deciding which of three lines was the same as a standard line. Each of the confederates in turn made a clearly erroneous judgment, and eventually 75% of real subjects agreed with this judgment at least once.

Why does such conformity occur? It might be a **normative influence**: people's judgments are affected if they want to be accepted by others (see impression management, Topic I3). Conformity depends on believing that others know what one's response is. If the subject is allowed to write the answers then conformity almost disappears. However, it also occurs when the individual is physically isolated from other people, being shown others' responses by signals. Then, conformity is ascribed to **informational influence**: others' judgments are used as evidence about reality. Both processes probably occur in most conformity situations.

A famous study in 1963 by Stanley Milgram showed that social influence can induce **obedience** to instructions that are apparently injurious to others. Subjects from the general population were given the task of 'teaching' another subject (actually an experimental confederate) in a learning task, by administering increasingly powerful electric shocks every time the other made a mistake. After each shock, the subject heard sounds of increasing distress over a loudspeaker. If the subject was reluctant to give a shock, the experimenter gave a simple instruction such as 'please go on'. *Every* subject continued to give shocks up to the level labeled 'intense shock', and 65% continued to the highest level, beyond a 'danger' label. This level of obedience was maintained in later studies in which the subject could see as well as hear the confederate, which were conducted in seedy surroundings, rather than in prestigious Yale University, or when the experimenter was not present in the room with the subject. It dropped by only 10% when the confederate was heard demanding to be released from the study. When the subject was given the task of holding the resisting confederate's hand on the electrodes, compliance dropped to 30%. When the experimenter was replaced by a 'fellow subject', lacking the former's apparent authority, 20% of subjects still went to the highest shock level. The underlying feature of these modifications seems to be increases and decreases in the 'psychological distance' between the subject and the consequences of his or her actions. Some subjects described a process of 'dehumanizing' of the person they were electrocuting. Milgram interpreted his results as showing a shift from an **autonomous state**, in which people see themselves as responsible for their actions, to an **agentic state**, in which they see themselves as the agents of others. The parallels between these studies and the actions of apparently ordinary people in situations such as warfare, or living and working under a totalitarian regime are obvious.

Crowd behavior

Under certain conditions, crowds of people may be observed acting in extreme ways, usually involving panic or violence. The characteristic feature is that people act in ways that that they would not do when alone, and this led to the proposal that people in crowds lose a sense of personal identity, experiencing **deindividuation**. This is most likely to occur when excitement is high, releasing normally inhibited impulses (which may be violent, panicky, or orgiastic, depending on the situation). Deindividuation is also supposed to depend on anonymity, although attempts to demonstrate this experimentally have been open to alternative explanations. Anonymity could lead to a more cognitive explanation for crowd behavior: the individual, feeling anonymous, might feel more likely to get away with the extreme behavior.

Group decision making

How does the behavior of a group as a whole differ from that of the individuals that compose it? In 1961, James Stoner tested the assumption that a decision made by a group would be conservative. Subjects considered a series of scenarios involving a choice between a risky, but potentially rewarding, and a safe, but less rewarding, course of action. They first decided individually what level of risk was acceptable when advising the central figure in each. Then they discussed the scenarios in groups, and reached a group consensus about the level of risk acceptable. The subjects finally again made individual judgments. Group decisions were riskier than the average of initial individual decisions, and these riskier decisions were carried over into the final, individual decisions. This phenomenon became known as the **risky shift**.

The original explanation for this result was the **diffusion of responsibility**: individuals could distance themselves from responsibility for the group decision. However, individual risky shifts were shown when subjects discussed the scenarios, but did not reach a group decision. It was also discovered that, for scenarios for which individuals initially made cautious judgments, the group decision was a shift to greater caution, and the phenomenon was relabeled **group polarization**. One explanation of group polarization is **informational influence** (see *Conformity and obedience*). A second possibility comes from **social comparison theory**, which suggests a process like **impression management** (see Topic I3).

Group decisions do not always produce polarization. In some circumstances, a group may reach unrealistic and potentially disastrous decisions by failing to consider information or arguments that do not fit with the group's initial view. This process was described in the 1970s by Irving Janis, who blamed serious errors of the US administration on this phenomenon, which he called **groupthink**. Groupthink is most likely to occur when a group is cohesive, has a strong directive leader, is insular (cut off from outside information), and lacks systematic procedures. All of these features have been identified in decisions such as the disastrous 1961 US invasion of Cuba in the Bay of Pigs episode, and the launch of the Challenger space shuttle in 1986 which exploded, killing seven people.

Leadership

In most groups, it is possible to identify a leader or leaders. A leader is a person who guides and facilitates group activities. What factors determine who will be a leader? Until the middle of the last century, the dominant view was the 'great man' theory, which suggested that leaders had personal characteristics that would make them rise to leadership in any group. This was generally replaced by a situational view, suggesting that the person who emerges as a leader depends on the nature of the group and the demands facing it. The modern view is a transactional one, in which the emerging leader's behavior changes depending on the situation.

Leaders' behavior has been examined to identify **leadership styles**. Kurt Lewin in the 1930s identified three such styles. The **authoritarian leader** is a dictator, imposing his or her will on members, and watching them closely. The **democratic leader** remains part of the group, discussing objectives with them. The **laissez-faire leader** leaves the group to its own devices. Lewin's research seemed to show that democratic leadership was most effective, in that these groups were overall more productive and had higher morale. Authoritarian groups were more productive *while the leader was watching them*, and laissez-faire groups did little work and had low morale.

J3 Intergroup relations

Key Notes

Social categorization

The groups we belong to help determine our behavior. Social comparison leads to value judgments about other groups, which can lead to negative stereotyping and prejudice. Downward social comparison enhances the individual's self-esteem, and leads to over-valuing of our own group and its members.

Social identity theory

We derive our social identity from the groups we belong to. Social identity theory sees social identity as an essential part of the self-concept.

Stereotypes

A stereotype is a belief that members of a group have a common characteristic or set of characteristics. Schemata (cognitive frameworks that we use to help us to understand the world) lead us to make predictions based on minimal information. One factor that maintains stereotypes is illusory correlation: noticing only instances that fit the stereotype.

Social representation theory

People within a group share a common world view, a social representation of reality, which also guides the individual's understanding of the world. Anchoring is the integration of a new phenomenon by linking it with things with which we are already familiar. Objectification usually involves making abstract ideas more concrete, the use of metaphor, or the linking of ideas with individuals (personification).

Related topics

Perceiving others (I3)
Self-perception (I4)
Groups (J2)

Social categorization

In Topic I3, we saw how our perception of others can be influenced by the groups of which we perceive them to be members. This process of **social categorization** is an important way in which we structure our social environment. Each of us is a member of numerous categories, or groups. Categories may be determined by a variety of features; for example, physical characteristic (gender, skin color), interest (hikers, opera lovers), occupation (teacher, accountant), relationship (mother, husband). Some categories involve real status differences (e.g. full vs. assistant professor), while processes of **social comparison** lead us to make comparative value judgments about other groups in relation to our own. This can lead to negative stereotyping and prejudice. Social comparison tends to be made with groups perceived as of lower status than our own. This practice operates to enhance the individual's **self-esteem**, because we identify with groups of which we are members.

Conversely, social comparison leads us to perceive the status of a group to which we belong as higher than that of similar groups. **Minimal group studies**, many of which were conducted in the 1970s, set up laboratory groups by arbitrarily assigning subjects to groups. Typically, when subjects were then asked to evaluate the performance of members of their own and other groups, they gave members of their own group higher scores. So, discrimination between in-group and out-group members occurs even in the absence of real differences between groups or group members. More recent research has shown that such discrimination is reduced by allowing participants to make less directly comparative judgments. For real groups, competition (e.g. for jobs, prizes) may be an important feature of comparisons, which can then lead to hostility.

Social identity theory

Henri Tajfel argued from the 1970s that we derive a view of ourselves, our **social identity**, from the groups we belong to. We often join particular groups because in so doing we enhance our self-esteem. The extent to which a group we belong to does this is an important determinant of how much we identify with that group. If we do not identify with the group, we will try to distance ourselves from it, or even to leave it. Social identity theory sees social identity as an essential part of the self-concept. We do not act differently in different social groupings because we are playing different roles (see Topic I3), but because we are expressing a particular aspect of our self.

Stereotypes

A **stereotype** is a belief that members of a group have a common characteristic or set of characteristics. It is shared by members of that group or, more commonly, members of another group. Actual or perceived group membership may be determined by a superficial feature such as skin color, place of residence, or national origin. The stereotype (racial, national, etc.) may then be applied to the individual. Since stereotypes are generally value laden, the result is prejudice and discrimination. The origins of stereotypes are not well understood, but their operation and perpetuation have been described within the loose theoretical framework of schema theory. A **schema** (plural *schemata* or *schemas*) is a cognitive framework that we use to help us to understand the world (see also Topic H2). An implicit personality theory could be viewed as a schema, so that believing that somebody is a warm person is part of a schema that includes a number of other positive personality traits. Schemata lead us to have expectations about what characteristics go together, and what consequences we can expect from particular acts. They reflect the implicit assumption that social objects and events have invariant properties, in much the same way as do physical objects. On this basis, we make predictions based on minimal information. This gives us an immediate way of responding to a new situation or person, based on similarities with situations or persons we have encountered before. Because it is based on minimal information, however, it is not likely to be an accurate prediction.

One factor that maintains stereotypes is **illusory correlation**. In the social, as in the physical, world, events occur in conjunction or in sequence. We do not notice or remember all of these co-occurrences, but we are more likely to notice those which accord with our schemata. We recall the brash American and the reserved Englishman, but not the quiet American or the pushy Englishman. As a consequence, we falsely notice a correlation between nationality and personality which leads us to believe that certain traits really are over-represented in particular national groups.

Social representation theory

An approach to understanding the behavior of people within groups emerged in Europe in the last two decades of the last century. Social representation theory, deriving largely from the work of Serge Moscovici, holds that people within a group or society are able to understand one another and to communicate effectively because they share a common world view; a **social representation** of reality. Not only do these social representations permit communication, but they also guide the individual's understanding of the world, so that our world view is a combination of individual and shared social representations.

Two key processes take place in using social representations to understand new or unfamiliar objects, events, or ideas. **Anchoring** is the integration of a new phenomenon by linking it with things with which we are already familiar. This may involve categorizing or labeling the new phenomenon with existing names. Once this has occurred, the new phenomenon is 'understood' in the sense that we can apply our existing knowledge to it. The second process is **objectification**, which usually involves making abstract ideas more concrete (e.g. thinking of the psychoanalytic view of the mind in terms of a spatial model). Objectification can also involve the use of metaphor (e.g. ideas associated with environmental concerns are thought of as 'green'), or the linking of ideas with individuals (**personification**). These processes are seen as different from the processes of social cognition that we looked at in Section I. Social representations apply social meaning to ideas and phenomena, and social representation theory is concerned with the social effects of anchoring and objectification, and how those processes operate in social groups.

K1 THE STUDY OF PERSONALITY

Key Notes

Defining personality	The study of personality is concerned with the entire person. Definitions of personality tend to be derived from particular views of how that study should be approached, and any attempt at a general definition will inevitably irritate those of a different persuasion. We adopt a definition that has stability, consistency, and internality as its principal components.
Structure and process	Personality structure is described in terms of components that (once they are fully formed) are considered stable and enduring. Personality processes are descriptions of motivational states, which give rise to behavior whose expression is mediated by that structure.
Growth and development	Most personality theories describe the adult personality and only the psychoanalytic theories pay close attention to the developmental processes that lead up to adulthood. However, the importance of both genetic and environmental factors in the formation of personality is acknowledged. In particular, it is evident that experience within the family and peer groups is of great importance.
Person–situation controversy	It has been suggested that people behave consistently because of features of the situation rather than features of their personality. This debate has focused attention on the importance of the interaction between the person and the situation as joint determinants of behavior.
Related topic	The nature of development (H1)

Defining personality

It is very difficult to construct a definition of personality that will satisfy even the majority of personality theorists and researchers, since there are many ideas about what constitutes the core of the concept; what should be included and what should be left out. The study of personality is concerned with the *total* person and how the various aspects of an individual's functioning are related to one another. But there would be some disagreement about whether a definition should include, for example, intelligence as a component of personality or whether it should be regarded as one the subordinate functions. Trait theorists may include it, psychodynamic theorists probably would not.

It is implicit in any definition that behavior is to some degree determined *by* personality, since we must make inferences *about* personality by observing behavior. We must also assume that personality is stable over time so that consistencies in behavior can be explained. Child's definition encapsulates these features. It states that personality encompasses the "more or less *stable internal* factors that make one person's behavior *consistent* from one time to another, and *different* from the behavior that another person might exhibit in comparable situations". Note that this definition could be read to include intelligence as a component of personality.

Structure and process

The concept of personality **structures** refers to the stable component parts. The concept of personality **processes** refers to the dynamic relationship between the components; the motivational functions that account for behavior. Personality theories differ considerably in the relative emphasis that they place on these two concepts.

A **trait** is a descriptor that expresses consistency of behavior. When we describe people as being 'honest', 'good-humored', 'kind', 'nervous' we are making a statement about how they generally behave across a range of circumstances. Some theorists regard traits as the structural elements of personality while others use more complex concepts.

Motivational processes are the motors that power behavior. They may loosely be classified into three types: pleasure, personal growth, and cognitive motives. **Pleasure** motives are further divided into tension–reduction motives (or drives) and incentive motives, which emphasize goals. **Personal growth** (or self-actualization) motives refer to the need to realize one's potential. **Cognitive motives** are those that underlie behavior that seeks to understand and predict the world. Some of these motives are rooted in human physiology while the origin of others is unclear. Motivational forces are translated into action by way of personality processes. At present, these processes are very much a matter for conjecture and, with the famous exception of Freud, most personality theorists have little to say on the matter.

Growth and development

Although we have emphasized that consistency of behavior is a defining factor of personality, it is self-evident that this is really an adult characteristic and that personality has to develop during the early years of life (Topic H1). The determinants of this development are split between **genetic** and **environmental** factors, and the issue of which is the more important is known as the **nature–nurture** controversy. Since genetic factors determine our complete biological structure, it is obvious that at some level they are determinants of all behavior. What are crucial for personality development are the complex **interactions** that occur between the individual and the environment and afford opportunities for learning. Arguably, the most important macroscopic aspects of the environment are the **family** and the **culture** within which it is situated. Differences in personality may be underpinned by genetic factors that set the possible outcomes (the **reaction range**) for development, but differences in specific outcomes arise from differences in experience. Siblings in the same family group may differ considerably, which emphasizes that **shared experience** may be less important for personality development than different experiences. Some differences of experience do occur within the family group, but many occur outside it, in the peer environment.

Most personality theorists are concerned with describing the adult personality and, as yet, only the psychodynamic theories contain a detailed account of it. However, theories of child development, while not explicitly addressing personality issues, do consider the way in which, for example, a child acquires a sense of right and wrong, and these should not be overlooked.

Person–situation controversy

Because people and situations interact there may be some difficulty in determining whether observed consistency in behavior arises because of the person (i.e. the personality) or the situation. This is referred to as the person–situation (or internal–external) controversy. The issue is whether someone really is 'the same person' from situation to situation, or whether they show some behavior patterns in certain contexts and not in others. Some theorists emphasize that situations modify how people behave, and challenge the idea of behavioral consistency as a defining characteristic of personality. There is some experimental evidence on this issue but it is not clear-cut. In the laboratory, however, behavior tends to be more determined by situations that:

- are novel, formal and public;
- have detailed instructions;
- allow little choice of behavior;
- are of brief duration.

Personality tends to be the more important determinant in situations that have the opposite characteristics.

However, current thinking emphasizes the importance of the interaction between person and situation. Three general types of interaction have been described which account for differences in behavior:

- **reactive** interaction, where individuals **interpret** their experiences differently;
- **proactive** interaction where individuals **choose** their experiences selectively;
- **evocative** interaction, where individuals **evoke** different reactions from a situation because of the way they themselves behave.

K2 THE TRAIT APPROACH

Key Notes

The nature of traits

A trait is a stable characteristic of a person that may be used to explain consistencies in the behavior of an individual, and the differences in behavioral consistencies between individuals. Traits are hypothetical entities that are generally revealed by statistical analysis of empirical data. Their nature, inter-relationship, and measurement are therefore linked to the methodology used to enumerate them.

Cattell's model

Cattell pioneered the use of factor analysis in personality theory. Starting with rating scales derived from a study of personality-related words in the English vocabulary, he obtained data from a large sample of ordinary people and derived 16 personality factors.

Eysenck's theory

Eysenck argued for the usefulness of second-order factors, which can be regarded as clusters of traits. He initially derived two such factors, introversion/extraversion and neuroticism, from data on patients suffering from neurotic disorders, and suggested an underlying physiological basis for them. Individual differences for these two factors were assumed by Eysenck to be mediated, in part, by genetic factors.

The five factor model

There is now broad agreement that a good description of the structure of personality can be provided by five factors: the 'big five'. Although there is a great deal of data to support this approach, it has no real theoretical basis.

Evaluation of the trait approach

Trait theories are supported by a good deal of empirical data and have provided useful tools for assessment purposes. In general, they are descriptive rather than explanatory, and do not offer any insight into the processes underlying individual differences. However, there is a suggestion that evolutionary theory supports the five factor model.

Related topic

Approaches to motivation (B1)

The nature of traits

The trait approach starts from the premise that adults exhibit broad predispositions which are revealed though consistencies in behavior, thoughts, and feelings. The task of a trait theory is to analyze these consistencies in order to discover a set of **traits** that are common across all individuals, but possessed by each person in varying strengths or amounts. This variation gives rise to individual differences in personality.

Trait theorists agree that the components of human personality are organized into a hierarchy. At the most fundamental level, behavior can be considered as **specific responses**. These can be linked together to form **habits** which, in turn, are grouped together to form **traits**. Traits may then be linked together to form second-order factors or **superfactors** (*Fig. 1*).

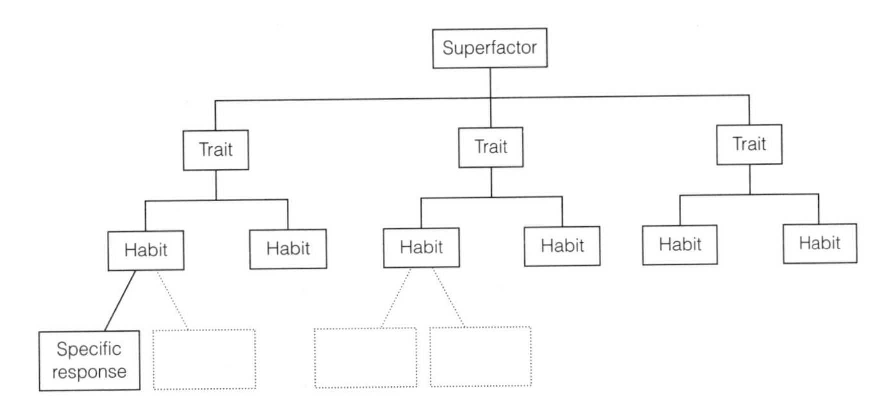

Fig. 1. Diagrammatic representation of the hierarchical organization of personality.

There are two distinctions to be made amongst traits. The first is between **ability, temperament**, and **dynamic** traits. Ability traits refer to skills and abilities; intelligence is an example of this. Temperament traits refer to the emotional aspects of personality, and dynamic traits to the motivational aspects. The second distinction is between **surface** and **source** traits. Surface traits are the expression of behaviors that covary but do not necessarily have a common cause. Source traits are the expression of behaviors that covary and form a natural personality structure or dimension. These structures are identified by applying **factor analysis** to samples of behavior, which are usually captured by self-report questionnaires. The factors are interpreted by considering those characteristics that the covarying behaviors have in common.

Cattell's model

Raymond Cattell based his model on the assumption that every important attribute of personality will be represented by at least one word in the English language; this is called the **lexical hypothesis**. He argued that if people rate the degree to which they possess these attributes then it should be possible to uncover underlying structures common to everyone. His questionnaire studies used large samples of normal people and led him to the conclusion that there are 16 source traits that describe human personality. These can be measured using the **sixteen personality factor (16 P.F.) questionnaire.** Cattell used special names for each of these **bipolar** factors, but the terms given in *Table 1* represent their broad meanings.

Eysenck's theory

H.J. Eysenck also used factor analysis as a method for examining the structure of personality. His work concentrated on second-order factors and initially he proposed two: neuroticism (emotionally stable–unstable) and a bipolar factor named introversion–extraversion. Eysenck proposed an underlying biological (and therefore genetically linked) basis for these factors based on the concept of **cortical arousal** (Topic B1). A third factor, psychoticism, was added to the theoretical framework at a later date.

A large body of research demonstrates that variation on the introversion–extraversion dimension is associated with variation in a wide range of other characteristics such as tolerance of pain and fatigue, social interaction patterns, and suggestibility. Variation in neuroticism scores is associated with differing

Table 1. Cattell's 16 personality factors derived from questionnaire data

Reserved	Outgoing
Less intelligent	More intelligent
Stable, ego-strength	Emotionality/Neuroticism
Humble	Assertive
Sober	Happy-go-lucky
Expedient	Conscientious
Shy	Adventuresome
Tough-minded	Tender-minded
Trusting	Suspicious
Practical	Imaginative
Forthright	Shrewd
Placid	Apprehensive
Conservative	Experimenting
Group-dependent	Self-sufficient
Undisciplined	Controlled
Relaxed	Tense

levels of anxiety, and certain attentional and cognitive biases. There may also be associations between extreme measures on these two dimensions and psychological ill-health. The less important psychoticism scale seems to measure **psychopathy** rather than a disposition towards psychosis.

The five factor model

Recent research based on the lexical hypothesis leads to the conclusion that the best description of personality in these terms requires five second-order factors; this is known as the '**big five**' theory. The meaning of the factors can be illustrated using trait adjectives (*Table 2*).

There is as yet no theoretical account of why *these* factors (rather than any others) should be the most important attributes of personality, although it has been suggested that each one corresponds to a fundamental aspect of human interaction, and that they have evolved in response to the demands of survival and reproductive success.

Evaluation of the trait approach

The trait approach is concerned with specifying those underlying attributes with regard to which people differ from one another. It has generated a lot of empirical research including work into the **predictive utility** of traits and their relationship with interpersonal behavior and psychopathology. Evidence for the importance of genetic contributions to the determination of personality is consistent with trait models of personality.

There are weaknesses, however. The trait approach is not explanatory, and there is always a danger of circularity whereby a trait is used to explain the behavior from which it was originally inferred. The assumption that consistencies in behavior indicate stable underlying personality traits ignores the importance of situational factors, and there is an absence of theory to integrate traits into the broader picture of human functioning and to account for personality development and change.

Table 2. The 'big five' trait factors and illustrative attributes

Characteristics of the high scorer	TRAIT SCALES	Characteristics of the low scorer
	NEUROTICISM (N)	
Worrying, nervous, emotional, insecure, inadequate, hypochondriacal	Assesses adjustment vs. emotional instability. Identifies individuals prone to psychological distress, unrealistic ideas, excessive cravings or urges, and maladaptive coping responses	Calm, relaxed, unemotional, hardy, secure, self-satisfied
	EXTRAVERSION (E)	
Sociable, active, talkative, person-orientated, optimistic, fun-loving, affectionate	Assesses quantity and intensity of interpersonal interaction; activity level; need for stimulation and capacity for joy	Reserved, sober, unexuberant, aloof, task-oriented, retiring, quiet
	OPENNESS (O)	
Curious, broad interests, earth, creative, original, imaginative, untraditional	Assesses proactive seeking and appreciation of experience for its own sake; toleration for and exploration of the original	Conventional, down-to- narrow interests, unartistic, unanalytical
	AGREEABLENESS (A)	
Soft-hearted, good-natured, trusting, helpful, forgiving, gullible, straightforward	Assesses the quality of one's interpersonal orientation along a continuum from compassion to antagonism in thoughts, feeling, and action	Cynical, rude, suspicious, uncooperative, vengeful, ruthless, irritable, manipulative
	CONSCIENTIOUSNESS (C)	
Organized, reliable, hard-working, self-disciplined, punctual, scrupulous, neat, ambitious, persevering	Assesses the individual's degree of organization, persistence and motivation in goal-directed behavior. Contrasts dependable, fastidious people with those who are lackadaisical and sloppy	Aimless, unreliable, lazy, careless, lax, negligent, weak-willed, hedonistic

From Costa, P.T. and McCrae, R.R. (1985) The NEO Personality Inventory Manual.

K3 THE PSYCHOANALYTIC APPROACH

Key Notes

The structure

Freud's theory of personality employs three interacting systems: the id, which is present at birth, and the ego and the superego which emerge in a developmental sequence. The biologically determined drives that comprise the id are channeled into action through the ego that is, in turn, moderated by the superego. The activities of these systems are not always accessible to conscious thought.

The processes

At the core of psychodynamic theory is the notion of the release of energy generated by the biological drives of the id. While this demands instant gratification, the ego must moderate the individual's activity because of the constraints of the outside world and the ethical principles embodied in the superego. There is always tension between these three systems and the resulting conflicts give rise to anxiety. The ego protects itself from these by means of defense mechanisms.

Personality development

As the child begins to differentiate the self from the rest of the world, its mode of thought changes from the fantasy of primary process to the more reality-based secondary process. Psychosexual development takes place in five stages, as the focus for the libido moves from the mouth to the anus and then to the genitals. If these transitions are not fully accomplished then fixation can occur, which gives rise to characteristic personality types. At the same time, the individual achieves identification with the same-sex parent through the resolution of the Oedipus or Electra complex.

Evaluation of the approach

The major problem of psychoanalytic theory is that the central terms are ambiguous. Although they have considerable descriptive power, it is difficult to relate them to specific behavioral observations and hence to formulate testable hypotheses. The theory can encompass the complexity of human personality but cannot predict what will happen given a specific set of circumstances. There is limited evidence in support of some components, such as the importance of unconscious processes, and against others, such as the Oedipus and Electra complexes. In general, however, its scientific usefulness is very limited.

Related topics

Anxiety disorders (L4) Treatment of psychopathology (L7)

The structure

The psychodynamic approach has its origins in the work of Sigmund Freud and makes use of the important assumption of **psychological determinism**, the doctrine that all psychological events have a cause. Freud developed a model of the mind that employed three general systems, the id, the ego, and the superego. A baby is born with only the **id**, which comprises mainly biologically determined drives. These are grouped into **life** instincts, which include sexual drives and tendencies towards self-preservation, and **death** instincts, which are primarily aggressive.

The energy for life instincts is called the **libido**, and that for the death instincts **thanatos.** Early infant behavior is directed towards the immediate satisfaction of these drives; this is the **pleasure principle**. As the infant develops and learns how to react with the environment, the **ego** becomes differentiated from the id. The ego works to satisfy the needs of the id, taking into account the opportunities and limitations of the outside world; this is the **reality principle**. In later childhood, the **superego** begins to develop. This incorporates the moral and ethical values transmitted to the child by the parents or other carers. It represents society's proscriptions, and is associated with punishment. Behavior that conflicts with it gives rise to guilt. Alongside it develops the **ego-ideal** which represents prescriptions for life and behavior and is associated with rewards. Behavior that conflicts with it gives rise to shame. The superego and the ego-ideal serve to place constraints on the ways in which the ego can satisfy the needs of the id. The activities of these three systems are not always accessible to conscious thought. Freud used three levels of consciousness: the **conscious**, the **pre-conscious**, and the **unconscious**. Thoughts at the conscious level are simply those that are the subject of attention at any time, while thoughts at the pre-conscious level are unattended but could be brought in to consciousness without much effort. Thoughts at the unconscious level, however, are very difficult or even impossible to bring into consciousness. The id functions entirely at the unconscious level while the other systems function at all three levels, although the most important functions of the ego are conscious. The three systems and the three levels of consciousness together specify Freud's structural model.

The processes

It is important to note that there is generally tension between the three systems. The id demands instant gratification, but this is often inconsistent with the boundaries set by the superego, and these conflicts give rise to **anxiety** (Topic L4). Freud conceived of the ego as a relatively weak structure that will make use of a variety of unconscious processes, or **defense mechanisms**, to protect itself from anxiety if the resolution of these conflicts becomes too difficult. These serve, for example, to distort perceptions of reality and exclude feelings from awareness. All defense mechanisms require constant effort to keep that which is threatening out of consciousness. The major primary defense mechanism is **repression** whereby unacceptable thoughts are banished from consciousness. **Reaction formation** occurs when an individual defends against the expression of an unacceptable impulse by expressing its opposite. **Projection** provides protection from guilt or anxiety about feelings towards another by ascribing ('projecting') those same feelings on to the other person. **Denial** of reality can be applied to internal states, for example a denial of anger, or to external events. **Rationalization** involves the perception of an action but a re-interpretation of its motivation so that it appears acceptable.

Personality development

The psychodynamic view emphasizes the importance of childhood experience in the development of personality. Thought processes change from **primary process** to **secondary process** during infant development. Reality and fantasy are intertwined in primary process thought, which is the language of the unconscious, but which may reveal elements of itself in dreams. Secondary process thought is the language of rational thought, of reality. Psychosexual development takes place as the focus of the libido moves from the mouth to the anus and then to the genitals. The resolution of the male **Oedipus complex** and the female **Electra complex**, leading to identification with the same-sex parent, to conclude the genital stage is particularly important. The emotional growth of the child depends crucially on the anxiety and gratification that occurs during these changes, and developmental failures can occur. In particular, individuals may fail to progress completely from one stage to the next, giving rise to **fixations**. Fixated individuals try to obtain the same types of satisfaction that were appropriate for an earlier stage of development. Fixation is said to give rise to personality types with characteristics related to the stage at which fixation occurred. These are the **oral** personality (narcissistic, no clear idea of others as separate entities, always asking for something), the **anal** personality (orderly, mean and obstinate), and the male **phallic** personality (overt displays of masculinity, exhibitionist) or female **hysterical** personality (flirtatious, idealizing romance and life in general). A related phenomenon is that of **regression** in which someone temporarily returns to an earlier mode of satisfaction, often as a result of stress (Topic L7).

Evaluation of the approach

The central terms of psychoanalytic theory are ambiguous and processes tend to be defined in terms of metaphors and analogies. Thus, although Freud conceived of his theory as a scientific project, it is difficult to relate the structures and processes to specific behavior and hence to formulate testable hypotheses. The theory can encompass the complexity of human personality in descriptive terms but cannot predict what will happen in any defined set of circumstances. That being said, psychoanalytic theory has generated a lot of research questions and indirect evidence has emerged to support, for example, the existence of repression. Given its all-embracing nature, one might have expected developments of psychoanalytic theory to accommodate our expanding knowledge of human behavior and psychobiology. This has not taken place, although some psychoanalysts have tried to extend the mainstream view. As a theory of personality, psychoanalytic theory is of limited usefulness.

K4 THE PHENOMENOLOGICAL APPROACH

Key Notes

Rogers' concept of the self

The phenomenological approach is based on individual conscious experience from which people develop their self-concept. The most important motivating force for action is the need for the person to develop to their full potential and become their ideal self. People strive for consistency between their behavior and their self concept, and to maintain their self-esteem.

The Q-sort

Rogers made use of this technique as a way of exploring the self-concept in a therapeutic setting. The Q-sort provides systematic ways of eliciting the attributes which a person thinks are most characteristic of themselves. The attributes are usually derived from interviews with the participant thus reflecting their experience.

Kelly's concept of the construct system

Kelly based his model on the idea that people construct their view of the world based on a network of personal evaluative dimensions called constructs. The construct system is dynamic; it determines the way in which we perceive people and events, and react to them.

The repertory grid technique

Only the individual can make their construct system explicit and this technique was devised to assist them. Constructs are elicited by contrasting elements of a set (which may be people, things, or events) in groups of three. The constructs are then applied to all elements of the set, and the resulting patterns can be used to explore the relationship between the person and significant people or events in their life.

Related topic

Self perception (I4)

Rogers' concept of the self

Carl Rogers' built his theory around the notion of the **self-concept.** Unlike the ego, the self-concept does not control behavior. Instead, it represents organized and consistent patterns of perceptions about the world and its meaning, which are unique to the individual (Topic I4). The self-concept that an individual would like to possess is called the **ideal self**. While in Freudian theory the main characteristics of personality are fixed early on in life, in Rogers' theory they are continually developing. This development is motivated by a need for self-fulfilment known as **self-actualization**. Rogers also emphasized the importance of **self-esteem** and the role of the child–parent relationship in its development. High levels of self-esteem are more likely to emerge when (i) parents accept the individuality of the child and show interest and affection; (ii) demands made on the child are clear, and rewards and punishments are fairly and consistently applied; and (iii) treatment within defined limits is democratic and the opinions of the child are recognized and respected.

The Q-sort

The nature of a person's self-concept can be investigated using the **Q-sort** technique. The individual is given a large number (100 or so) of statements of personality characteristics and is asked to sort them into ordered piles according to how characteristic each one is of themselves. A typical sort would have 11 piles ranging from 'most characteristic of me' at one end to 'least characteristic of me' at the other. A 'forced-normal' distribution is imposed, with the person being told how many cards may be placed in each pile. The statements used in a Q-sort may be derived from clinical interviews with the participant or from other sources; there is no single standardized list. The technique can also be applied to the investigation of the ideal self with the phrases such as 'most like I would wish to be' and 'least like I would wish to be' substituted at the end points.

According to Rogers, all behavior and experience is evaluated in terms of the self-concept. Individuals strive to behave in a manner **consistent** with their self-concept, and to elicit behavior from others that is **congruent** with it. Discrepancies threaten the structure of the self and may give rise to defensive processes similar to those suggested by Freud. **Distortion** or **denial** of experience can lead to anxiety, and it is the relationship of experience to the self-concept that provides the therapeutic key.

Kelly's concept of the construct system

George Kelly's personal construct theory does not make assumptions about internal structures such as the ego or the self. It is based on the cognitive principle that individuals derive organizing principles called **constructs** from their interaction with the world and use them to represent it. Since every individual has a unique and personal view of reality, their set of constructs will also be **personal**. It is this construct set which underpins the characteristic ways of reacting and feeling which the outside observer calls 'personality'. Constructs express contrasts and are said to have two **poles**. The poles do not have to be direct opposites, but often are.

People can only perceive and interpret events within the limits of their constructs, but have the freedom to revise their construct system in whole or in part if the old one is no longer useful. **Core** constructs are those which are central to the person's functioning. Alteration of these has strong implications for the whole system. **Peripheral** constructs can be modified without change to the core structure.

Kelly used the metaphor of **man-the-scientist** to describe the way in which constructs are used as hypotheses to anticipate future events. Subsequent observations may confirm or refute a hypothesis, and the well-adjusted response is to modify the construct system. Core constructs will require stronger contrary evidence than peripheral constructs before modification occurs. Problems can arise when the individual denies the validity of the observations rather than reject the hypothesis, or when they are very selective about how they sample their data.

The repertory grid technique

Kelly found therapeutic value in exploring an individual's construct system to reveal where it needed modification due to evident inconsistencies, but the individual was unable or unwilling to do this without assistance. The tool for this exploration is the **repertory grid** technique. This is simply a procedure that helps an individual to make their constructs explicit. Constructs about a set of people, situations, events, or even objects are elicited by comparing triads of elements from that set. For any given triad, the individual must name a way in which two of the elements are similar to each other and different from the third. This attribute is by definition a construct since it is one of the ways in which the individual categorizes the elements of the set. Further triads elicit additional constructs until the individual can think of no more. All elements of the set are then rated on all constructs (*Table 1*). Patterns and inconsistencies may be evident on inspection, but there are analytic methods available that assist interpretation.

Table 1. Portion of a completed repertory grid

	Construct				
Element	Likes me/ doesn't like me	Tolerant/ short-tempered	Open/ evasive	Shares interests/ nothing in common	Intelligent/ slow-witted
Brother	*✓	✓	*✓	✓	✖
Father	*✖	✓	✖	*✖	*✖
Mother	*✓	✓	✖	*✖	✓
Work colleague 1	✓	*✖	✓	✓	*✓
Work colleague 2	✓	*✓	*✖	✖	✖
Work manager	✖	*✖	*✖	✖	✓
Friend 1	✓	✓	✓	*✓	*✓

Elements coded with a ✓ on a construct were allocated to the first of the two poles, and those with a ✖ to the second. Element marked with a * constituted the original triad which elicited the construct.

K5 Personality assessment

Key Notes

The nature and purpose of personality assessment

Personality assessment is useful for clinical work and for research. It is also increasingly used in occupational settings. Most assessments consist of paper and pencil questionnaires (often called inventories) with multiple choice or rating scale answers. The reliability and validity of personality assessments is an important issue.

Trait-based methods

These are all personality inventories which derive their validity either from the theory that generated them or from criterion keying, which ensures they discriminate between two or more groups of interest. Trait-based inventories are easy to use and there is a large body of research that assists interpretation.

Psychodynamic methods

Projective tests are devices that assist participants, with the aid of a skilled therapist, to uncover sources of conflict by telling stories in response to ambiguous pictures or interpreting ambiguous figures. Although there are standard interpretative techniques, the utility of the methods depends heavily on the therapist.

Phenomenological methods

The Q-sort and the repertory grid technique are both devices that assist participants in making explicit the way in which they internalize their experience. The interpretation of the outcomes is a matter for discussion between the test administrator and the participant. Although the methods are standard, the tests themselves are, by their nature, tailored to the individual.

Related topic

The measurement of intelligence (K6)

The nature and purpose of personality assessment

Personality assessments are generally used in three areas: clinical practice, occupational selection, and research into personality itself. In common with other kinds of psychological testing, instruments or procedures used to assess personality should have demonstrable **reliability** and **validity**. Two kinds of validity are of particular concern: **criterion** (or **empirical**) validity and **construct** validity. Tests that are constructed to discriminate between two or more groups have to show criterion validity. Thus, an occupational psychologist may give advice on selection and recruitment of employees based on tests which yield different scores for people who are successful in a particular job compared with those who are not so successful. There does not have to be any theoretical basis for the selection of items in such a test. Criterion validity is purely a matter of how the test performs against some external benchmark.

Construct validity is concerned with how well a test measures or describes the attribute in question. This is a more subtle matter because there are generally no external and independent standards against which to check it. In general, construct validity is derived from the theory underpinning the test and from research findings that accumulate during its development. There is necessarily a close relationship between research into personality and the methods used to assess it, and data may lead to the modification of a theory as well as the refinement of a test. It is not surprising, therefore, that the major methods of assessment are closely linked to particular theoretical positions (Topic K6).

Trait-based methods

A good example of a criterion-based personality assessment is the Minnesota Multiphasic Personality Inventory (MMPI). Its scales were developed empirically by criterion keying 550 items to eight specific psychiatric diagnoses using clinical groups and normal controls. Two additional scales (masculinity–femininity and introversion–extraversion) were added later and were keyed to normal controls. Three further scales check that the inventory has been honestly and consistently filled in (*Table 1*).

Table 1. The MMPI scales

Scale name	Abbreviation	Interpretation of high scores
Hypochondriasis	Hs	Abnormal concern with bodily functions
Depression	D	Depressed, pessimistic
Hysteria	Hs	Unconsciously avoids conflict by denial of problems
Psychopathic deviancy	Pd	Lack of conformity to social norms; emotional shallowness
Masculinity–femininity	Mf	Feminine orientation (males); masculine orientation (females)
Paranoia	Pa	Suspicious; delusions of grandeur or persecution
Psychasthenia	Pt	Compulsions, abnormal fears, anxiety or guilt
Schizophrenia	Sc	Thought disorders; withdrawn
Hypomania	Ma	Impulsive, inappropriately excited, over-activity
Social introversion–extraversion	Si	Introverted, little interest in people, insecure
Lie	L	Overly good self-report
Validity	F	Carelessness in completing test; deliberate malingering
Correction	K	Defensiveness in admitting problems

The MMPI yields a score on each of the scales and these are presented as a **personality profile**. The inventory is widely used in research on normal as well as clinical populations, and has recently been extensively revised with additional scales.

Cattell constructed the sixteen personality factor (16 P.F.) questionnaire (see Topic K2) by assembling questions which best represent each factor and presenting them as a self-report inventory. Scores on the 16 P.F. can be interpreted by profile matching with norms for selected groups, and computerized scoring with interpretative comments is available. The 16 P.F. test has been widely used in counseling and research despite its modest reliability. There are also serious doubts about its validity since recent factor-analytic studies have consistently failed to yield the 16 factors it is supposed to measure.

The Eysenck personality questionnaire yields measures of neuroticism (N) extroversion/introversion (E), and psychoticism (P); it also includes a lie score (L). This test and its predecessors, the Maudsley Personality Inventory and the Eysenck personality inventory, have been extensively used in personality research and clinical practice.

The NEO-personality inventory has been developed from the five factor theory with each of the primary factors being differentiated into six or more **facets**. The resulting scales have good reliability, and their construct validity is demonstrated by high correlations with other inventories based on the factor-analytic approach.

Psychodynamic methods

Psychodynamic methods are based on the principle of **projection**. The test material is intentionally ambiguous, and respondents must draw on their personal experience in constructing their responses, thus providing the therapist with information which might otherwise remain inaccessible.

The two best known projective tests are the Rorschach Psychodiagnostic Technique and the Thematic Apperception Test. The Rorschach test consists of 10 bilaterally symmetrical figures that resemble inkblots, hence its popular name. Respondents are invited to report what they see in the figure, or what it might represent. There are some elaborate scoring schemes, but responses tend to be judged on their content and this is, of course, highly subjective. The Rorschach technique has generated a lot of research, notwithstanding its low reliability and validity. The Holtzman inkblot technique uses similar stimulus material to the Rorschach, but the test has been constructed and standardized like a personality inventory and has good reliability; its validity is still a matter for research.

The Thematic Apperception Test utilizes black and white pictures of people in ambiguous situations. The respondent is asked to tell a complete story about each of a set of pictures appropriate to their age and sex. From these stories, the experienced therapist can gain information about the dominant emotions and pressures in the respondent's life. As with the Rorschach test, interpreting the outcome of the test is very subjective, but there are systematic scoring procedures and norms. Variations of the test exist for children, the elderly, and some minority ethnic groups.

Phenomenological methods

Assessment methods that use standard materials run contrary to the phenomenological approach, and the most common tool is the interview. However, the Q-sort and the repertory grid technique are both used as structured investigative tools, since the core material in each case originates from the participant. Every Q-sort or repertory grid is, in principle, different, and there is no question of comparing an individual against sets of norms. However, it is possible to replicate either technique, using the same material with the same individual, in order to monitor the progress of a therapeutic intervention.

K6 THE MEASUREMENT OF INTELLIGENCE

Key Notes

Reliability

Any useful test of intelligence must provide consistent and replicable results so that the performances of individuals can be compared with each other. There are three principal methods of assessing reliability which each yield a measure of consistency called a reliability coefficient.

Validity

The validity of an intelligence test must be assessed in relation to a specific definition of intelligence and the purpose for which the test was designed. There are, accordingly, a range of techniques available. There is often a danger of circularity when reasoning about validity, since it may be difficult to tell whether poor validity results from an indifferent test or an inappropriate definition of intelligence.

Standardization

Standardization is part of the development process for any intelligence test. Firstly, the test is administered to a representative sample of people drawn from the population that the test is designed to assess. Then, the statistical properties of the scoring are adjusted to standardize the mean score at 100 and the standard deviation at about 16. The final standard scores form the norms for the test and individual performance is evaluated against them. Some tests have sub-scales or even sub-tests that have been separately standardized for specialist assessment purposes.

Related topic

Personality assessment (K5)

In practical terms, intelligence, like personality, is defined by the test that is used to assess it. However, there has, historically, been more emphasis on measurement and on comparing individuals with their peers in the domain of intelligence; and procedures for measuring test reliability and validity, and for establishing standardization norms have been thoroughly explored.

Reliability

The reliability of a test is the extent to which it provides consistent and replicable results, and it can be measured in three ways. If a group of individuals take the same test on two occasions separated by a few weeks or months then the correlation between the two sets of scores is known as **test–retest** reliability. Splitting a test into two (for example by grouping all odd-numbered items into one sub-test and all even-numbered items into another) and then correlating the scores from the two halves gives a measure called **split half** reliability. Constructing two tests to be equivalent to each other and correlating the results from each (as in the case of form L and form M of the Stanford–Binet test) results in a measure called **parallel form** reliability. This is the least common measure because of the cost of developing and standardizing the test. Reliability correlation coefficients in excess of 0.85 are generally reported for each of these methods.

Validity

The validity of a test is the extent to which it measures what it is supposed to measure. This can be assessed in three principal ways. **Content validity** involves examining the items to be tested to ensure that they seem appropriate. When this is simply done by inspection of the items it is known as **face validity**. Although this may seem superficial, a test with items which *appear* unrelated to the attribute being measured is unlikely to be trusted by either the person administering the test or the person being tested, no matter what research evidence might be produced to support it. A more rigorous procedure involves a statistical technique called factor analysis, which helps identify those factors or abilities that a test actually measures. A test that is designed to measure verbal ability and which yields only one factor to which all the items contribute significantly would have good **factorial validity**.

Criterion validity is measured by comparing the results of the test against some external criterion such as performance in an examination. Sometimes a test may be used to *predict* performance (for example as part of a selection procedure), in which case the degree to which that prediction is successful is a measure of **predictive** validity. The difficulty with this procedure is that performance of the criterion task will inevitably be affected by factors other than intelligence, and this in turn will reduce the correlation between the test score and criterion performance making it difficult to estimate the 'true' value.

Construct validity is always grounded in a specific theory of intelligence. If a test produces the results predicted by a theory then it is said to have good construct validity. For example, if a particular theory required that intelligence score increase with age, and this is so in the case of a particular test, then the test has construct validity with respect to that theory. Unfortunately, there is some ambiguity when predicted results fail to occur. Either the theory is wrong or the test is a poor one, but it may not be clear which is the correct explanation.

Standardization

Binet and Simon developed the first modern intelligence test for the purpose of identifying children who would benefit from remedial education. Binet prepared tables of the **average** performance on the test for children of any particular age, and the score of an individual child was then located in these tables (called **age-based norms)** to give their **mental age**. Thus, if a 6-year-old child scored as highly as an *average* 7-year-old child, that child would have a mental age of 7. The child's mental age was then divided by their chronological age and multiplied by 100 to give the **intelligence quotient** (**IQ**). A child who scored at the average level for their age would always have an IQ of 100.

Intelligence tests for adults were developed along similar lines to those pioneered by Binet; a well-known example is the **Wechsler Adult Intelligence Scale** (**WAIS**). This comprises 11 **sub-tests,** some of which may be separately standardized for specialized purposes. The sub-tests are summed to form two **sub-scores**, the performance score and the verbal score. These in turn are summed to form the full-scale score, which is compared with norms to give the IQ measure (*Fig. 1*).

Some of the items on the WAIS depend heavily on white American culture (e.g. the comprehension sub-scale includes items on the meaning of proverbs) and fluency in English (the vocabulary test requires definitions of words).

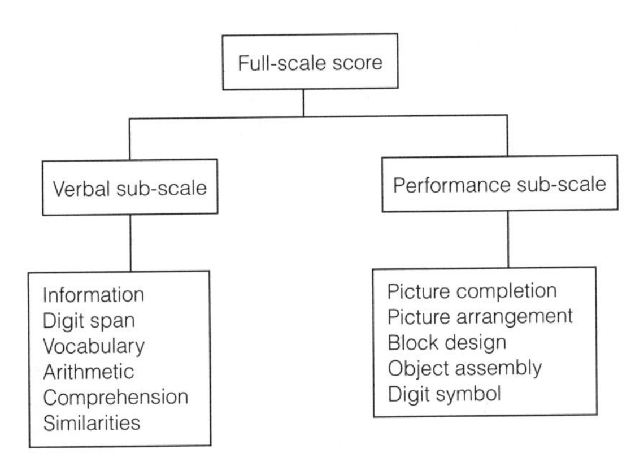

Fig. 1. WAIS sub-tests and sub-scales.

People from other cultural and linguistic backgrounds are obviously disadvantaged if assessed by means of tests that have such a **cultural bias**.

Modern tests are standardized by administering them to a large sample of people (the **norm group**) which is representative of the population for whom the test is intended. Tests are scored in such a way that scores are normally distributed with a mean of 100 and a standard deviation of 16 or thereabouts. Using the properties of the normal distribution, an individual score can be interpreted in terms of the percentage of the population who would be expected to obtain that score or less. In the above example, a person who scores 116 (one standard deviation above the mean) would be classed as more intelligent than 84% of their norm group. Binet's concept of IQ as the ratio of mental age to chronological age is no longer in technical use.

K7 THEORETICAL PERSPECTIVES ON INTELLIGENCE

Key Notes

Definitions

The principal difference between definitions of intelligence lies in their scope. Earlier theories were derived primarily from data on academic abilities that were generally restricted to verbal and arithmetic items together with some tests of abstract reasoning. Later theories extended the definition to include culturally valued abilities and a range of cognitive processes. Not everyone would agree with these definitions, but they have widened the scope of research into intelligence considerably.

Early theories

Both Spearman and Thurstone derived theories of intelligence from the factor analysis of intelligence test data. The central difference between them lay in Spearman's notion of general intelligence (*g*), which he believed underpinned all test performance, while Thurstone preferred to argue in terms of discrete primary abilities. Although each used a restricted, if implicit, definition of intelligence by modern standards, they both had a profound influence on psychometric testing.

Crystallized and fluid intelligence

Cattell and Horn distinguished between abilities that require the use of previously acquired knowledge and abilities that require mental flexibility in novel situations. This approach has been incorporated into at least one major intelligence test. It is possible that Spearman's *g* underlies both types of ability, but note that crystalline intelligence increases with age while fluid intelligence peaks in early adulthood.

Multiple intelligences

When intelligence is defined within a cultural, rather than scholastic, framework it is evident that what is regarded as intelligent behavior varies from one society to another. Gardner suggested, on the basis of a wide range of data, that there are seven varieties of intelligence, each with their origin in a different part of the brain.

Information processing approach

An alternative to the view of intelligence as the possession of a range of abilities is that it arises from a variety of cognitive processes. These are involved in the acquisition of knowledge, problem solving, personal and social relationships, and the overall planning and control of behavior. Attempts to analyze intelligence in these terms are at present only descriptive, but offer promising new research directions.

Related topics Mental representation (F4) Problem solving (F5)

Definitions

The definition of intelligence is as problematic as the definition of personality. Early workers, such as Binet, conceived of intelligence simply as 'mental ability' but, more recently, writers have focused on the capacity for abstract reasoning or the ability to adapt to novel situations or solve problems (Topic F5). The capacity to gain new knowledge or acquire new skills has also been stressed. Sternberg's definition ("... mental activity directed towards purposive adaptation to, and selection and shaping of, real world environments relevant to one's life") captures current thinking and highlights the point that intelligence is inseparable from culture. The shift from a narrower to a broader definition of intelligence follows changes in theoretical approach.

Early theories

Spearman's two factor theory

Spearman observed that performance on the sub-scales of intelligence tests was almost always positively correlated, but that these correlations were not perfect. He proposed that the reason for this is that performance on intelligence tests is partly determined by **general intelligence** (or *g*) and partly by **specific abilities** that are needed to complete different tasks (*Fig. 1*). So, for example, a person's score on a vocabulary test would be partly determined by their general intelligence and partly by the degree to which they possessed a specific verbal ability; hence, performance depends on two factors. The degree to which a sub-test correlates with the others indicates the extent to which it draws upon *g* for its successful completion. The degree to which it *does not* correlate with the others indicates the extent to which it draws upon a specific ability. Spearman argued that since *g* was important in almost all tasks, it was the most important aspect of intelligence. There have been attempts to devise tests which measure *g* alone, probably the best known being **Raven's Progressive Matrices**.

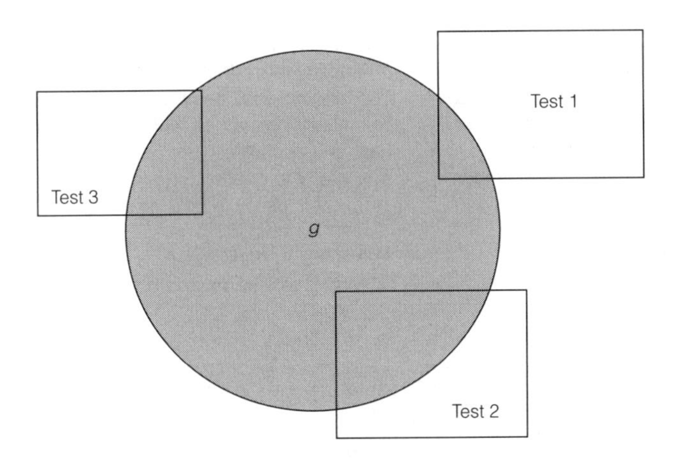

Fig. 1. Spearman's two factor model of general intelligence (g).

Thurstone's primary mental abilities

Thurstone took a different view of the intercorrelations between tests from that of Spearman (*Fig. 2*). He proposed seven **primary mental abilities** that could, in principle, be measured independently of each other (*Table 1*).

One important consequence of Thurstone's model has been the development of **multiple aptitude batteries**, which yield a **profile** of abilities, rather than a single, overall, measure. The Chicago Test of Primary Mental Abilities was based directly on Thurstone's work, but there have been many others, such as the Differential Aptitude Test developed for specific use in areas such as personnel selection, educational testing, and counseling.

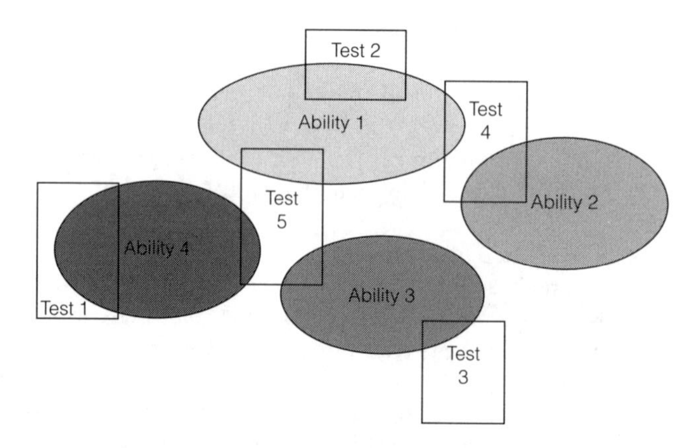

Fig. 2. General multi-factor model. Note: test 1 is a good measure of ability 4, while tests 2 and 3 are weaker measures of their respective abilities. Scores on these tests should not be correlated. Tests 4 and 5 test more than one ability, and scores will intercorrelate with other tests.

Table 1. Thurstone's primary mental abilities

Ability	Description
Verbal comprehension	The ability to understand the meanings of words; measured by vocabulary tests
Number	The ability to work with numbers
Space	The ability to perceive spatial relationships and to visualize changing positions between objects
Associative memory	Principally concerned with rote memory
Perceptual speed	Fast and accurate comprehension of visual patterns
General reasoning	The ability to deal with novel problems. It has been suggested that this ability is best measured by arithmetic reasoning tests
Word fluency	The ability to think of words rapidly. Measured by tests such as solving anagrams and producing word which rhyme

Crystallized and fluid intelligence

Cattell and Horn suggested a position intermediate between the Spearman and the Thurstone theories. They made a distinction between abilities that involve the application of previously acquired knowledge and skills (known as **crystallized intelligence**) and the ability to deal with new problems that require novel approaches and mental flexibility (known as **fluid intelligence**). There is a relatively high correlation between measures of crystallized and fluid intelligence, which might be taken to indicate that they are both manifestations of *g*. However, they differ in that crystalline intelligence seems to increase with age while fluid intelligence peaks in early adulthood and then declines with age. A recent revision of the Stanford–Binet test incorporates this distinction within a hierarchical structure in which the overall score is a measure of general intelligence (*Fig. 3*).

Multiple intelligences

Most approaches to the study of intelligence are based on the concept of mental competence and, consequently, research tends to be restricted to the domain of education. Gardner proposes an alternative view: that intelligence takes a

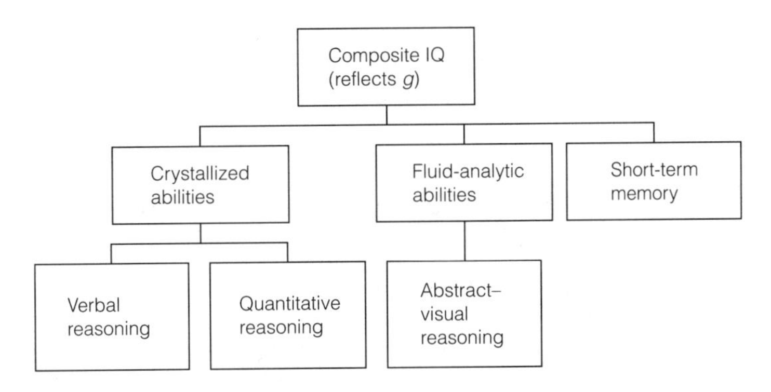

Fig. 3. Hierarchical structure of the revised Stanford–Binet test.

variety of forms depending upon the demands made upon the individual to adapt to their environment. Thus, the prize-winning physicist, the Olympic athlete, and the street-wise gang leader each exhibit a different form of intelligence that is highly effective in their environment. He proposes seven types of intelligence (*Table 2*). The first three are similar to abilities proposed by Thurstone while the others are quite distinctive.

Table 2. Gardner's seven varieties of intelligence

Type of intelligence	Characteristics
Linguistic	The ability to use language fluently and to reason verbally
Mathematical	The ability to work with numbers and solve abstract problems
Visual–spatial	The ability to comprehend visual patterns and to visualise the relationship between objects
Musical	The ability to perceive pitch and rhythm
Bodily–kinesthetic	Control of one's own body motion and the ability to manipulate objects with precision
Interpersonal	The ability to understand other people and interact with them in a sensitive way
Intrapersonal	The ability to understand oneself

Gardner suggests that each variety of intelligence arises from the function of distinct, though interacting, modules within the brain. He cites evidence from a range of sources including the study of brain-injured patients (where one type of intelligence may be impaired but others unharmed), and from the study of people who have exceptional abilities (e.g. musicians, gymnasts) in some areas but are quite ordinary in respect of others. He also considers the manifestation of intelligence in different cultures. His analysis has proved controversial since, amongst other things, it raises the theoretical issue of whether we need to maintain a distinction between intelligence and talent in the study of human abilities.

Information processing approach

Theories of intelligence generally describe individual differences in terms of abilities which people possess to a greater or lesser degree. An alternative perspective is to analyze differences in terms of **cognitive processes**, and proponents of this view focus on such issues as the internal representation of information (Topic F4), and the nature and selection of strategies for processing information.

Sternberg's approach has been to consider intelligence in terms of its relationship with the individual's inner world, with their external world, and with their experience. His analysis examines each of these relationships in terms of a **sub-theory** (*Table 3*).

Table 3. Outline structure of Sternberg's triarchic theory

Sub-theory	Domain
Contextual	The individual's external world. Contextual processes are concerned with the demands made on the individual by their culture and environment. Includes academic problem-solving abilities and personal–social intelligence
Componential	The individual's internal world. The most important processes (called **meta-components**) concern the planning and regulation of behavior. Others deal with the acquisition of knowledge and the execution of strategies devised by meta-components
Experiential	The processes of interaction between intelligence and experience, and between the internal and the external world. Includes the ability to carry out familiar tasks automatically, and to deal effectively with novel situations

Much of what Spearman and Thurstone would have described as intelligence arises from the processes of the componential sub-theory, and the distinction between fluid and crystallized intelligence can be recognized in the processes of the experiential sub-theory. Processes underpinning five of Gardner's seven intelligences can also be located within the sub-theories, but Sternberg does not include musical and bodily–kinesthetic abilities within his scheme. The triarchic theory provides a detailed framework that remains, for the most part, descriptive, and the lack of detail about the relationships between the sub-theories is a shortcoming. Modern intelligence tests do not provide effective assessment of much that Sternberg considers important, and new measurement methods are needed to assess its predictive power.

K8 HEREDITY, THE ENVIRONMENT AND INTELLIGENCE

Key Notes

The 'nature–nurture' debate

Some decades ago, opinion was divided on whether intelligence was primarily inherited or acquired. We now know that both factors are almost equally important determinants of measured intelligence, although different conclusions may apply in the case of broader definitions of what intelligence actually is.

Genetic specification

Individuals have a genetic specification, the genotype, which cannot be directly observed. As they develop, this genetic specification interacts with the environment to produce the observable phenotype. Any assessment of abilities is concerned with the phenotype.

Twin and adoption studies

An important way of gathering data on the relative effects of inherited and environmental factors is to study people with differing degrees of genetic relationship who are raised in environments of differing degrees of similarity. Such studies have often suffered from methodological flaws and should not be taken as providing precise estimates of the heritability of intelligence. Taken overall, however, they do support the conclusion that genetic and environmental factors are of roughly equal importance.

Environmental enrichment

The interaction of genetic and environmental factors gives rise to a range of possible phenotypes known as the reaction range. There is some evidence that an environment which offers good opportunities for interaction and stimulation early in life will lead to enhanced development within the reaction range.

Related topic

The nature of development (H1)

The 'nature–nurture' debate

The study and measurement of intelligence has always had a political dimension, and this is particularly evident in the debate concerning the relative importance of environmental and inherited factors in determining intelligence and measured IQ. Controversy has been fueled by arguments based on the assumption that intelligence is determined almost entirely by one factor or the other. For example, there is a correlation between measured intelligence and socio-economic status such that people from disadvantaged backgrounds tend to have a lower measured IQ than those who are better off. In the early part of the 20th century, some scientists and politicians used this fact to argue that there was little point in trying to educate individuals with low IQ, because their disadvantages stemmed from inherited (and therefore immutable) factors. If it were not so, they would have bettered themselves. The opposing argument was that the tests themselves were culturally biased towards the better off, and that measured IQ was predominately determined by environmental factors such as early home background and schooling. Neither of these extreme positions is helpful (Topic H1).

Genetic specification

The genetic specification of an individual is known as their **genotype**. Environmental factors interact with the genetic specification to produce a range of possibilities known as the **reaction range**. The observable pattern of characteristics of an individual that result from such interactions is called the **phenotype**. It is important to remember that the assessment of intelligence concerns phenotypes, since genotypes are not directly observable.

Twin and adoption studies

One of the earliest approaches to studying the degree to which intelligence is due to inherited factors was to study groups of people who differed in the degree of their genetic similarity. The study of twins has been particularly important. Identical (or **monozygotic**) twins develop from a single fertilized egg and have identical genotypes. In Western society, they tend to be brought up in near-identical environments and treated in a very similar fashion in their early years by family and friends, who may have difficulty telling them apart. Fraternal (or **dizygotic**) twins develop from two eggs fertilized at the same time; they have half their genes in common. Fraternal twins also tend to be treated in a similar fashion during their early years, but their genotypes are no more similar than any pair of siblings and, of course, they will not look alike nor necessarily be of the same sex. If intelligence were entirely determined by genetic factors then the correlation between the IQ scores of pairs of identical twins would be +1.00, while the corresponding value for pairs of fraternal twins would be much lower. In fact, the correlation for identical twins is about 0.86, while that for fraternal twins is about 0.57.

Sometimes, through unfortunate circumstances, twins are separated at birth, adopted, and raised in different environments. When identical twins develop in different environments, the correlation between their IQ scores is lower than that for identical twins reared together, and a similar difference appears when fraternal twins are reared apart. Adoption studies also provide data on children who are reared together, but are genetically unrelated. These indicate a correlation of about +0.30 during childhood that drops to nearly zero in late adolescence, thus demonstrating the diminishing effect of a *shared* environment over time.

Data from twin and adoption studies must be treated with caution because many studies suffered from methodological problems. However, the correlations serve at least to illustrate the modification of the genotype by the environment (*Table 1*). Despite these reservations, the current view is that the

Table 1. Correlations between IQ scores from twin and adoption studies

Relationship	Median correlation
Identical twins reared together	+0.86
Identical twins reared apart	+0.75
Nonidentical twins reared together	+0.57
Siblings reared together	+0.45
Siblings reared apart	+0.21
Child and biological mother separated by adoption	+0.31
Child and unrelated adoptive mother	+0.17

environment and heredity each contribute about 50% to the determination of individual differences in intelligence.

Environmental enrichment

Environmental enrichment refers to the process of providing as stimulating an environment as possible in order that a child's intellect may develop toward the upper bound of their reaction range. Evidence from the Head Start programme in America, and other sources, show that such interventions are most effective when they begin early in a child's life. Although gains in measured IQ are often small, Head Start children tend to achieve higher academic standards and remain in education for longer.

L1 NORMALITY AND ABNORMALITY

Key Notes

Defining abnormality

Psychopathology or abnormal psychology is the study of abnormal behavior and mental function. Most definitions of abnormality include deviance, distress, dysfunction, and dangerousness. None of these alone is sufficient to define mental abnormality.

Models of abnormality

Views of the origin of psychopathology have historically included demonic possession and medical models (of illness caused by physical and/or psychological disturbances). Modern views of psychopathology tend to be multifactorial, including sociocultural and family influences, as well as somatic and personal factors.

Classification of mental disorders

Classifying mental disorders helps clinicians and researchers to communicate, permits diagnosis, helps to identify causes, and can help to identify potential treatments. Modern classification systems, including the American Psychiatric Association's *Diagnostic and Statistical Manual of Mental Disorders* (DSM-IV), are derived from Kraepelin's 19th century classification.

Related topic

The study of personality (K1)

Defining abnormality

Psychopathology or **abnormal psychology** is the study of abnormal behavior and mental function. However, there is no agreement about what constitutes abnormality in this context, and definition is made more difficult because psychological or mental disorders are so varied. Most definitions of abnormality include the following concepts, constituting 'the four Ds':

- deviance
- distress
- dysfunction
- dangerousness

Abnormal function is **deviant**, but deviant from what? We might consider behavior to be deviant if it is unusual: it is not normal in the sense of not statistically usual. However, this is not a useful approach; most of us cannot compose a symphony or paint a masterpiece, yet it is not helpful to view those who can do so as abnormal. Further, what is usual behavior depends on sociocultural norms. For much of the last millennium in Europe it may not have been considered unusual for people to claim to be receiving instructions from a supernatural authority. But that would only have been so if the authority was the Christian God. Today, this might be regarded as a sign of abnormality.

Suffering **distress** is not alone a sufficient criterion for abnormality. We would all expect to be distressed by traumatic events such as the sudden death of a close relative. However, prolonged distress following an event that most people recover from quickly might be considered deviant, and hence abnormal. Even here the boundaries are unclear. For example, **post-traumatic stress disorder** (see Topic L4) is now considered to be a very likely consequence of traumatic events, but is viewed as a disorder that requires treatment. Also, deviant behavior is not necessarily accompanied by distress.

Abnormal behavior may be **dysfunctional**; that is, it might interfere with an individual's ability to function adequately. Thus, for example, mental illness might prevent people from performing a job, forming relationships, or looking after themselves. Yet dysfunction is not necessarily a sign of mental illness. While many instances of self-starvation are clearly a sign of abnormality (see Topic L5), death through hunger strike in the service of a cause might be considered a heroic act.

Sometimes, behavior is considered abnormal if it is **dangerous** to other people. **Insanity** (a legal rather than psychological term) may be used as a defense in murder trials; the defendant being considered incapable of realizing the nature of his or her actions at the time of the killing. However, being dangerous is not of itself a sign of mental abnormality.

It is impossible to provide a generally applicable definition of abnormal behavior. There is no clear boundary between eccentricity and mental illness. In 1961, Thomas Szasz went further, claiming that there is no such thing as 'mental illness'; that it is constructed by society to justify a variety of ways of controlling people who either have problems with living or who make the rest of society uncomfortable. However, most now agree that there are many instances of behavior and mental process that it is useful to consider 'abnormal', and which psychiatrists, clinical psychologists, and others can attempt to help.

Models of abnormality

The earliest explanation for mental disturbance was probably that it is caused by **demonic possession**. This is explicit in early Egyptian and Chinese writings, and in the Bible. Many archaeologists believe that this view was held in the Stone Age. Skulls from this period have been found with holes drilled into them, presumably to release evil spirits. This view has long been discarded by psychology and the medical sciences.

From around 500 BC, Greek and Roman philosophers and physicians viewed a number of mental disorders as being diseases caused by physical dysfunction. This **somatogenic** (physical causation) **medical model** of psychiatric illness was eclipsed in the Middle Ages by a return to belief in demonic possession. Sufferers were generally viewed as being in league with the devil, and were often subjected to physical 'exorcism' using various forms of torture. By the 16th century in many European countries, and later on other continents, mentally ill people were confined to asylums, the most famous being the Bethlehem Hospital in London (known as 'Bedlam'). Asylum inmates were largely restrained and confined, often in chains, and were not treated as patients.

The medical model started to become dominant again at the end of the 18th century. Philippe Pinel in France started to treat asylum inmates as patients in need of care. Similar **moral treatments** (treating patients with respect and giving them dignity) were quickly adopted in other countries. However, while these enlightened physicians were able to claim considerable success, releasing many inmates back into society, the general approach until the mid-19th century was still one of confinement. In the second half of the 19th century, the somatogenic medical model became dominant. It was demonstrated that **general paresis**, a syndrome with physical and mental symptoms, was a late manifestation of syphilis. This led to the search for somatic causes for other mental conditions, a search that well into the 20th century had borne little fruit.

While some were trying to find physical causes of psychological abnormality, others were developing **psychogenic** versions of the medical model, viewing mental problems as illnesses with psychological rather than physical causes. (Some writers limit the term 'medical model' to conceptions of psychological abnormality as having physical causes.) In Vienna in the late 19th century, Joseph Breuer returned to a technique developed 100 years earlier by Friedrich Mesmer, using **hypnotism** (originally known as **mesmerism**) to treat patients with hysterical disorders. Breuer was joined by Sigmund Freud who started, at the end of the 19th century, to develop his theory of the mind and of the origins of abnormality: **psychoanalysis** (see Topic K3).

Today, there are a wide range of approaches to explaining mental abnormality. In one form or another, the medical model is still dominant. Some conditions are considered to be primarily organic (somatic) in origin, and others psychological in origin. Many consider it to be most helpful to view all conditions as having multiple causality, which may include sociocultural and family influences, as well as somatic and personal factors. Each approach to abnormal behavior has an associated method of treatment (see Topic L7).

Classification of mental disorders

The medical model implies that there are a number of distinct mental illnesses. If that were so, it would be useful for several reasons to classify illnesses:

- communication: clinicians can more easily discuss the states they see with others (the classification provides **diagnoses**);
- it allows researchers to ensure that they are studying the same states as others;
- it can help to identify causes;
- it can help to identify potential treatments.

In 1883, the German psychiatrist Emil Kraepelin devised a classification system that is the basis of those used today. As a firm believer in a somatogenic medical model, Kraepelin advocated that patients be interviewed just as would those suffering from physical illnesses. The aim of the clinical interview was to identify **symptoms** (what the person complains of) and **signs** (behavioral or other indications of abnormal function), and to establish the problem's **onset** and **course** (how it has developed). Together, these should permit the clinician to diagnose the person as suffering from a particular **syndrome** or illness. Today, the categories of illness and the relationships between them, identified by Kraepelin, form the basis of the widely used classification systems. These are the World Health Organization's *International Classification of Diseases*, and the American Psychiatric Association's (APA) *Diagnostic and Statistical Manual of Mental Disorders*. As conceptions of psychological abnormality change, so classification systems have to be changed. The latest version of the APA manual is known as DSM-IV, and is the fifth version since its inception in 1952. Table 1 shows the outline of the DSM-IV classification system. We will refer to this throughout this Section L.

One of the main changes in the DSM classification has been a move towards basing diagnosis on specific criteria, and away from the assumption that most disorders fell into one of three large groups. These were **organic brain syndromes**, such as the **dementias**, caused by brain dysfunctions; **neuroses**, such as phobias and obsessive–compulsive disorders, caused by anxiety; and **psychoses**, such as schizophrenia, involving thought and mood disturbances amounting to a loss of contact with reality.

Table 1. The DSM-IV[a] classification of mental disorders, with examples of each category

Diagnostic category	Example disorders
Disorders usually first diagnosed in infancy, childhood, or adolescence	Mental retardation Learning disorders Attention deficit and disruptive behavior disorders
Delirium, dementia, and amnestic and other cognitive disorders	Substance withdrawal delirium Dementia of the Alzheimer's type
Mental disorders due to a general medical condition	Catatonic disorder due to general medical condition
Substance-related disorders	Alcohol dependence Caffeine intoxication Hallucinogen-induced psychotic disorder
Schizophrenia and other psychotic disorders	Schizophrenia, paranoid type Substance-induced psychotic disorder
Mood disorders	Major depressive disorder Bipolar disorders
Anxiety disorders	Panic disorder Specific phobias
Somatoform disorders	Somatization disorder Hypochondriasis Body dysmorphic disorder
Factitious disorders	Characterized by intentional production of physical or psychological symptoms, including self-injury
Dissociative disorders	Dissociative amnesia Dissociative fugue
Sexual and gender identity disorders	Sexual dysfunctions (including hypoactive sexual desire, male erectile disorder, premature ejaculation) Paraphilias (including pedophilia, exhibitionism, masochism) Gender identity disorder (cross-gender identification)
Eating disorders	Anorexia nervosa Bulimia nervosa
Sleep disorders	Primary insomnia Primary hypersomnia
Impulse control disorders not elsewhere classified	Kleptomania Pathological gambling
Adjustment disorders	With depressed mood With anxiety (in response to identifiable stressors)
Personality disorders	Paranoid personality disorder Antisocial personality disorder

[a]*Diagnostic and Statistical Manual of Mental Disorders,* 4th Edn, 1994. American Psychiatric Association, Washington, DC.

L2 SCHIZOPHRENIA AND OTHER PSYCHOSES

Key Notes

The nature and incidence of schizophrenia	Psychosis is characterized by loss of contact with reality, which may involve hallucinations (false sensory perceptions, usually hearing voices), delusions (false beliefs), or withdrawal from the social and physical world. The term psychosis is now largely restricted to schizophrenia.
Signs and symptoms of schizophrenia	The symptoms of schizophrenia include delusions of persecution, of reference, of grandeur, and of control. A person whose life is directed by delusions has paranoid type schizophrenia. Catatonic type schizophrenia is characterized by immobility or posturing. Disorganized type schizophrenia is characterized by inappropriate affect or blunted affect.
Explaining schizophrenia	Schizophrenia is multifactorial in origin. It has a genetic component, and is associated with an increase in activity in dopaminergic circuits in the brain that link perception with emotion and memory, or with enlarged cerebral ventricles. Schizophrenia is influenced by sociocultural factors. Families' interaction styles have also been implicated in schizophrenia.

Related topics Normality and abnormality (L1) Disorders of personality, identity and memory (L6)

The nature and incidence of schizophrenia

Psychosis is characterized by loss of contact with reality. The psychotic person may have such extreme impairment or distortion of perception and thought that he or she is unable to function adaptively. Psychotic people may have **hallucinations** (false sensory perceptions, usually hearing voices) or **delusions** (false beliefs), or may shut themselves off from the social and physical world. The term psychosis is used much less widely than previously (see Topic L1). It is largely restricted to schizophrenia (which we concentrate on here), **schizoaffective disorder** (in which schizophrenic symptoms are accompanied by disturbances of mood), **delusional disorder** (nonbizarre delusions without other schizophrenic signs), and **substance-induced psychotic disorder** (hallucinations or delusions following use of psychotropic, i.e. mind-changing, drugs).

Schizophrenia (literally 'split mind') is a term introduced by Kraepelin (see Topic L1) to describe a particular psychotic condition (or group of conditions). The term causes some confusion, as schizophrenia does not involve **multiple personality**, which is classed as a **dissociative disorder** (see Topic L6). About one person in every hundred on earth meets the criteria for schizophrenia. About the same proportion of men and women are schizophrenic. The incidence is nearly five times as high in the lowest socioeconomic groups as it is in the highest (see *Fig. 1*), and is some three times higher in separated or divorced persons, and twice as high in single people, as in married people. In the UK and in the USA, the incidence is much higher in people of Afro-Caribbean origin

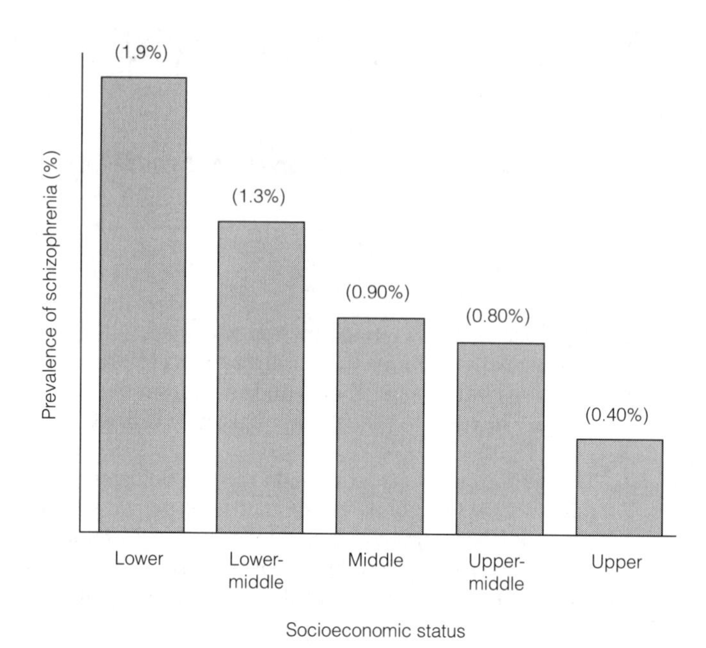

Fig. 1. Incidence of schizophrenia in different socioeconomic groups. Data from Keith, S.J., Regier, D.A. and Rae, D.S. (1991) Schizophrenic disorders. In: Robins, L.N. and Regier, D.A. (Eds) Psychiatric Disorders in America. *Free Press, New York.*

than in white races. This may be due to higher rates of divorce and higher levels of poverty in Afro-Caribbean groups. The direction of cause and effect in the relationships between incidence and socioeconomic and marital status is impossible to determine. The stress of poverty or marital breakdown may contribute to the onset of schizophrenia, but on the other hand schizophrenic behavior may be a factor causing poverty (e.g. through difficulty keeping jobs) and marital discord.

Signs and symptoms of schizophrenia

There is an enormous variation in the signs, symptoms, and course of schizophrenia. A person may have one overwhelming delusion (e.g. that he is Christ) or many delusions, which may form an internally consistent and persistent system, or may be fragmented and variable. The person may feel uplifted by or confused by the delusions. Delusions are sometimes classified as:

- **delusions of persecution** (or influence): a belief that one is the victim of plots;
- **delusions of reference**: believing that the actions of others (e.g. radio announcers) or world events are directed at oneself;
- **delusions of grandeur**: a belief that one is of great importance;
- **delusions of control**: a belief that one's actions are controlled by another.

A person whose life is directed by delusions, not necessarily of persecution, is said to have **paranoid type schizophrenia**. Most hallucinations are auditory, and are frequently voices, which may support delusions (e.g. may be giving the person instructions or shouting abuse). Some people with schizophrenia show **poverty of speech (alogia)**, characterized by brief and content-free replies, which may be accompanied by **blocking of thought** (having fewer and less varied thoughts than most people). On the other hand, some patients show **positive formal thought disorders**, including disorganized speech, following unexpected chains of thought, using invented words, and **perseveration** (the repetition of words or phrases).

Other sufferers show **psychomotor symptoms**, including loss of spontaneous movement, or the adoption of odd postures or movements. In an extreme form, this may be a total lack of response to the environment: **catatonia**. One form of illness, **catatonic type schizophrenia**, is characterized by such immobility or posturing, often with extreme negativity, **echolalia** (repetition of what is heard), or **echopraxia** (repetition of observed actions).

A third variety of illness, **disorganized type schizophrenia**, is characterized by **inappropriate affect** (e.g. laughter when hearing tragic news) or **blunted affect**, together with incoherent speech and confused behavior.

Many sufferers cannot be assigned to one of these diagnostic categories, and are classified as **undifferentiated type schizophrenia**. This diagnosis is used today for people who meet the DSM-IV criteria for schizophrenia, but who do not satisfy the criteria for any of the three main types. Some clinicians favor a simpler classification into **Type I schizophrenia** (predominantly the 'positive symptoms' of delusions, hallucinations, and disorganized speech) and **Type II schizophrenia** (with the 'negative symptoms' of flat affect, poverty of speech, and reduced voluntary movement). As we will see below, this classification seems to fit better with certain brain dysfunctions.

Explaining schizophrenia

Many theorists accept that schizophrenia is multifactorial in origin. It is now clear that there are neurophysiological or neuroanatomical malfunctions that provide the underlying organic pathology. There is a **genetic** component in schizophrenia, shown by the following types of data:

- **relationship**: the closer one's relationship to a schizophrenic person, the more likely one is to be schizophrenic oneself, ranging from 2% for first cousins to 48% for identical twins;
- **adoption**: the genetic basis of this effect is confirmed by findings that blood relatives of adopted schizophrenia sufferers were four to five times as likely to be sufferers themselves as were members of the adopting families;
- **molecular genetics** (see Topic A1): there have been reports that schizophrenia sufferers and their relatives have structures on chromosome 5 that are different to those of nonsufferers. However, this has not yet been firmly established.

One possible route by which genetic factors could predispose a person to schizophrenia is by producing organic pathology. Two likely disorders have been identified. According to the **dopamine hypothesis**, the underlying pathology is an increase in activity in dopaminergic circuits in the brain (see Topic A2). This was suggested first by the observation that antipsychotic drugs (see Topic L7) eventually produce tremors like those in Parkinson's disease. Since Parkinson's disease is known to be due to insufficient dopamine in motor neural circuits, the drugs were thought to have their effect by decreasing the activity of dopaminergic circuits. This is confirmed by experimental evidence that injecting schizophrenia sufferers with drugs that increase dopamine activity increases their symptoms, and that amphetamines, which enhance dopamine activity, sometimes lead to **amphetamine psychosis**. These psychotic effects are mainly of 'positive symptoms', and the early antipsychotic drugs are mainly effective on these symptoms. This has led some to believe that this biochemical abnormality underlies Type I schizophrenia (see above). How an excess of dopamine activity causes these symptoms is unclear. The neurons affected extend from the midbrain into the limbic system (see Topic A3), areas that probably function to link perception with emotion and memory. It may be that hyperactivity causes a state in which the brain is unable to ignore sensory inputs or internally generated thoughts, leading to a sort of 'cognitive overload'.

Postmortem examinations of the brains of people with Type II schizophrenia, and more recently techniques such as PET and MRI (see Topic A4), have revealed that they frequently have enlarged cerebral ventricles (see Topic A3), particularly in the right hemisphere, associated with reduced blood flow nearby. This might reflect underactivity or even atrophy of adjacent brain areas. Exactly how these defects might cause negative schizophrenic symptoms is not clear.

While there is almost certainly underlying organic pathology, schizophrenia is influenced by sociocultural factors. In particular, many believe that 'schizophrenic' is a convenient label for behavior in others that makes us uncomfortable. More importantly, once the label is assigned, the individual is treated differently in such a way as to provide a 'self-fulfilling prophecy'. This was demonstrated in the 1970s in a study by Rosenhan, who got eight 'normal' people to report to different hospitals reporting hallucinations. All were diagnosed as psychotic and admitted as patients, at which point they tried to behave normally. For many this became progressively more difficult, as they were depersonalized by staff, and had normal behavior described as if it were abnormal (e.g. note taking as 'engaging in writing behavior'). They felt powerless, and became bored, listless and apathetic; in effect, showing negative schizophrenic symptoms. Sociocultural pressures cannot be the whole explanation of schizophrenia, however, since schizophrenic symptoms do not differ between very different cultures.

Families have also been implicated in schizophrenia. Psychoanalysts in the mid-20th century talked about the **schizophrenogenic** (schizophrenia-causing) **mother**, who was supposed to be overprotective but rejecting; cold and domineering. In the 1970s, Gregory Bateson proposed the **double-bind hypothesis**. This was that the parents of schizophrenic patients behave towards them in self-contradictory ways (double-binds); for example, making positive statements with negative nonverbal signals. Repeated exposure to these ambiguous messages may lead to the development of abnormal ways of coping with the social environment; for example, ignoring people altogether, or focusing only on verbal or only on nonverbal messages. However, this theory cannot account for many aspects of schizophrenia, and has received very little empirical support.

A more recent family approach that has received more support focuses on how emotional communication within the family affects the course of schizophrenia. Schizophrenic patients returning home to live with relatives are more likely to relapse if they have a relative who is high in **expressed emotion (EE)**. This reveals itself in interviews as high levels of critical comment, hostility, and/or emotional overinvolvement. Relatives can be taught to modify their emotional interaction with the patient, and this reduces the likelihood of relapse. Again, this does not suggest a cause of schizophrenia, and EE has also been shown to influence the course of a variety of other conditions.

L3 MOOD DISORDERS

Key Notes

Unipolar depression

Major depressive disorder occurs in twice as many women as men. It is most frequent in separated and divorced persons. Symptoms are persistent sadness, together with a variety of somatic, cognitive, emotional, and motivational disturbances. Suicide is a serious risk. Seasonal affective disorder is a distinct condition that occurs through the winter months.

Causes of unipolar depression

Depression is triggered by stressful events. Beck's cognitive theory of depression argues that at the heart of unipolar depression is persistent negative thinking. Seligman's learned helplessness theory holds that people become depressed when they perceive themselves faced with an uncontrollable situation. The catecholamine hypothesis proposes that reduced brain norepinephrine causes depression.

Bipolar disorders

In bipolar disorders, the individual swings between serious depression and mania. Between 0.4% and 1.3% of the population suffer; equally often men and women.

Causes of bipolar disorders

There is a genetic factor in bipolar disorders. Norepinephrine is low in depressed episodes and higher than normal in manic episodes. Lithium reduces norepinephrine activity, yet is effective on both manic and depressive states. Serotonin is also implicated, but is reduced in both mania and depression. Lithium interacts with the transport of sodium ions across membranes that is essential to proper neural function.

Related topic

Normality and abnormality (L1)

Unipolar depression

The mood we all experience from time to time which we call depression should not be confused with the extreme, debilitating clinical states, sometimes with psychotic features, also called depression. Most cases of clinical depression are diagnosed in DSM-IV (see Topic L1) as **major depressive disorder** (sometimes called **unipolar disorder**, in contrast to bipolar syndromes which are looked at below). In any 1 year, 5–10% of the population of developed countries will suffer from this, and a further 3–5% will suffer from a less extreme form of the illness. The incidence increased through most of the 20th century. The average age of onset of major depression is 27 years, and twice as many women as men are diagnosed with the condition. It is more frequent in separated and divorced persons than in those of other marital status (see *Fig. 1*).

While persistent sadness (depressed mood) is the characteristic symptom of major depression, others include loss of feeling of affection for others, fatigue, varied and often vague physical symptoms, sleep disturbances, psychomotor disturbances (decreased, or sometimes increased, activity levels), reduced drive,

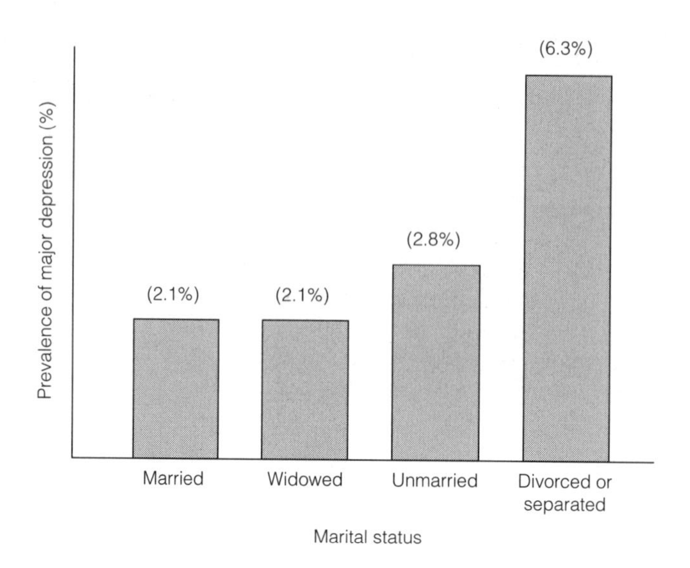

Fig. 1. Incidence of major depressive illness in any 1 year in adults in the USA. Data from Weissman, M.M. et al. *(1991) Affective disorders. In: Robins, L.N. and Regier, D.A. (Eds)* Psychiatric Disorders in America. *Free Press, New York.*

poor appetite, reduced ability to concentrate, feelings of worthlessness and self-blame, and frequent thoughts of death. Sufferers are at great risk of **suicide**, but less often when in the deepest despair than when showing signs of recovery. Women attempt suicide about three times as often as men, but men are four times as likely to succeed.

Seasonal affective disorder (SAD) is a condition that occurs through the winter months in a small number of people, and is characterized by depression, fatigue, and oversleeping, combined with increased appetite and consequent weight gain. It is believed that this condition is associated with a disturbance of the process by which melatonin acts on the suprachiasmatic nucleus to synchronize diurnal rhythms (see Topic B6).

Causes of unipolar depression

Episodes of depression often seem to be triggered by stressful events, and those subject to high levels of stress are more likely to develop depression. Clinicians used to distinguish between **reactive depression**, triggered by external events, and **endogenous depression**, with no such trigger. This distinction has given way to a search for the combination of internal and external factors that result in any individual's depressive episode. The dominant views now are cognitive and neurochemical in nature. Starting in the 1960s, Aaron Beck developed a **cognitive theory of depression**, which argues that at the heart of unipolar depression is persistent **negative thinking**. This involves **maladaptive attitudes** (e.g. that one's general worth is adversely affected by every minor failure), a **cognitive triad** (interpreting the self, one's experiences, and one's future negatively), **errors in thinking** (e.g. personalization: viewing oneself as the cause of bad external events, like bad weather), and **automatic thoughts** (persistent and uncontrollable thoughts reminding one of one's inadequacy). This negative thinking is so pervasive that all aspects of life and activity are adversely affected.

Another cognitive view is that of Martin Seligman. He conducted studies of animals in situations from which they could not escape, observing that they eventually become passive and stop trying. He called this state **learned helplessness**, and argued that people become depressed when they perceive themselves as faced with uncontrollable situations. However, helplessness does not necessarily lead to depression, and it has been argued that people become depressed in these circumstances because they have an **attributional style** (see Topic I2) which causes them to interpret bad external events as due to their own failings. A criticism of these cognitive approaches is that they confuse cause and effect; the cognitive characteristics are symptoms of depression rather than causes of it. Beck has argued that there is a positive feedback process operating: the cognitive factors produce depression, which enhances the cognitive difficulties, and so on. Further, there is evidence that people with depressive cognitive or attributional styles are more likely to become depressed following stress.

The cognitive approaches do not explain why some people become clinically depressed and others do not. Even if we accept that there are individual differences in cognitive styles, how do they originate? There is evidence from studies of twins and other relatives, and from adoption studies that there are genetic factors involved in unipolar depression. Biochemical research has shown that depression is associated with low levels of the neurotransmitter norepinephrine (see Topic A2), leading to the **catecholamine hypothesis** of depression. Evidence for this includes the finding that **reserpine**, a drug used to treat hypertension, lowers norepinephrine levels and sometimes causes depression. Further, antidepressant drugs called **monoamine oxidase inhibitors** work by increasing the availability of norepinephrine in synapses, while another class of antidepressants, **tricyclics**, also increase norepinephrine levels in the brain. Similar lines of argument have led to the proposal that it is the neurotransmitter **serotonin** that is important. The newest class of antidepressants, including Prozac (see Topic L7), act by increasing the amount of serotonin in serotonergic synapses. Many believe that neural circuits based on each of these transmitters are involved in unipolar disorders.

Bipolar disorders

In the bipolar disorders, the individual swings between extremes of serious depression and **mania** (the conditions were formerly called manic–depressive illness). Manic episodes are periods of persistent elevated or irritable mood, characterized by inflated self-esteem, decreased sleep, being extremely talkative, distractibility, 'flights of fancy', agitation, or engaging in pleasurable activities that might have serious physical or social consequences. The rate at which a person swings from one extreme to another can vary from a few hours to several months. A person may have only one manic or depressed episode. Bipolar disorders are less frequent than unipolar disorders, amounting to between 0.4% and 1.3% of the population, and are equally frequent in men and women. The diagnosis of bipolar disorders depends on the severity of the manic episode, the nature of the most recent episode (manic or depressed), and whether the states are separated in time or occur mixed on the same day.

Causes of bipolar disorders

As with unipolar illness, it is established that there is a genetic factor in the likelihood of developing bipolar disorders. As well as the usual relative and adoption studies, careful analysis of families with high rates of bipolar disorder have suggested links with genes on the X chromosome (see Topic A1) responsible for other disorders, including red–green color blindness. The catecholamine hypothesis has been extended to bipolar illness. Norepinephrine is low in depressed episodes, and higher than normal in manic episodes. Reserpine (see above) sometimes decreases manic symptoms. However, the most effective treatment for bipolar illness, **lithium**, also reduces norepinephrine activity, yet is effective on both manic and depressive states. Serotonin is also implicated, but is *reduced* in both mania and depression. Administering the drug **tryptophan**, which increases serotonin, to manic patients can alleviate their symptoms, suggesting that the low level of serotonin is a causal factor. Lithium is also known to interact with the transport of sodium ions across membranes that is essential to proper neural function (see Topic A2). Abnormalities in sodium transport have been found in people with bipolar disorders, and the therapeutic effect of lithium may be based on its interaction with this mechanism. It is difficult to reconcile the facts about neurotransmitter function, and current research aims to elucidate the roles of serotonin and norepinephrine, and of sodium transport defects, in bipolar disorders.

L4 ANXIETY DISORDERS

Key Notes

Panic

Panic attacks are short periods of intense fear, and panic disorder consists of repeated panic attacks. They seem to have a basis in overactivity of noradrenergic circuits between the midbrain and the amygdala. Panic-prone persons may tend to amplify bodily sensations and interpret them as signs of impending disaster.

Phobias

Phobic disorders are intense and irrational fears of specific objects or situations. They may result from learned associations between fear and particular objects. The preparedness theory holds that phobias are most likely of certain classes of (dangerous) natural stimuli, for which we have evolved a predisposition.

Generalized anxiety disorder

Generalized anxiety disorder is unfocused, prolonged anxiety and worry. It has been explained as excessive existential anxiety, generalization of conditioned fear to a very wide range of situations, modeling, or as due to the adoption of a pessimistic cognitive style. There may be insufficient GABA (γ-aminobutyric acid) activity in neurons connecting midbrain centers to the amygdala to inhibit fear responses.

Obsessive–compulsive disorder

Obsessions are persistent thoughts or images that intrude into consciousness. Compulsions are repetitive behaviors that a person feels compelled to perform. Obsessive–compulsive disorder may start with strategies that dispel anxiety generated by unwanted thoughts. An underlying cause may be low levels of the neurotransmitter serotonin, permitting excessive activity in the frontal cortex.

Post-traumatic stress disorder

Post-traumatic stress disorder (PTSD) is persistent anxiety, intrusive thoughts, flashbacks, recurrent dreams, and feelings of re-enactment of an intensely traumatic event. PTSD may be more likely in people exposed to childhood traumas, or who lack a strong social support network.

Related topic

Normality and abnormality (L1)

Panic

Fear or anxiety experienced in physically or socially dangerous situations is normal, and even adaptive. When they are excessively prolonged, are unusually intense in relation to events, or occur with no precipitating event, they might be considered pathological. A number of anxiety disorders are recognized, and a feature of many of them is **panic attacks**. A panic attack is a period of intense fear or discomfort commencing suddenly, and peaking within 10 min. It will have many of the following symptoms: palpitations or fast heart rate, sweating, trembling, sensations of difficulty breathing, chest discomfort or pain, nausea, dizziness or light-headedness, feelings of unreality, fear of losing control or dying, numbness or tingling, chills or hot flushes. If panic attacks occur repeatedly, with the person worrying between attacks that another will occur, a diagnosis of **panic disorder** might be made. This can be with or without **agoraphobia** (fear and avoidance of places from which escape might be difficult, or in which the person fears they might suffer a panic attack).

Panic attacks respond to some antidepressant drugs known to reduce norepinephrine activity (see Topics A2 and L2). Stimulation in monkeys of a midbrain structure, the **locus coeruleus**, which communicates with the **amygdala** in the limbic system via noradrenergic neurons (see Topics A2 and A3), produces a state like panic, and destruction of this region prevents fear responses. The amygdala is known to be integral to the learning and triggering of emotions, particularly fear. Drugs which specifically increase norepinephrine activity trigger panic attacks in people with panic disorder, while drugs that decrease norepinephrine reduce panic symptoms. Taken together, these findings suggest that panic attacks have, at their physiological core, overactivity of noradrenergic neural circuits that stimulate fear response mechanisms in the amygdala.

In the last few years, cognitive theorists have suggested that panic-prone persons have a tendency to amplify bodily sensations, and to interpret them as signs of impending disaster rather than as responses to discrete stimuli such as an argument. This proposal is not incompatible with the physiological view just described, which might indicate a pathway by which this cognitive process occurs.

Phobias

Phobic disorders are intense and irrational fears of specific objects or situations. The object could be almost anything, but common examples are high places (acrophobia), spiders (arachnophobia), and enclosed spaces (claustrophobia). While most of us might be apprehensive when faced with many of these classes of object, in phobias the fear is pervasive and debilitating, including the symptoms of panic attacks. It may prevent a person leaving the house for fear of encountering the phobic object. Agoraphobia, which we mentioned in the preceding section, is a more complex case, with more general fears.

It is widely believed that specific phobias result from learned associations between the physical sensations of fear and particular objects which become the phobic object. Later, the fear sensations can **generalize** (see Topic E1) to other stimuli. One problem with this view is that phobias for a lot of dangerous objects (e.g. knives, cookers) are extremely rare. The **preparedness theory** suggests that we have evolved a readiness to respond with fear to certain classes of (dangerous) natural stimuli, making it more likely that phobias will develop to those.

Generalized anxiety disorder

When anxiety and worry are prolonged (for at least 6 months), but are not focused on particular events ('free-floating anxiety'), and are accompanied by difficulty concentrating, fatigue, feelings of inadequacy, insomnia, or irritability, it may be diagnosed as **generalized anxiety disorder**. Up to 6% of the population may suffer from this in any year, and it is not as disabling as other anxiety disorders. Most sufferers, however, will at some time experience one of the acute anxiety disorders, or sometimes depression. Little is known of the causality of this disorder, and none of the proposed explanations is satisfactory. Some argue that it is an excessive existential anxiety, or concern with one's reason for being. Others have suggested that the conditioned fear of phobias generalizes to a very wide range of situations, or that anxiety is learned by observing the reactions of others (**modeling**). Cognitive theorists argue that sufferers adopt a particular cognitive style characterized by pessimism, and the belief that unpleasant events are unpredictable. This will include automatic thoughts emphasizing the likelihood of failure and social embarrassment (rather than the self-doubt we saw in depression). There is little evidence of a genetic component in generalized anxiety disorder; however, a physiological correlate has been identified. A class of drugs used to relieve anxiety, **benzodiazepines** (e.g. Valium, Librium), work by attaching to receptors in the limbic system that are normally receptors for the inhibitory neurotransmitter **GABA** (γ-aminobutyric acid; see Topic A2). It is thought that these normally function to control fear responses, but that in generalized anxiety disorder, for some reason, there is insufficient GABA activity to inhibit fear responses.

Obsessive–compulsive disorder

Obsessions are persistent thoughts or images that intrude into consciousness. They are repetitive, uncontrollable, feel 'alien' to the person experiencing them (although they are recognized as coming from within the person), are not about 'normal' concerns, and cause anxiety. Attempts to dispel them or ignore them fail, and make the sufferer more anxious. **Compulsions** are repetitive behaviors or mental acts that a person feels compelled to perform. Failing to perform them, or trying not to perform them, leads to anxiety, so they seem to function to reduce anxiety. Common compulsions include hand-washing, checking (e.g. that doors are locked or electric appliances switched off), and counting. When extreme, obsessions or compulsions (usually both together) are diagnosed as **obsessive–compulsive disorder**. The condition can be extremely disabling, as both obsessions and compulsions can make it impossible to pursue a normal life.

Cognitive–behavioral theorists have tried to explain obsessive–compulsive disorder as growing out of the sort of repetitive but unwanted thoughts (e.g. of sexual acts, of being in contact with germs, of hurting others) that we all occasionally experience. People who are unable to dismiss such thoughts will start to worry about them, and the possibility that they might act them out. To attempt to dispel these thoughts, such a person might engage in mental or behavioral strategies designed to reduce the anxiety caused by the intrusive thoughts. If they succeed, these thoughts or behaviors are rewarded, and become more likely to recur, eventually becoming automatic obsessions or compulsions. Whether this provides a satisfactory explanation is unclear. A physiological explanation is suggested by the finding that sufferers from obsessive–compulsive disorder have low levels of activity of the neurotransmitter **serotonin** (see Topic A2). Another finding is of excessive activity in areas of the frontal cortex (see Topic A4) and midbrain structures that control the passage of impulses to the thalamus (see Topic A3), which in turn essentially channels sensory information to the point of awareness. Such activity could allow more impulses to reach awareness. Serotonin is involved in these circuits, but just how this would result in obsessive–compulsive disorder is not known.

Post-traumatic stress disorder

Shortly after a traumatic event, such as combat, rape, a natural disaster, or a serious accident, during which a person has experienced intense fear, most people will experience anxiety, and may have recurrent intrusive thoughts about the event. In some people, these persist long after, and are now considered to constitute **post-traumatic stress disorder (PTSD)**. Sufferers experience vivid reliving of the event ('flashbacks'), may have recurrent dreams about the event, may feel or act as if the event were recurring, and will be distressed by stimuli that remind them of the event. During each of these types of occurrence, they may experience heightened physiological arousal and its effects on behavior (e.g. irritability, lack of concentration, insomnia). There is little understanding of why some people develop this persistent pattern of responding after severe trauma, while others do not. Some recent research has suggested that vulnerability develops in people exposed to childhood events such as assault, mental disorder in the family, or parents divorcing. However, how this leads to susceptibility to PTSD is unclear. Other possible factors include the lack of strong social support networks, and personality styles; for example, those who respond to stress with positive attitudes ('hardiness') are unlikely to develop PTSD.

L5 EATING DISORDERS

Key Notes

Anorexia nervosa

Anorexia nervosa is a refusal to maintain body weight within the normal range. It is much more common in females. When weight control is achieved by self-starvation, the illness is described as restricting type anorexia nervosa. In binge-eating/purging type anorexia nervosa, sufferers engage in binge eating and/or purging.

Bulimia nervosa

Bulimia nervosa is characterized by binge eating, with vomiting or laxative use. Most sufferers are women. Bulimia nervosa often starts after a period of dieting, and bingeing episodes are often triggered by stress. The binges usually start with feelings of anxiety, which are relieved by the uncontrollable binge, followed by guilt and depression. Purging temporarily relieves the guilt and depression, rewarding the whole binge–purge cycle.

Explanations of eating disorders

Recent explanations of eating disorders are multifactorial. Sociocultural factors include the thin 'ideal' female figure depicted in the media. Characteristics of the families of eating disorder patients have been noted, including achievement pressure which can engender perfectionism, and disturbed patterns of interaction. The development of an eating disorder is a way of gaining control.

Related topics

Hunger and eating (B3) Normality and abnormality (L1)

Anorexia nervosa

Anorexia nervosa involves a refusal to maintain body weight within the normal range, fear of gaining weight, low self-esteem coupled with distorted perception (overestimation) of body size, and loss of menstrual periods. The name, which literally means *loss of appetite of nervous origin*, is a poor one, because sufferers are usually preoccupied with food. It is vastly more common in females, about 1% of whom will have the illness at some time. Many more will show some signs of the disorder. It can start in childhood, but most commonly between 14 and 18 years of age. Weight loss is mainly achieved by self-starvation and, when confined to that, the illness is described as **restricting type anorexia nervosa**. Up to half of sufferers engage in binge eating and/or purging (self-induced vomiting or use of laxatives or enemas). These cases may be diagnosed as **binge-eating/purging type anorexia nervosa**. The extreme weight loss in anorexia nervosa leads to a range of physical problems, and can be fatal.

Bulimia nervosa

The main characteristic of **bulimia nervosa** is binge eating, accompanied by compensatory calorie-losing behavior which may allow the maintenance of a normal body weight. The compensatory behaviors are most commonly vomiting or laxative use, but may include periods of fasting and strenuous exercise. The vast majority of those with bulimia nervosa are women, and the most frequent age at onset is between 15 and 19 years. Up to 6% of young women will suffer from bulimia nervosa, although more will occasionally purge, and up to 50% admit to one or more episodes of binge eating. The onset of bulimia nervosa often follows a period of dieting, and bingeing episodes, at least initially, are often triggered by stress or weight concern. Bingeing and purging are usually carried out in secret. The binges in bulimia nervosa usually start with feelings of tension or anxiety, which are relieved by the consumption of huge quantities of food in the course of an hour or more. The person usually feels unable to control the eating, and the relief of tension is followed by guilt and depression. Purging follows, apparently as a way of relieving these feelings, and of avoiding weight gain. However, neither vomiting nor laxative and enema use prevent the absorption of calories, and vomiting interferes with physiological appetite control mechanisms (see Topic B3), making the person hungrier. The purging temporarily relieves the guilt and depression, rewarding the whole binge–purge cycle. However, the person with bulimia nervosa develops low self-esteem, particularly feeling powerless and disgusted with the self.

Explanations of eating disorders

There have been an enormous variety of theories for the etiology of eating disorders. Psychodynamic theories were dominant through much of the 20th century. They suggested, for example, that anorexia nervosa results from unresolved 'oral conflicts'. More recently, a multifactorial view has found favor, recognizing the importance of sociocultural, family, personal, and physiological factors that predispose the individual, precipitate the illness, and perpetuate its course. A lot of attention has been paid to the thinness of the 'ideal' female figure, as depicted in the media and, particularly, by fashion models. The emergence of this ideal through the second half of the 20th century paralleled an increase in the incidence of anorexia nervosa. Attempts by young women or girls to attain this figure usually involve dieting, and this predisposes the development of anorexia and bulimia nervosa.

The families of eating disorder patients often show a history of pressure on the child towards academic and personal achievement. Personal achievement may include physical appearance and thinness. This can encourage the development of **perfectionism** in the child, which is thought to be another predisposing factor. Some researchers and clinicians have drawn attention to disturbed patterns of interaction in the families of eating disorder patients, particularly concerning the expression of emotion. Salvador Minuchin has described an **enmeshed family pattern**, in which members are overinvolved with one another, yet do not communicate well, and this tends to produce dependency in members. Similarly, Hilde Bruch argued that disturbed family interactions (especially between mother and daughter) lead to 'ego deficiencies', including a poor sense of autonomy and control. It might be that the development of an eating disorder is a way in which the young person can gain control over herself.

Many studies have reported particular physiological changes or abnormalities in anorexic persons. However, such changes might be a result rather than a cause of anorexia. There is no clear evidence for an underlying physical cause in eating disorders.

L6 Disorders of personality, identity, and memory

Key Notes

Personality disorders

Personality disorders are fixed, long-term patterns of behavior and thinking that lead to personal or social difficulties, and sometimes to danger to the self or others. They show a wide range of symptoms, which overlap considerably between different disorders.

Dissociative disorders

Dissociative disorders involve the separation of parts of an individual's identity. They include dissociative amnesia, when the person suddenly becomes unable to recall central information about themselves. Dissociative identity disorder is the occurrence in the same person of two or more distinct personalities.

Memory disorders

Amnestic disorders are memory problems with an organic origin, characterized by anterograde amnesia (reduced ability to learn new information). They result from dysfunction of the brain mechanisms involved in memory formation and recall.

Dementias

Dementias are progressive cognitive deficits, the occurrence of which increases dramatically with age. Alzheimer's disease is caused by deterioration and protein deposits in various brain regions.

Related topics

Memory loss (F3)
The study of personality (K1)
Normality and abnormality (L1)
Schizophrenia and other psychoses (L2)

Personality disorders

Personality disorders are fixed, long-term patterns of behavior and thinking that lead to personal or social difficulties, and sometimes to danger to the self or others. They differ from most of the other disorders we have looked at by beginning in childhood or adolescence, and remaining relatively unchanged until late in life. The other disorders mostly have periods of remission, and change in intensity and in nature over time. DSM-IV (see Topic L1) recognizes 10 personality disorders, which are shown, with their major features, in *Fig. 1*. However, as the figure shows, there is considerable overlap of these features, and many clinicians doubt the utility of this classification. We will look at four disorders that have relationships with other conditions.

People with **paranoid personality disorder** are distrustful and suspicious of others, which causes them to avoid forming close relationships. They tend to find hidden meanings in everyday actions, and are cautious, vigilant, and quick to respond to threats, real or imagined. They are likely to think themselves blameless, and so are very sensitive to criticism (real or imagined). Although they may occasionally have episodes in which they seem to lose contact with

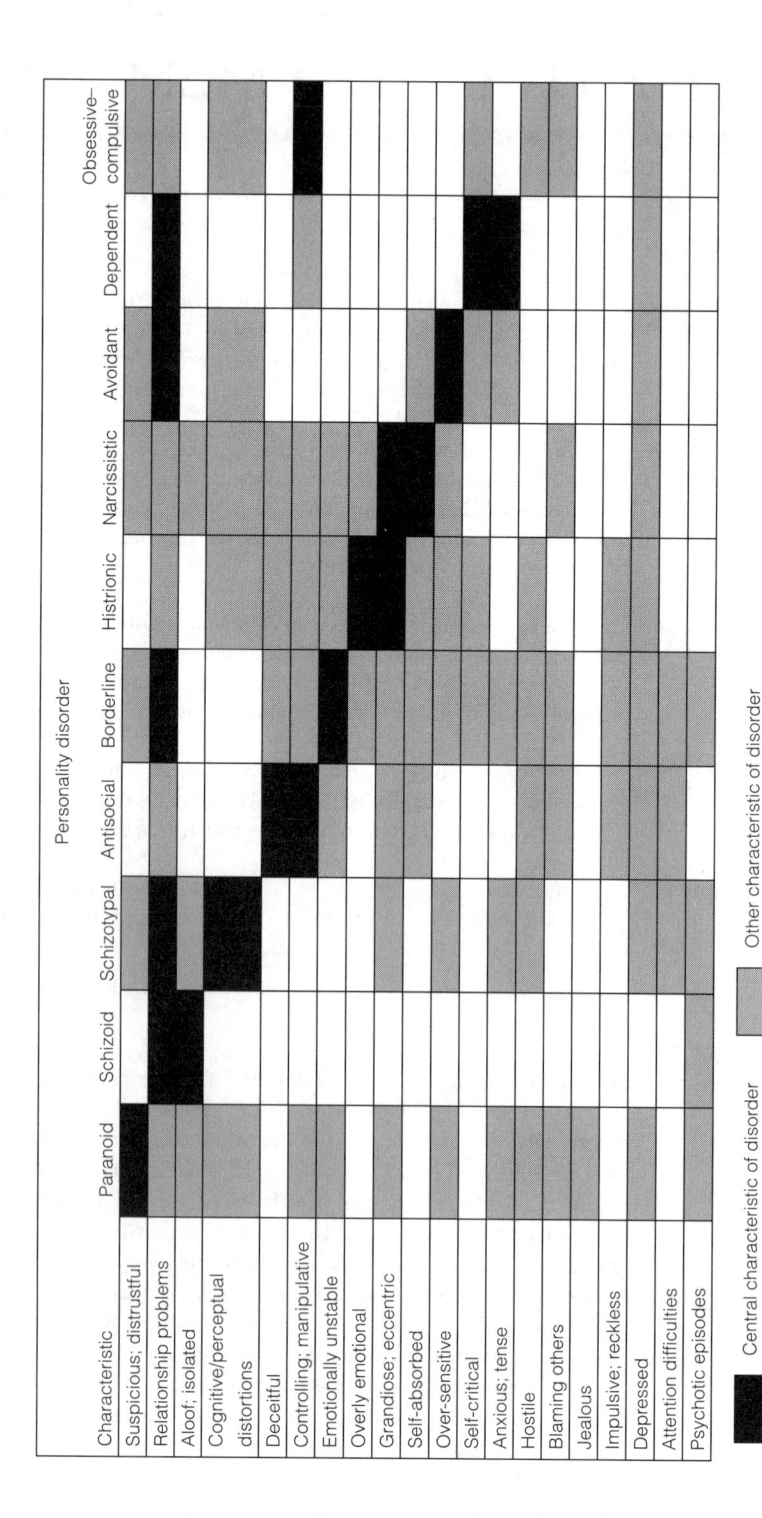

Characteristic	Personality disorder: Paranoid	Schizoid	Schizotypal	Antisocial	Borderline	Histrionic	Narcissistic	Avoidant	Dependent	Obsessive–compulsive
Suspicious; distrustful	■		▒		▒		▒	▒		▒
Relationship problems	▒	■	■	▒	■	▒	▒	■	■	▒
Aloof; isolated	▒	■	▒				▒			
Cognitive/perceptual distortions	▒		■			▒	▒	▒		▒
Deceitful				■	▒	▒	▒			
Controlling; manipulative	▒			■	▒	▒	▒		▒	■
Emotionally unstable	▒			▒	■	▒	▒			
Overly emotional					▒	■	▒			
Grandiose; eccentric	▒		▒	▒	▒	■	■			
Self-absorbed				▒	▒	▒	■	▒		
Over-sensitive	▒		▒		▒	▒	▒	■		
Self-critical					▒	▒		▒	■	▒
Anxious; tense	▒		▒		▒			▒	■	
Hostile	▒		▒	▒	▒	▒				▒
Blaming others	▒			▒	▒		▒			▒
Jealous	▒									
Impulsive; reckless				▒	▒	▒				
Depressed	▒		▒	▒	▒	▒	▒	▒	▒	▒
Attention difficulties			▒	▒	▒					
Psychotic episodes	▒	▒	▒		▒					

■ Central characteristic of disorder　▒ Other characteristic of disorder

Fig. 1. Characteristics of personality disorders, showing overlap between the features of different disorders.

reality, usually after stressful events, these are brief. The paranoid beliefs, although persistent, are not bizarre or delusional, like those in schizophrenia.

Schizotypal personality disorder is characterized by widespread peculiarity of cognitive processes, and difficulties in social and personal relations. While their unusual thinking is often delusional, their symptoms are not as severe as those of schizophrenia, and do not show the time course of psychosis. Instead, they have the steady, persistent quality of all the personality disorders.

People with **antisocial personality disorder** are sometimes called **psychopaths** or **sociopaths**. They show a persistent pattern of irresponsible, reckless, and antisocial behavior. They are manipulative, often deceitful, and like to control situations and other people. They are frequently imprisoned for offenses involving violence, and have a high rate of alcoholism.

Obsessive–compulsive personality disorder is characterized by a life ruled by orderliness, perfection, and control. This preoccupation can interfere with task performance, so that the person may be unable to complete a project, being unable to perform processes to their own satisfaction. They are likely to be overconscientious and inflexible, devoted to work at the expense of leisure and relationships, reluctant to delegate tasks, and unable to discard worn-out objects. This disorder should not be confused with obsessive–compulsive anxiety disorder (see Topic L4), as the personality disorder does not show the true obsessions and compulsions of the anxiety disorder. That is, whereas in the anxiety disorder there are specific, ritualized thoughts and acts which are distressing to the sufferer, in the personality disorder the obsessive–compulsive style is more generalized, and is accepted by the individual as an appropriate way of life.

Dissociative disorders

Dissociative disorders involve the separation of parts of an individual's identity. They include **dissociative amnesia**, which is popularly known as 'losing one's memory'. The individual suddenly becomes unable to recall central information about themselves, such as their name, family status, and occupation. It is usually precipitated by stressful events. At the end of the 20th century, a debate raged over the alleged recall of sexual abuse in childhood, often by adults undergoing psychotherapy. Proponents of the view that real 'repressed childhood memories' were being recalled pointed to dissociative amnesia as a well-documented example of the complete forgetting of such important information. Those who called this 'false memory syndrome' argued that the alleged memories were suggested or implanted by therapists. While childhood sexual abuse remains an enormous problem, many people were wrongly accused on the basis of such 'memories'. When a person suffering from dissociative amnesia disappears to another location, sometimes adopting a new identity, the condition is known as **dissociative fugue**.

Dissociative identity disorder (or **multiple personality disorder**) is the occurrence in the same person of two or more distinct personalities. One personality is usually dominant, but other, often radically different, personalities emerge suddenly, often following stress. Sometimes the personalities are aware of one another; sometimes not. The subpersonality, by for example engaging in activities that would be alien to the dominant personality, may leave effects that are distressing. The condition is much more frequent in women than in men, and almost all sufferers have a history of childhood abuse.

Little is known of the causation of dissociative disorders, although there are a number of theories. Psychodynamic theory sees them as extreme forms of repression (see Topic K3) of childhood trauma. Behavioral theorists view memory loss as a learned response to stress. Others believe that dissociation is a state akin to self-hypnosis, the memory effects being similar to induced posthypnotic amnesia.

Memory disorders

Some memory problems have a clear organic origin, and are called **amnestic disorders**. They are characterized by **anterograde amnesia**, which is a reduced ability to learn new information. They may also involve **retrograde amnesia**, which is a loss of memories that had already been established (such as in dissociative amnesia). The amnestic disorders result from dysfunction of the brain mechanisms involved in memory formation and recall (see Topic F3). This may result from head injury, circulatory problems, or from chronic alcoholism, when it is called **Korsakoff's syndrome**. In its early stages, the progression of Korsakoff's syndrome can be halted by large doses of the vitamin **thiamin**, but if untreated it becomes a profound memory loss. Deficiency of dietary thiamin accompanying high alcohol intake causes deterioration of neurons in forebrain structures involved in the formation of memories, the hippocampus, thalamus, and hypothalamus (see Topic A3).

Dementias

Memory loss together with other cognitive failures, including aphasia (see Topic A4), difficulty performing everyday tasks, and personality changes, are characteristics of **dementias**. The incidence of dementia increases markedly with age, from less than 5% before the age of 65 to 30% in the over 80s. About half of all cases of dementia are **Alzheimer's disease**. Brains of sufferers show changes in a number of locations, including shrinkage or death of neurons in the cerebral cortex and the forebrain structures mentioned above, and deposits of proteins between cells in the cortex and the amygdala. A number of possible causes have been suggested, but research has identified a gene responsible for many cases, which causes production of the protein that is deposited in the brain.

L7 TREATMENT OF PSYCHOPATHOLOGY

Key Notes

Psychodynamic therapies

Psychodynamic therapies derive from psychoanalysis. Illness is caused by defense mechanisms that prevent awareness of conflicts. Treatment uses free association which leads to insight and conflict resolution. Some other dynamic therapies are more directive, and might deal with immediate symptoms.

Humanistic and existential therapies

Client-centered therapy is nondirective, and aims to help individuals achieve their full potential by examining and understanding themselves and their life. Existential therapy builds up a therapeutic relationship between client and therapists, who work together to give meaning to the client's life.

Behavioral therapies

Behavior therapy considers that symptoms are acquired through maladaptive learning. Techniques used include systematic desensitization, based on relaxation to reduce anxiety during treatment, aversion therapy, which punishes inappropriate behaviors, and operant methods, such as token economies, which reward appropriate behaviors.

Cognitive therapies

Rational–emotive therapy is a directive method of showing the patient that their assumptions and cognitive processes are unsuitable. Cognitive therapy helps clients to recognize the negative thought processes that underlie their problems. Cognitive–behavioral approaches consider cognitions to be learned responses that are susceptible to modification by conditioning techniques, or skills that can be modified by training.

Somatic therapies

The first antipsychotic drugs, phenothiazines, remain the most effective treatment of schizophrenia. Monoamine oxidase inhibitors, tricyclic antidepressants, and the selective serotonin reuptake inhibitors have been used to treat depression. Bipolar disorders are treated with lithium salts. Anxiety disorders are controlled by anxiolytic drugs, such as the benzodiazepines. Electroconvulsive therapy was once widely used to treat depression. Psychosurgery (including frontal lobotomy) involves operating in the brain to change mental states or behavior. All these treatments have side-effects of differing degrees of severity.

Related topic

Normality and abnormality (L1)

Psychodynamic therapies

Psychodynamic therapies derive from Sigmund Freud's theory of **psychoanalysis**, and the therapy of the same name (see Topic K3). According to psychoanalysis, a patient's illness is caused by unconscious defense mechanisms. These prevent awareness of conflicts and traumatic memories from childhood, and are shown in psychological and bodily symptoms, and sometimes maladaptive behavior. Symptoms do not constitute the illness, but are merely the way it presents itself. If symptoms are removed, they will be replaced by other problems. The central therapeutic technique of psychoanalysis is **free association**. The patient talks about whatever comes into their head, generally uninterrupted by the psychoanalyst, and eventually becomes aware of the unconscious elements. Once they surface, the patient can achieve **insight** into the conflicts, and resolve them. Freud also introduced **dream analysis** as a therapeutic method, arguing that in dreams the unconscious elements revealed themselves in symbolic form to the trained analyst. Everyday slips, choice of words, or small mannerisms were all interpreted as revealing the nature of underlying conflicts. Psychoanalysis is an extremely time-consuming process; some patients spend many years in analysis, attending a number of sessions each week.

A range of later dynamic therapies are all based on the notion of unconscious conflicts. However, they modify the classic technique in different ways. Some are more directive, with the therapist interviewing and discussing, rather than simply listening to the patient. Other dynamic psychotherapies are more likely to deal with immediate issues (symptoms), and will get the client to address relationship and cultural issues, instead of allowing these to continue until (hopefully) the underlying conflicts are resolved.

Humanistic and existential therapies

These therapies, like dynamic therapies, treat the individual as a whole person. However, instead of seeing mental illness as the result of unconscious impulses, they view disorders as resulting from self-deception, or a failure of the person to recognize his or her needs, thoughts, and emotions. The aim of Carl Rogers' **client-centered therapy** is to help individuals to achieve their full potential by examining and understanding themselves and their life. The techniques of client-centered therapy are nondirective: the therapist will not advise or direct the client, but will help them to reach their own understanding and hence resolution of their situation.

Existential therapy sees neuroses as resulting from an inability to understand life and the self. Central to the therapeutic process is a relationship built up between client and therapist, who work together to give meaning to the client's life. There are no specific therapeutic techniques, and methods used vary between existential therapists. The central feature, however, is the therapeutic relationship.

Behavioral therapies

Behavioral approaches are problem based; that is, they focus on symptoms, and do not consider that there are underlying difficulties that need to be addressed before the patient can be helped. **Behavior therapy** considers that the symptoms shown by a patient *are* the illness, and that they are acquired through maladaptive learning. Therapy consists of unlearning these symptoms and/or learning more adaptive responses. This is most easily seen and applied in the case of phobias, which are viewed as anxiety responses associated with inappropriate stimuli. Simple exposure to the stimulus will not extinguish the response, since that will produce anxiety which reinforces the stimulus–anxiety link. Joseph Wolpe introduced the technique of **systematic desensitization** to treat phobias. The principle is that anxiety is prevented from occurring by introducing a competing response, usually muscular relaxation. The patient is helped in constructing an **anxiety hierarchy**, placing situations in increasing order of fearfulness. The patient does not have to confront the hierarchy items physically but, after relaxation, imagines each of them in turn until no anxiety is felt. If necessary, further relaxation instructions are given until the patient is able to confront the top item in the hierarchy without anxiety. Behavior therapists claim a very high transfer of success from imagined to real situations.

Behavior therapists also use **aversion therapy** to treat inappropriate or excessive attraction to people or objects; for example, excessive alcohol consumption, smoking, or sexual 'deviations'. Presentation of the stimulus, either in reality, in pictures, or in imagination, is accompanied by an aversive stimulus such as electric shock or a drug-induced nausea. A third behavioral technique to control behavior is the use of **operant techniques** (see Topic E2). For example, disruptive behavior or failure to perform appropriate behaviors might be treated in an institutional setting by a system of rewards and withholding of rewards: a **token economy**. This is said to work by operant conditioning: rewarded behaviors are more likely to recur, and will replace nonrewarded behaviors.

Cognitive therapies

As we have seen, cognitive theorists suppose that many psychological problems have a basis in inappropriate thoughts or assumptions. **Cognitive therapies** use methods to change these cognitive processes to more realistic or adaptive ones. As different cognitive distortions have been proposed, so a variety of approaches to cognitive therapy exist. **Rational–emotive therapy**, devised by

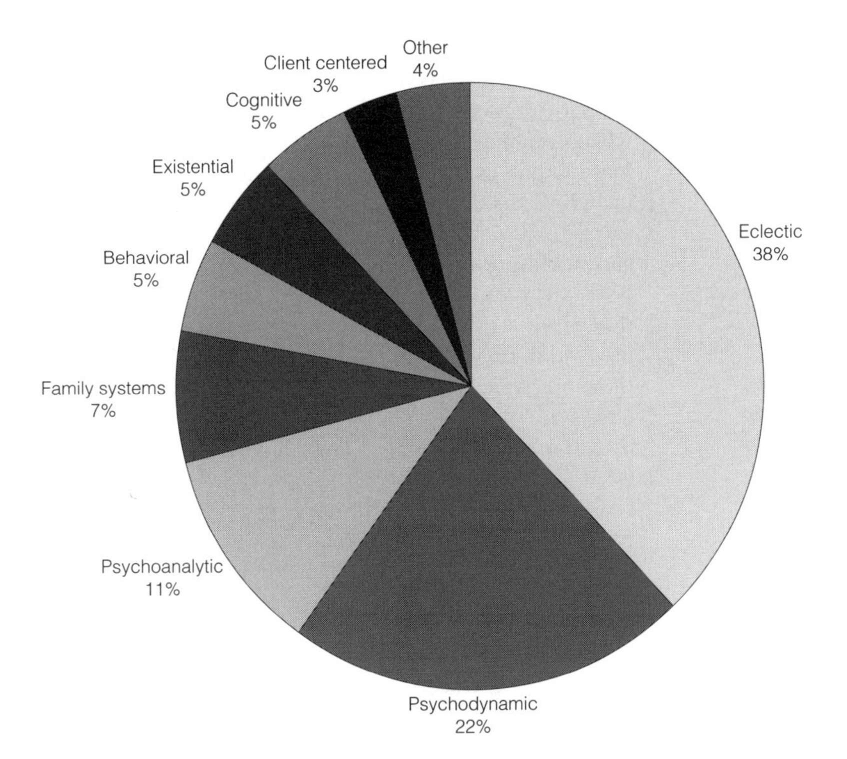

Fig. 1. Approach to treatment taken by contemporary mental health care professionals. The largest group ('Eclectic') do not subscribe to any one theoretical school. Amongst psychologists, a larger percentage use cognitive and behavioral techniques, while amongst psychiatrists, more use dynamic approaches. Adapted from Comer, R.J. (1995) Abnormal Psychology. *W.H. Freeman & Co., New York.*

Albert Ellis in the 1960s, is a directive method of showing the patient that their assumptions and cognitive processes are unsuitable. This is done by directly challenging statements made by patients, and getting them to practice using more appropriate cognitive strategies. Aaron Beck, also starting in the 1960s, developed what he called **cognitive therapy**. This is similar to rational–emotive therapy, and has been most widely used in treating depression. Therapists help clients to recognize the negative thought processes that Beck argues underlie depression (see Topic L3).

In the last quarter of the 20th century, **cognitive–behavioral** therapies were devised, and have become amongst the most widespread approaches. These consider cognitions to be learned responses that are susceptible to modification by conditioning techniques, or skills that can be modified by training. One technique used is **self-instruction training**, in which clients are taught to recognize when they are engaging in maladaptive thought processes (e.g. "I can't cope with this"), and to counteract them by giving themselves more suitable advice (e.g. "easy does it – I can handle this").

Somatic therapies

Many drugs have been used for treating psychopathology. As we have seen in other topics in this section, they were often introduced with no clear idea of how they worked, but later investigations of their actions have helped the understanding of causation. Drugs which alter mental state are called **psychotropic drugs**. The first drugs to be used, in the 1950s, became known as **antipsychotic drugs**. The **phenothiazines** (e.g. chlorpromazine) were originally used to treat allergic reactions, but it was noticed that they also calmed people. Administered to psychotic patients, they reduce the psychotic symptoms. Together with related drugs, they remain the most effective treatment of schizophrenia, best when combined with psychotherapy and community support. Unfortunately, these drugs can have serious side-effects which mimic neurological disease, including shaking and other movement disorders. Recently, a different type of antipsychotic drug (e.g. clozapine) has been introduced, which does not have such widespread side-effects.

Two classes of drug have been used since the 1950s to treat depression, **monoamine oxidase inhibitors** (MAOI, originally tried as a cure for tuberculosis) and **tricyclic antidepressants** (introduced to treat schizophrenia), and these are effective in the majority of cases (Topic L3). Both groups have side-effects, ranging from sleepiness to heart irregularities, high blood pressure, and liver damage, and may interact badly with other drugs and even with food (MAOI drugs). In the 1980s, a new class of drug was developed, the **selective serotonin reuptake inhibitors** (e.g. Prozac), which act on mechanisms prolonging the action of serotonergic synapses. These are far safer, but do have side-effects: up to 30% of recipients report reduced sexual desire. Bipolar disorders are treated with **lithium** salts (see Topic L3).

Anxiety disorders are controlled by **anxiolytic drugs**, such as the benzodiazepines (see Topic L4), also introduced in the 1950s. By the 1980s, these were the most widely prescribed drugs in the USA. They are very effective but induce physical dependence if used for long periods, and they are now mainly used for short-term control of acute anxiety.

Other forms of somatic therapy include **electroconvulsive therapy (ECT)**. This was introduced in the 1930s as a treatment for schizophrenia, but in the second half of the 20th century was very widely used to treat depression, helping about 70% of patients, including those who had not responded to drug treatments. Today, it is used less frequently owing largely to the introduction of more effective drugs, and the spread of cognitive therapies which are very effective in depression. One final form of somatic treatment is **psychosurgery**. Psychosurgery involves operating in the brain to change mental states or behavior. It has its origins in prehistory in the practice of **trephining** (see Topic L1): making holes in the skull of a person who is behaving oddly. In the modern era, it stems from the introduction in the 1930s of **frontal lobotomy**, a procedure that severed the connections between the frontal cortex and the thalamus. This was performed to relieve psychotic symptoms (or as a permanent means of keeping disturbed patients quiet), and it (and later less drastic surgery) was widely carried out before the introduction of antipsychotic drugs in the 1950s. By that time, it became clear that lobotomy was not as effective as originally claimed, and also had drastic side-effects, including seizures, listlessness, and even death. Today, much more focused and precisely located lesions are being tried to relieve dysfunction in specific brain sites, and are reserved for severe cases (including intractable pain) where other available methods have been tried and have failed.

M1 HEALTH BELIEF MODELS

Key Notes

The purpose of health belief models

Health belief models are attempts to structure what we know about the variables which affect our behavior in health-related domains, and the processes that link them. They can be used to inform health interventions, and are valuable tools in the research process. All health belief models share the common problem of finding valid and reliable ways of measuring the variables that they employ.

The basic health belief model

This is a cognitive model that describes the way individuals respond to a health threat in terms of a rational weighing up of the costs and benefits of any suggested course of action. A particular weakness of this model is the omission of any reference to social factors.

The protection motivation model

This model is similar to the basic health belief model with the addition of the variable of self-efficacy. This is the individual's perception of their ability to do something given their personal and social resources. Self-efficacy has proved to be an important factor in predicting health behaviors.

The theory of planned behavior

This is a social psychological model that has proved very useful in predicting some types of behavioral change in the health domain. Its strengths are the inclusion of social and emotional factors as important determinants of behavioral change, and the introduction of behavioral intent as a precursor to action.

The health action process model

This is an adaptation of the theory of planned behavior which imposes a temporal structure on the change process by dividing it into a motivational stage, which culminates in the intention to act; and an action stage, which deals with action plans and the maintenance of change.

The transtheoretical model

This model provides a description of five stages of change derived from psychotherapeutic observations. It provides a useful tool for understanding where a particular client is in the cycle of change, and this enables interventions to be designed appropriately.

Related topics

Attitudes (I1)
Treatment of psychopathology (L7)

The purpose of health belief models

Infectious diseases and social deprivation are no longer the major determinants of ill health and premature death in Western society. New risks have emerged which are strongly linked to our personal behavior, and health psychology is concerned with any behavior that has implications for our health. These include **health-promoting** behavior (such as taking exercise, eating a healthy diet), **health-protecting** behavior (such as having vaccinations, wearing a helmet when riding a bicycle), and **health-impairing** behavior (such as smoking).

Health belief models are attempts to understand the personal and social influences that are related to these behaviors and they have three main purposes: to make explicit our understanding, to predict outcomes, and where appropriate to help design interventions such as health promotion material.

Testing the models through research requires measurement of behaviors, and this can be problematic. Designing a suitable scale to measure, for example, the perceived reward for performing a particular action is not trivial, and the validity of such measures is often open to doubt. Comparison between studies is very difficult when measures are made by different means.

Six broad classes of factors are thought to predict health behavior: **genetic** (known inherited dispositions towards disease); **social** (such as norms, social learning); **emotional** (stress, fear); **perception of symptoms** (pain, irregular heart beat, etc.); **the beliefs of the patient**; and **the beliefs of the health professionals.** Most research has focused on the beliefs of the individual and the ways in which they interact with the other factors to generate, and modify, health behaviors.

The basic health belief model

This basic model assumes that health behavior is a function of five variables (*Fig. 1*). If, in the example shown in *Fig. 1*, an individual perceives they have high susceptibility to coronary heart disease due to family history, then the model predicts that behavior change is likely to occur.

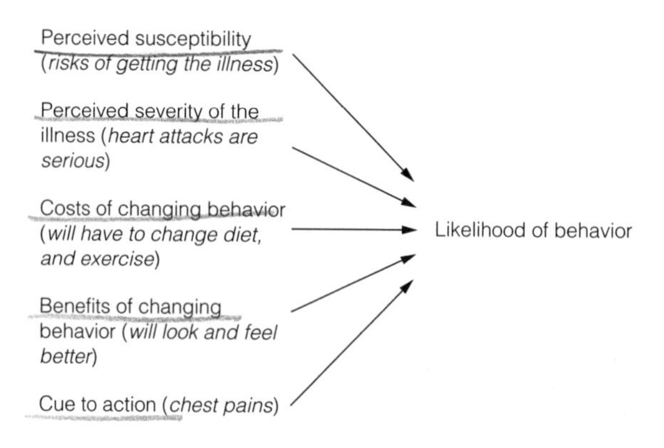

Fig. 1. The basic health belief model.

Although there has been some research support for the model, there have also been criticisms. In particular, it takes no account of individual fears, social factors, or other barriers which prevent action (Topic I1). The model also assumes a degree of rational decision making in weighing up costs and benefits.

The protection motivation model

This model can be seen as a development of the basic model and is particularly designed to understand how people react to fear-inducing communications. Motivation to undertake protective behavior is a function of six variables: the **severity** of the threat; the perceived **vulnerability** to the threat; the **ability** to perform the necessary coping behavior (self-efficacy); the **effectiveness** of the behavior in reducing the threat; the **costs** of carrying out the protective behavior; and the **rewards** associated with the **maladaptive** behavior. Self-efficacy is central to this model. If self-efficacy is low, then the individual is very unlikely to carry out the behavior. Similarly, if the costs and rewards are high, the probability of the behavior occurring is reduced. Like the basic model, this model also assumes a degree of rational information processing and has no role for social factors. However, there has been some research support for it. For example, work on breast self-examination confirmed that the perceived effectiveness of the behavior, the severity of the disease, and individual self-efficacy are the best predictors of the intention to carry out the examination.

The theory of planned behavior

The theory of planned behavior is a general model concerned with the relationship between attitudes and behavior. The model emphasizes that an **intention**, the plan to carry out a particular behavior, is the most important determinant of that behavior. An intention to act is the result of two parallel cognitive processes, one concerned with beliefs about the behavior and the other with social norms. Self-efficacy operates as a third factor (*Fig. 2*). This model has been applied to behaviors such as contraceptive use, smoking initiation and cessation, and participation in exercise; it has demonstrated good predictive ability.

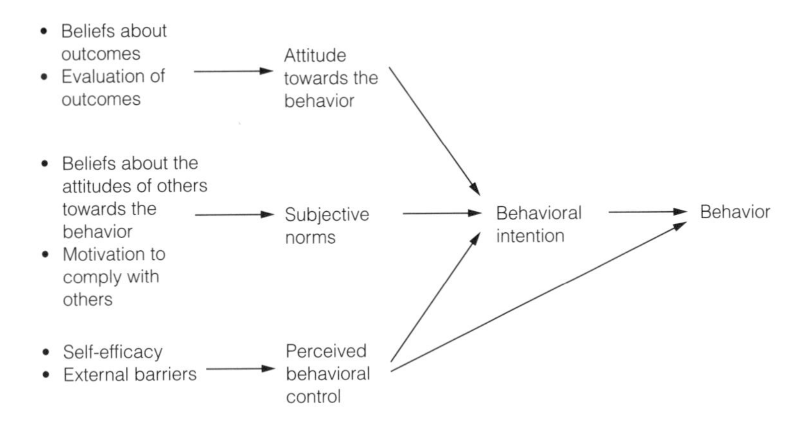

Fig. 2. Components of the theory of planned behavior.

The health action process model

This model is similar to the theory of planned behavior, but includes a **temporal dimension**, so that changes are described as taking place at a **motivational** stage followed by an **action** stage. The appraisal of threat, outcome, and self-efficacy occurs in the motivational stage, and leads to the intention to act or not. The action stage is an integration of cognitive and situational factors (*Fig. 3*). In this scheme, the appraisal of threat-and-outcome expectancies comes first, then the judgement of self-efficacy. If this leads to an intention to act, the individual will consider in more detail what to do in order to achieve the goal. This stage is directly influenced by perceptions of self-efficacy and situational factors, but not by outcome expectancies. Once a course of action has been started it must be maintained; behavior which receives social support is more likely to be sustained.

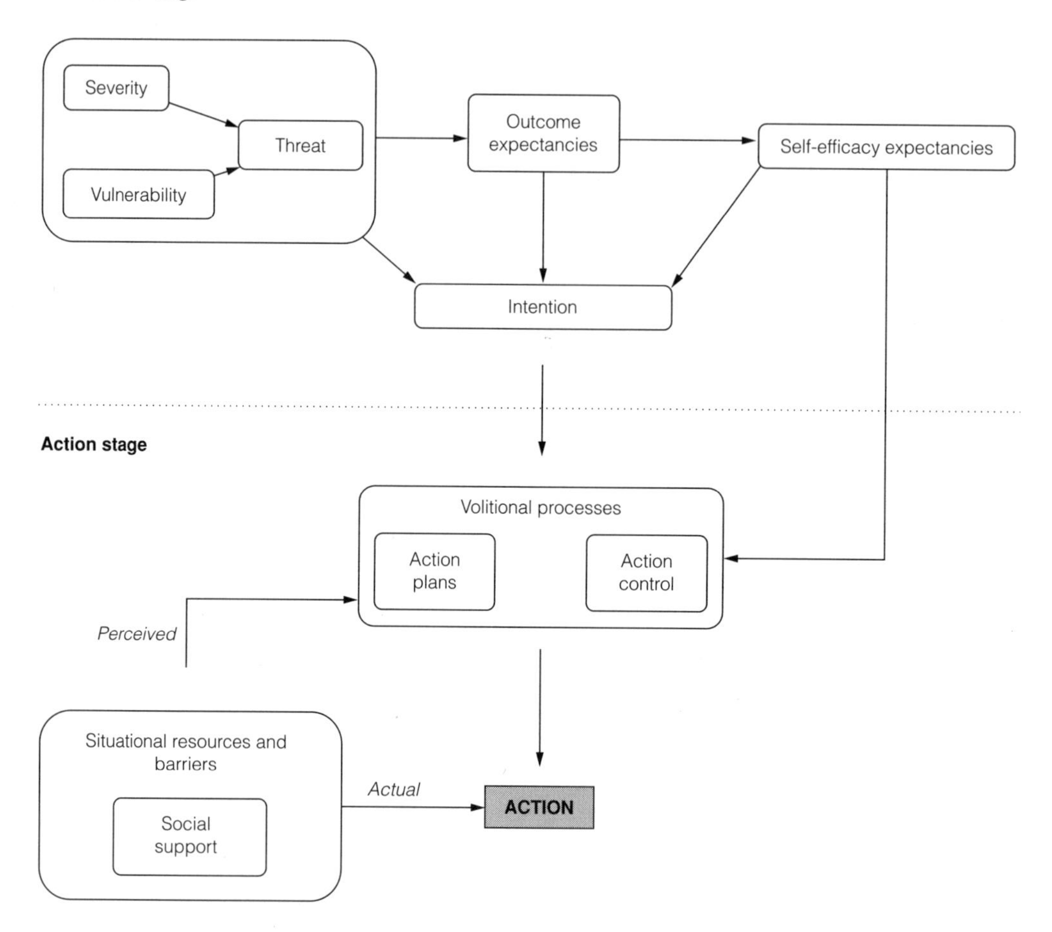

Fig. 3. The health action process model.

The transtheoretical model

This model was based on data collected from patients in psychotherapy. Five stages are suggested:

- **precontemplation**: not intending to make any changes;
- **contemplation:** considering a change, but not actively doing anything;
- **preparation**: making small changes;
- **action**: actively undertaking new behavior to overcome the problem;
- **maintenance**: sustaining the changes over time.

These stages do not take place in the linear fashion assumed by other models, and an individual may move onward and back between stages several times before progressing further. Since relapse is a common feature of some behavior change, such as smoking cessation and losing weight, the model has been described as a 'revolving door' (*Fig. 4*). This model has been tested in the context of several health-related behaviors, and data support the existence of the five stages of change. The model predicts that the best results are obtained if interventions are tailored to match the client's stage of change, instead of treating everyone in the same way (Topic L7).

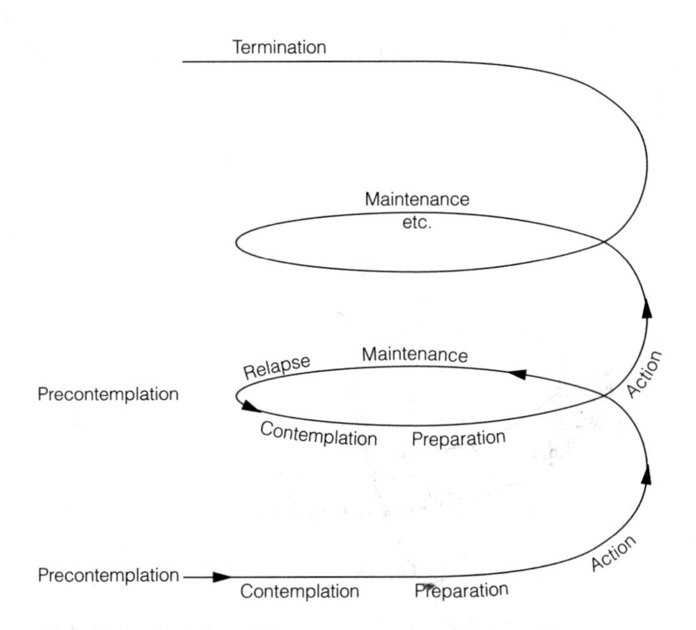

Fig. 4 The transtheoretical 'revolving door' model of stages of change. Redrawn from Prochaska, J.O. and Norcross, J.C. (1994) Systems of Psychotherapy: A Transtheoretical Analysis *(3rd Edn). Reprinted with permission of Wadsworth, a division of Thomson Learning.*

M2 SMOKING AND ALCOHOL ABUSE

Key Notes

Two approaches to addiction

Addictive behaviors are sometimes regarded as diseases that have their origins in physical or psychological abnormalities. The psychological view is that they are generally acquired and maintained through social learning processes that depend on expectations and rewards which are socially transmitted.

The stages of substance use

Four stages of substance use can be distinguished. When someone is trying to change their behavior, it is particularly useful to consider ways in which relapses can be anticipated and avoided.

Promoting cessation

Self-help groups and health promotion campaigns can help people modify their own behavior, but the most effective methods involve programs that are tailored to the individual. These may include cognitive strategies and behavior modification techniques based on learning theory.

Related topics

Associative learning (E1)
The psychoanalytic approach (K3)
Treatment of psychopathology (L7)

Two approaches to addiction

Early workers regarded alcoholism as a **disease** that was *caused by* alcohol. This is sometimes called the **1st disease hypothesis**. Focus later shifted to the individual as the source of the problem and the **2nd disease hypothesis** was developed. Explanations centered on physical or psychological **abnormalities**, or acquired **dependency**. All three have in common the ideas that addiction is an illness that is irreversible, that you either have it or do not, and that it must be treated through abstinence. Suggested physical abnormalities included a genetic predisposition to becoming an alcoholic, and endocrinological differences in the way in which alcohol is metabolized. Psychological abnormalities tended to be based on the Freudian concept of oral fixation (Topic K3). The acquired dependency explanation of alcoholism was based on four stages of response to the addictive substance: acquired tissue tolerance, adaptive cell metabolism, withdrawal and craving, and loss of control. The disease approach is focused on alcoholism for the historical reason that alcohol has been regarded as a health risk for much longer than has tobacco.

An alternative approach is to consider addictive behaviors as being acquired according to the mechanisms of **social learning theory.** These are generally forms of **classical** conditioning, which depend on associations between internal and external cues; or **operant** conditioning, where the **rewards** and **punishments** are of a social nature (**social reinforcements**), and the **contingencies** are determined by the dynamics of the social system within which the individual operates (Topic E1). A person will try out a new behavior in the **expectation** of a certain outcome. If that expectation is fulfilled, behavior is reinforced and the probability of it being repeated is increased. In the case of alcohol and tobacco use, the expected outcomes may be of both an internal (what it will make me feel like) and an external (what people will think of me) nature, but the expectations themselves are **transmitted** by social means. This may involve direct observation of other people's behavior and the rewards they appear to obtain, or indirect observation such as conversation or the mass media. In some cases, the behavior is learnt simply by way of this kind of observation, a process called **observational learning** or **modeling**. If the social processes involve others who are held in high esteem then their power is enhanced.

The stages of substance use

Four stages of substance use have been described: **initiation**, **maintenance**, **cessation**, and **relapse**. **Initiation** depends on expectations and reinforcement. In the case of smoking, the expectation is that it will be a sociable activity and will calm the nerves; that is, act as a stress-reducing agent. In the case of alcohol, the expectation is that it will ease tension and reduce anxiety. If these expectations are reinforced, there is a high probability that the behavior will become established. The **maintenance** of the behavior requires only intermittent reinforcement, and this is likely to be provided by the pharmacological properties of alcohol and nicotine, as well as social conditions. Extinction without external intervention is extremely unlikely.

The disease hypothesis predicts that **cessation** requires abstinence; all that is required is that the person stops the undesirable behavior. In contrast, health belief models indicate that the individual must have strong beliefs, a cue to action, suitable social support, and a belief in their own self-efficacy before they are likely to stop smoking or drinking. Learning theory predicts that successful quitting will depend on reinforcement being taken away from the old behaviors and transferred to the new ones.

Although many people are successful at stopping smoking or changing their drinking behavior, **relapse** rates are high. The pattern of relapses is similar across a range of substances (*Fig. 1*).

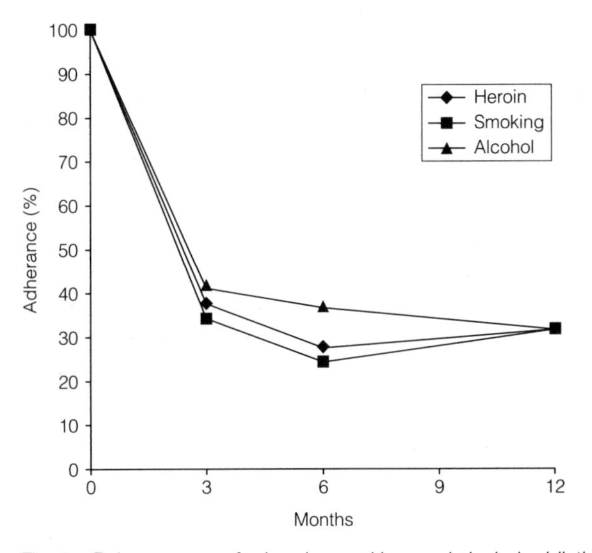

Fig. 1 Relapse curves for heroin, smoking, and alcohol addiction. From Hunt, W.A. et al. *(1971) Relapse rates in addiction programs.* J. Clin. Psychol. ***27****, p. 456. Reprinted by permission of John Wiley & Sons, Inc.*

The **relapse prevention model** emphasizes the importance of coping behavior and negative outcome expectancies in **high-risk** situations, which usually involve social pressure, negative emotions, or interpersonal conflict. **Lapses** reduce self-efficacy and raise positive outcome expectancies. The lapse may remain an isolated event, or the individual may return to the old behavior patterns (*Fig. 2*).

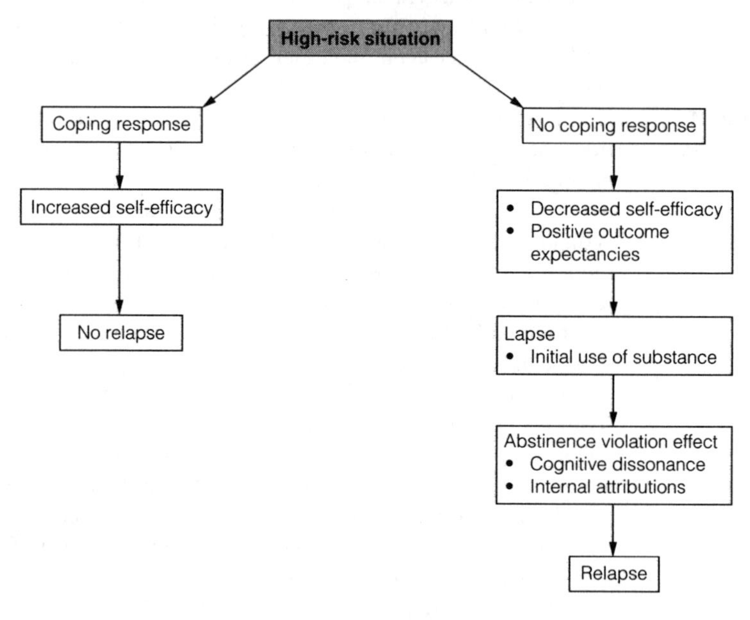

Fig. 2. The relapse process. From Marlatt, G.A. and Gordon, J.R. (1985) Relapse Prevention. *Published by Guilford Press, New York.*

The transition to relapse depends on the handling of the **abstinence violation effect**. Internal attributions for failure that lower perceptions of self-efficacy ("It's all my fault, I'll never manage to quit") increase the chances of a relapse. External attributions (e.g., blaming the lapse on the presence of others) reduce the chances of a relapse.

Promoting cessation

Self-help groups can provide valuable support for someone who wishes to give up smoking or change their drinking habits, and public health programs can help change health beliefs and provide a cue to act. However, the most effective intervention is an individual program of behavior change. In the case of smoking cessation, this is likely to include the following five stages:

- establishing motivation;
- identifying triggers to smoke;
- cutting down;
- stopping;
- relapse prevention.

There are several alternative strategies, based on learning theory. In **aversion therapy**, the individual is given a drug which makes them vomit when they drink alcohol. After a small number of pairings, the smell of alcohol will elicit the conditioned response without the drug (Topic L7). However, it does not address the reasons for the problem behavior. **Contracting** requires the individual to make a contract with their therapist, or a friend or partner, which establishes a set of rewards and punishments for their behavior. This is often successful in encouraging cessation, but relapses are common after the contract has expired. **Cue exposure** enables the individual to develop more appropriate behavior in the presence of environmental triggers. This technique is often used in the 'cutting down' stage mentioned above.

M3 Stress

Key Notes

Stress as a physiological response

The **general adaptation syndrome** describes the response of the body to a stressful threat. Repeated exposure to stressors can lead to physiological changes that contribute to the development of chronic illnesses.

Stress as a response to life events

The need to adapt to changes in our life can be a major source of stress. Measurements of the frequency and intensity of these events, both large and small, have revealed a relationship between them and health status.

The transactional view of stress

Viewing stress as a passive response to events around us ignores the fact that most of the time we are active participators in those events. The transactional view is that stress involves appraisal of threats and resources, and the development of effective coping strategies to deploy resources effectively.

The health impact of stress

Aside from those problems that are mediated by physiological effects, stress increases the frequency of other behaviors that compromise health including smoking, alcohol and caffeine consumption, and eating badly. In addition, stress increases the risk of accidents.

Stress management

It is possible to teach coping processes that enable individuals to deal effectively with stress. The main techniques involve relaxation to deal with physical effects, and cognitive appraisal to monitor stress-inducing cognitions and replace them with adaptive alternatives.

Related topics

Approaches to motivation (B1)   Treatment of psychopathology (L7)

Stress as a physiological response

Stress results from an interaction between physiological, psychological, and environmental processes. It is common to refer to stress as a **response** to a **stressor**. The first modern account of stress was given by Selye, whose concept of the general adaptation syndrome (GAS) was based on Cannon's conception of 'flight or fight'. The GAS has three stages:

(1) **alarm**: increase in activity when exposed to the threat. Sympathetic nervous system processes drive this;
(2) **resistance**: coping and reversing the effects of the alarm. Hormonal changes mediate this;
(3) **exhaustion**: occurs if resistance is no longer possible and adaptive resources are depleted.

Repeated exposure to stressors may harm the individual indirectly, since repeated activation of the sympathetic nervous system can suppress the activity of the **immune system** and may also contribute to the development of chronic conditions such **as coronary heart disease**, **arthritis**, **kidney disease**, and certain

gastrointestinal disorders. Selye called these **'diseases of adaptation'** since they are the result of continual challenge by stressful circumstances (Topic B1).

Stress as a response to life events

Stress and stress-related changes may be viewed as a **psychological response**, or adjustment, to **life changes**. The schedule of recent events (SRE) is an attempt to understand the psychological impact of major life events. Individuals are given a score according to how many events on a checklist of 43 positive and negative items they have experienced in (typically) the 6 months prior to completing the inventory. This score is used as an index of the intensity of their stress response. Although crude, some links have been found between this score and health status. Later modifications (SSRE) assign **weighted** scores to reflect the **severity** of events (*Table 1*).

Table 1. Extract from the social readjustment rating scale

Number	Event	Score
1	Death of spouse	100
2	Divorce	73
3	Marital separation	65
7	Getting married	50
12	Pregnancy	40
22	Change in responsibilities at work	29
30	Trouble with boss	23
42	Christmas	12
43	Minor violation of the law	11

An attempt to measure stress on a day-to-day basis uses minor events or **hassles**; these are defined as the 'irritating, frustrating, and distressing demands that to some degree characterize everyday transactions with the environment.' Examples of hassles are given in *Table 2*. The hassles scale yields scores which reflect both the frequency and intensity of stressful incidents.

Table 2. Sample items from the hassles scale

Item
Losing things
Troublesome neighbors
Nonfamily members living in your house
Trouble-making decisions
Too many responsibilities
Concerns about owing money

The life events approach has been criticized on four grounds. First, it is the individual's **interpretation** of their life events that determines whether they are stressful or harmless. Important attributes include how **desirable** an event was, how much **control** they had over the event and how much **adjustment** was required as a consequence.

Second, life events tend to be judged some time after they have happened, often when the individual has become ill. Thus, what is observed is the relationship between **self-reported** stress and **treatment-seeking** behavior. It has been suggested that both these reflect the stable personality attribute of **negative**

affectivity, which appears to correlate highly with symptom reporting, but is unrelated to objective health indicators. The implication of this is that the relationship between life event measures and health may be an artifact. Some **prospective studies** have supported the predictive nature of the scale, while others are less conclusive.

Third, life experiences may interact with each other, and the manner of the interaction (e.g. additive, multiplicative) in an individual case would be difficult to assess. The life events approach is based on the assumption of additivity.

Fourth, some life events on the scale (such as personal injury or illness, sexual difficulties) also reflect health status. These would artificially inflate the correlation between stress and health measures.

The transactional view of stress

The **transactional** view describes stress in terms of the interplay between the person and environmental variables, making use of the concepts of primary and secondary **cognitive appraisal**, and **coping**. An individual employs primary appraisal to sum up a situation in terms of its significance for their well-being, and secondary appraisal to evaluate their personal resources to deal with its demands. When the individual perceives that their resources are inadequate, they experience stress and must act to bring the situation under control and/or deal with their emotional reactions to it. The effort and uncertainty involved in achieving effective control may increase the total stress of the situation, and it has been suggested that people act so as to minimize the **total amount of stress** to which they are exposed; this is called a **minimax** model.

The health impact of stress

There are effects of stress on health apart from those that are mediated by physiological processes. In general, stress results in an increase in behaviors which have negative effects on health. Stress appears to a be a factor in determining when adolescents begin to **smoke**, and is also implicated in **relapse** after a period of nonsmoking. Smokers who are exposed to increases in stress report an increased urge to smoke and that smoking reduces the experience of stress. Indeed, smoking is one of the most commonly reported ways of dealing with stress. There is a complex link between stress and **alcohol** use. People may consume alcohol for its mood-changing properties, but stress seems to reduce the subjective impact and thus people under stress have to drink more to get the same effect. In reality, alcohol use renders the individual less able to cope. Stress is also associated with increased **caffeine** consumption, poor attention to **diet**, and reduced **exercise**.

Stress leads to an increased risk of **accidents** in the home, at work, and on the roads, but its role as a causal agent is complex and not fully understood. Behavioral consequences of stress may include poor concentration, slowed reaction times, and poor physical coordination and in dangerous or unforgiving conditions these can result in actions having unintended outcomes. Emotionally threatening situations are often dealt with by distortion or misperception, and this may in turn lead to the misjudgment of risk.

Stress management

Individuals manage stress by employing **coping processes**. These are described as either **problem focused** or **emotion focused** since the former are directed at the management of the problem and the latter at reducing the distress. Problem-focused behavior includes **monitoring** stress levels, **structuring** the resources available and **planning** their use, improving social skills in order to gain increased social support, and to manage stressors through negotiation and improved communication. **Avoidance** and **denial** are common emotion-focused behaviors. Both involve the distortion of reality, which is not necessarily maladaptive. Humor can be an effective emotion-focused strategy, as can exercise and the use of relaxation techniques. The **effectiveness** of coping processes is a good predictor of the health consequences of stressors.

Stress management programs are designed to teach coping strategies. There are two broad approaches, one based on **relaxation** training and the other on **cognitive** techniques. The goal of relaxation skills is to be able to reduce physical tension, and other symptoms of over-arousal, throughout the day. A useful secondary effect is that they often increase a person's **perceived control**. The whole process typically involves **learning** the skills, learning how to **monitor** tension, and **integrating** the practices into daily life. **Cognitive** techniques teach the individual to replace stress-inducing cognitions with more adaptive alternatives (Topic L7).

M4 CORONARY HEART DISEASE

Key Notes

The nature of coronary heart disease

Coronary heart disease (CHD) is a generic term covering diseases that affect the blood supply to the heart. The two most important diseases are angina and heart attack. These diseases typically have multiple causes, some of which have a strong psychological component.

Type A behavior

Type A behavior is in fact a cluster of behavioral traits that have been implicated in the etiology of CHD. Prominent characteristics include time urgency, rapid speech, free-floating hostility, and impatience and general restlessness. The degree to which an individual exhibits these behaviors is often called their Type A profile.

Assessing type A behavior

The structured interview is a technique designed to assess type A behavior by direct questioning and observation. The interview must be scored by at least three people, which makes it expensive to complete. The Jenkins activity survey is a self-report inventory that is convenient to administer, but is not a very good measure of some components of the profile. The hostility component of the type A pattern is usually measured by separate personality scales.

Changing the type A pattern

Changing type A behavior can reduce the risk of CHD. Popular techniques include anxiety management training and visual–motor behavioral rehearsal. Patients are taught to generate and practice new and adaptive responses and are provided with self-help exercises.

Essential hypertension

This is the term for raised blood pressure that occurs as a disease in itself (hence 'essential') rather than as a symptom of other diseases. Essential hypertension is a risk factor for CHD and for other diseases such as strokes. Blood pressure can be reduced by means of medication, but some drugs have unpleasant side-effects, and psychological interventions can be more acceptable.

Related topic

Personality assessment (K5)

The nature of coronary heart disease

The coronary arteries provide oxygen for the heart muscle itself. If fatty deposits narrow these arteries, the heart muscle may fail to get sufficient oxygen. This condition is called **ischemia**, and can result in intense pain called **angina.** If the blockage of blood flow causes actual damage to the heart tissue, the result is a heart attack or **myocardial infarction**. Angina and myocardial infarction are the two main forms of **coronary heart disease (CHD)**.

Major factors are implicated in the etiology of CHD; these include elevated serum cholesterol, hypertension, inactivity, obesity, smoking, age, diabetes, and family history. However, each of these separately accounts for a tiny fraction of the risk. Taken together, even the best combination will fail to identify, prospectively, most new cases.

Type A behavior

A specific pattern of behavior that is highly relevant to the development of CHD is known as **type A** behavior. The type A profile includes time urgency, rapid speech, free-floating hostility, impatience and general restlessness, and 'polyphasic activity'; that is, thinking about more than one thing at a time or engaging in more than one task at a time. The anger and hostility components of the profile are now thought to be the most important in predicting the development of CHD.

Assessing type A behavior

There are two main instruments for measuring the extent of Type A behavior: the **structured interview** and the **Jenkins activity survey**. The structured interview does not assume that people have a lot of insight into their behavior. The protocol not only contains questions that assess a person's behavior, but also questions that *provoke* Type A behaviors. The style of speaking, reactions to pauses by the interviewer, and nonverbal behaviors are also noted. Interviewers have to be carefully trained to administer this interview, and several independent raters usually score the responses using tape recordings of the interview.

The Jenkins activity survey is a **self-report questionnaire** with questions that assess the elements of the profile. It is not a good measure of hostility, impatience, and competitiveness; and although it is much more convenient to administer it is not as good a predictor of CHD as the structured interview.

Reviewing research on the relationship between the type A behavior pattern and CHD is highly problematical because of the range of methodologies that have been used, the mix of cross-sectional and prospective studies, and the range of definitions of the type A construct. **Meta-analyses** of published work have come to conflicting conclusions, but the most stringent studies confirm that the hostility component of the type A profile is the best predictor of CHD.

Measuring hostility is not without problems, but the trait hostility scale of the revised **Minnesota multiphasic personality inventory**, or the Cook and Medley **hostility inventory** which was derived from it, are frequently used (Topic K5). High hostility scores are associated with increased risk of CHD, but the degree to which hostility is expressed is important and high '**anger-out**' scores are more predictive. Extreme responses include getting angry when a person's belongings are disturbed or when they cannot find something.

Changing the type A pattern

Given the difficulties of defining the type A pattern adequately, and the inconsistent data on its relationship with CHD, it is perhaps surprising that there is good evidence to show that changing the behavior can result in reductions in the risk level for CHD. To be effective, interventions have to be quick and show rapid results since Type A people are impatient. One such program includes **anxiety management training** and **visual–motor behavioral rehearsal**. In the former, patients learn to recognize the early signs of stress build-up, and to employ relaxation techniques to reduce stress reactions. In the latter, patients learn to generate and practice new and adaptive responses using imagery as a substitute for real-life situations. In one study, 83% of treated participants showed improvements in CHD risk factors compared with controls. Other research has confirmed that the inclusion of a cognitive–behavioral component alongside psychological counseling greatly improves effectiveness. It is also useful to provide patients with self-help 'drills'. These are short and simple exercises for each day of the week, which help patients maintain their techniques. Examples include 'practice smiling', 'recall pleasant memories for 10 minutes', and 'eat more slowly today'.

Changing behaviors associated with increased risk of CHD remains problematical whether the intervention is at the level of the individual or society at large. The three main factors that are under individual and voluntary control are smoking, diet, and exercise. Health belief models indicate that change will only occur when individuals see themselves as vulnerable to disease, and also see the benefit of change (expressed in the most general terms) as outweighing the costs. It is difficult to convince people that their behavior in, say, adolescence will seriously affect their health 30 or 40 years later. Given the perceived lack of vulnerability to disease, the costs of changing are rarely seen to outweigh the benefits.

Essential hypertension

Hypertension is an elevation of the arterial blood pressure. Most hypertension is classified as **essential hypertension** meaning that it is a primary disease, rather than part of some other disease process (e.g. arteriosclerosis) or the side-effect of medication such as oral contraceptives. Hypertension is a risk factor for CHD as well as being associated with retinal damage, strokes, and kidney disease. Lowering blood pressure by means of medication reduces many of the problems but there are significant side-effects.

Despite the serious nature of hypertension there is no clear pattern of symptoms that identify it, and many patients have no symptoms at all. They then have to accept that they will feel *worse* with medication than without, and hypertensive patients are notoriously bad at complying with medical advice. Nondrug treatments for hypertension can be very effective. A range of relaxation techniques is available, and **autogenic** training using **biofeedback** has also proved useful. In the **Menninger technique,** for example, thermal biofeedback is used to teach patients how to warm their hands and thereby reduce peripheral resistance to blood flow, one of the main sources of high blood pressure. They are also taught yoga breathing techniques for relaxation. As patients master the exercises, they are able to take lower doses of medication.

M5 Doctor–patient interaction

Key Notes

Doctor–patient communication

Communication between doctor and patient is not always effective. It can be compromised when the doctor tries to convey information by using language that the patient does not fully understand, and when the patient has unrealistic beliefs about their health.

Patient satisfaction

Patient satisfaction is strongly influenced by the doctor's communication style. Informal, affiliative styles lead to greater patient satisfaction than do formal, business-like styles.

Compliance

Compliance (or adherence) refers to the extent to which patients follow the medical advice they are given. It is difficult to measure compliance, but there is no doubt that low compliance is a widespread problem. Compliance is strongly influenced by the effectiveness of doctor–patient communication and by patient satisfaction.

Related topic Communication (J1)

Doctor–patient communication

Good communication between doctor and patient is essential for effective healthcare. It is important that doctors talk to their patients using language that they understand, that they tell them everything they need (or want) to know, and that they check this information has been assimilated. It has been suggested that there are three principal reasons why doctor–patient communication may fail: (1) the doctor assumes some basic medical knowledge which the patient does not in fact have; (2) patients have misconceptions which interfere with proper comprehension; (3) the doctor tries to convey material which is too difficult for the patients to understand, or uses language which is ambiguous. Evidence in support of the first two points includes the finding that 50% of participants in one study did not know where the kidneys, heart, or lungs were located; and that people frequently have erroneous ideas about common aspects of diet and medication, and a poor understanding of both the causes and the seriousness of illness. A good example of the third type of problem arises when doctors give instructions about taking medicines. For example, some patients who are told to take medication 'four times daily' will assume that this means 'every 6 hours', and will set their alarm clock to take one dose in the early hours of the morning. Others will assume the instruction to mean that the medication should be taken at equally spaced intervals during waking hours.

Research on the uptake of information has been carried out using exit interviews at doctors' surgeries. Data show that patients' knowledge of the treatment they were to have was seriously deficient. In one study, half of the patients did not know for how long they should be taking their medication and 20% did not know what the treatment was for. An important reason for the lack of comprehension was the failure of the doctor to give clear and explicit information. Even when patients do understand what they are told they tend to forget a lot of it. Instructions and advice are more likely to be forgotten than anything else, but patients will remember: (1) what they are told first, (2) what they consider to be the most important, and (3) items which the doctor has been at pains to repeat. The age and intelligence of the patient have little bearing on what is forgotten after a consultation (Topic J1).

Patient satisfaction

Effective doctor–patient communication has a strong influence on whether the patient is satisfied with the outcome of a consultation and, in turn, on the degree to which they adhere to the advice they are given. A doctor's communication style is an important predictor of patient satisfaction. Friendly and informal (**affiliative**) styles are associated with higher patient satisfaction than are formal, business-like interactions. Furthermore, research with mothers who rated their satisfaction with the way doctors treated their children has demonstrated the importance of communicator variables, especially nonverbal cues such as eye contact and posture. The more a mother perceives her doctor as warm and caring, and as understanding her concern, the more satisfied she will be with the doctor. Conversely, if a doctor *fails* to provide information which the mother expected then satisfaction is reduced. Such information might include the likely progression of symptoms during recovery or advice on avoiding a recurrence of the illness.

Compliance

The degree to which patients follow the advice they are given is referred to as **compliance**, although the term **adherence** is often used since it reflects the voluntary nature of the behavior. The problem of low adherence is widespread, and it can often cause a serious breakdown in the treatment process. It also has serious financial implications for health care services; many millions of pounds are wasted because of noncompliance to prescribed drugs.

Patient satisfaction and doctor–patient communication are both major predictors of adherence; they have been linked in Ley's cognitive hypothesis model (*Fig. 1*). This scheme places emphasis on the transfer of information from doctor to patient, and predicts that enhanced patient knowledge and satisfaction will lead to increased compliance. Research on the model suggests that it is also necessary to take account of patient health beliefs, their perceived social support, and the life-style disruption caused by their adherence.

Unfortunately, it is difficult to measure adherence reliably. Most research indicates that patients do not report their own nonadherence accurately, which

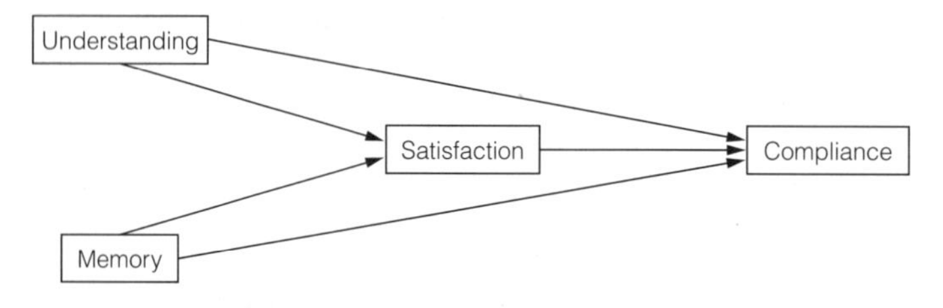

Fig. 1. Ley's cognitive hypothesis model of compliance.

is not surprising since it means admitting a negative behavior to the doctor. Measuring traces of medication in the blood or urine is not very reliable either; nor is it a practical way of monitoring individual patients. Attempts to improve adherence confirm that patients are more likely to follow the advice of doctors who are good communicators. They are also more likely to remember the information if it is properly organized, and using visual aids (including printed material) to complement spoken instructions can improve recall by a factor of six.

FURTHER READING

Section A

Carlson, N.R. (1999) *Foundations of Physiological Psychology*, 4th edn. Allyn and Bacon, Needham Heights, MA.

Rose, S. (1999) Précis of lifelines: biology, freedom, determinism. *Behav. Brain Sci.*, **22**, 871–921. (With peer commentaries.)

Robertson, I.H. and Murray, J.M.J. (1999) Rehabilitation of brain damage: brain plasticity and principles of guided recovery. *Psychol. Bull.*, **125**, 544–575.

Thompson, R.F. (2000) *The Brain: A Neuroscience Primer*, 3rd edn. Worth, New York.

Section B

Beck, R.C. (2000) *Motivation: Theories and Principles*, 4th edn. Prentice-Hall, Upper Saddle River, NJ.

Mazur, A. and Booth, A. (1998) Testosterone and dominance in men. *Behav. Brain Sci.*, **21**, 353–363. (With peer commentaries.)

Rolls, E.T. (2000) Précis of the brain and emotion. *Behav. Brain Sci.*, **23**, 177–234. (With peer commentaries.)

Wagner, H.L. (1999) *The Psychobiology of Human Motivation*. Routledge, London.

Woods, S.C., Schwartz, M.W., Baskin, D.G. and Seeley, R.J. (2000) Food intake and the regulation of body weight. *Annu. Rev. Psychol.*, **51**, 255–277.

Section C

Bowmaker, J.K. and Dartnell, H.J.A. (1980) Visual pigment of rods and cones in a human retina. *J. Physiol.*, **298**, 501–511.

Kalat, J.W. (1997) *Biological Psychology*, 6th edn. Brooks/Cole, Pacific Grove, CA.

Melzac, R. and Wall, P.D. (1965) Pain mechanisms: A new theory. *Science*, **150,** 971–979.

Norwich, K.H. (1984) On the theory of Weber fractions. *Perception Psychophys.*, **42**, 286–298.

Stevens, S.S. and Galanter, E. (1957) Rating scales and category scales for a dozen perceptual continua. *J. Exp. Psychol.*, **57**, 377–411.

Taylor, W., Pearson, J., Mair, A. and Burns, W. (1965) Study of noise and hearing in jute weaving. *J. Acoustical Soc. Am.*, **38**, 113–120.

Section D

Beck, J. (1966) Effects of orientation and of shape similarity on perceptual grouping. *Perception Psychophys.*, **1,** 300–302.

Cherry, E.C. (1953) Some experiments on the recognition of speech with one and two ears. *J. Acoustical Soc. Am.*, **25,** 975–979.

Coren, S., Ward, L.M. and Enns, J.T. (1994) *Sensation and Perception*, 4th edn. Harcourt Brace, Fort Worth, FL.

Gibson, J.J. (1968) What gives rise to the perception of motion? *Psychol. Rev.*, **75,** 335–346.

Gordon, I.E. (1990) *Theories of Visual Perception*. John Wiley & Sons, Chichester.

Gregory, R.L. Visual illusions. *Scientific American*, **219,** 66–76.

Kanizsa, J. (1976) Subjective contours. *Scientific American*, **234,** 48–52.

Moray, N. (1959) Attention in dichotic listening: Affective cues and the influence of instructions. *Quart. J. Exp. Psychol.*, **11**, 56–60.

Rock, I. and Palmer, S. (1990) The legacy of Gestalt Psychology. *Scientific American*, **263,** 84–90.

Schiffman, H.R. (1996) *Sensation and Perception: An Integrated Approach*, 4th edn. John Wiley & Sons, New York.

Triesman, A.M. and Gormican, S. (1988) Feature analysis in early vision: evidence from search asymmetries. *Psychol. Rev.*, **95,** 15–48.

Section F

Baddeley, A. (1997) *Human Memory: Theory and Practice*, revised Edn. Psychology Press, Hove, UK.

Collins, A.M. and Loftus, E.F. (1975) A spreading activation theory of semantic processing. *Psychol. Rev.*, **82**, 407–428.

Evans, J. St.B. T., Newstead, S.E. and Byrne, R.M.J. (1993) *Human Reasoning: The Psychology of Deduction*. Lawrence Erlbaum Associates, Hove.

Gick, M.L. and Holyoak, K.J. (1983) Schema induction and analogical transfer. *Cognitive Psychol.*, **15,** 1–38.

Korial, A., Goldsmith, M. and Pansky, A. (2000) Toward a psychology of memory accuracy. *Annu. Rev. Psychol.*, **51**, 481–537.

Kosslyn, S.M., Ball, T.M. and Reisser, B.J. (1978) Visual images preserve metric spatial information: Evidence from studies of image scanning. *J. Exp. Psychol.: Human Perception and Performance*, **4**, 1–20.

Lindsay, D.S. and Briere, J. (1997) The controversy regarding recovered memories of childhood sexual abuse. *J. Interpersonal Violence*, **12**, 631–647.

Pavio, A. (1969) Mental imagery in associative learning and memory. *Psychol. Rev.*, **76**(3), 241–263.

Pinker, S, and Prince, A. (1988) On language and connectionism: Analysis of a parallel distributed processing model of language acquisition. *Cognition*, **29**(1), 73–194.

Shepard, R.N. and Metzler, J. (1971) Mental rotation of three-dimensional objects. *Science*, **171**, 701–703.

Sternberg, R.J. (1996) *Cognitive Psychology*. Harcourt Brace, Fort Worth, FL.

Tversky, A. and Kahneman, D. (1981) The framing of decision and the psychology of choice. *Science*, **211**, 453–458.

Section H

Berk, L. (1998) *Development Through the Lifespan*. Allyn and Bacon, Needham Heights, MA.

Bjorklund, D.F. (1997) The role of immaturity in human development. *Psychol. Bull.*, **122**, 153–169.

Flavell, J.H. (1999) Cognitive development: Children's knowledge about the mind. *Annu. Rev. Psychol.*, **50**, 21–45.

Littard, A. (1998) Ethnopsychologies: Cultural variations in theories of mind. *Psychol. Bull.*, **13**, 3–32.

Rice, E.P. (1998) *Human Development*, 3rd edn. Prentice Hall, Upper Saddle River, NJ.

Section I

Aronson, E., Wilson, T.D. and Akert, R.M. (1999) *Social Psychology*, 3rd edn. Longman, New York.

Baron, R.A. and Byrne, D. (2000) *Social Psychology*, 9th edn. Allyn and Bacon, Needham Heights, MA.

Macrae, C.N. and Bodenhausen, G.V. (2000) Social cognition: Thinking categorically about others. *Annu. Rev. Psychol.*, **51**, 93–120.

Petty, R.E., Wegener, D.T. and Fabrigar, L.R. (1997) Attitudes and attitude change. *Annu. Rev. Psychol.*, **48**, 609–647.

Section J

Alexander, M.G., Brewer, M.B. and Herrmann, R.K. (1999) Images and affect: A functional analysis of out-group stereotypes. *J. Personality Social Psychol.*, **77**, 78–93.

Aronson, E., Wilson, T.D. and Akert, R.M. (1999) *Social Psychology*, 3rd edn. Longman, New York.

Baron, R.A. and Byrne, D. (2000) *Social Psychology*, 9th edn. Allyn and Bacon, Needham Heights, MA.

Postmes, T. and Spears, R. (1998) Deindividuation and antinormative behavior: A meta-analysis. *Psychol. Bull.*, **123**, 238–259.

Smith, E.R., Murphy, J. and Coats, S. (1999) Attachment to groups: Theory and measurement. *J. Personality Social Psychol.*, **77**, 94–110.

Section K

Anastasi, A (1990) *Psychological Testing*, 6th edn. Macmillan, New York.

Butcher, J.N. and Rouse, S.V. (1996) Personality: individual differences and clinical assessment. *Annu. Rev. Psychol.*, **47**, 87–111.

Costa, P.T. and McCrae, R.R. (1992) Four ways five factors are basic. *Personality and Individual Differences*, **13**, 653–655.

Moskowitz, D.W. (1982) Coherence and cross-situational generality in personality: a new analysis of old problems. *J. Personality Social Psychol.*, **43**, 754–768.

Neisser, U., Boodoo, G., Bouchard, T.J. Jr and Boykin, A.W. (1996) Intelligence: Knowns and unknowns. *American Psychologist*, **51**(2), 77–101.

Pervin, L.A. and John, O.P. (1997) *Personality: Theory and Research*, 7th edn. Wiley, New York.

Section L

Davidson, G.C. and Neale, J.M. (1998) *Abnormal Psychology*, 7th edn. John Wiley & Sons, New York/London.

Hartung, C.M. and Widiger, T.A. (1998) Gender differences in the diagnosis of mental disorders: Conclusions and controversies of the DSM-IV. *Psychol. Bull.*, **123**, 260–278.

Holmes, D.S. (1997) *Abnormal Psychology*, 3rd edn. Longman, New York.

Wampold, B.E., Mondin, G.W., Moody, M., Stich, F., Benson, K. and Allin, H-N. (1997) A meta-analysis of outcome studies comparing bona fide psychotherapies. *Psychol. Bull.*, **122**, 203–215. (With commentaries.)

Widiger, T.A. and Sankis, L.M. (2000) Adult psychopathology: Issues and controversies. *Annu. Rev. Psychol.*, **51**, 377–404.

Section M

Bandura, A. (1977) Self-efficacy: toward a unifying theory of behavioural change. *Psychol. Rev.*, **84,** 191–215.

Booth-Kewley, S. and Friedman, H.S. (1987) Psychological predictors of heart disease: a quantitative review. *Psychol. Bull.*, **101,** 343–362.

Di Clemente, C.C., Prochaska, J.O., Fairhurst, S.K. Velicer, W.F., Velasquez, M.M. and Rossi, J.S. (1991) The process of smoking cessation: An analysis of precontemplation, contemplation and preparation for stages of change. *J. Consulting Clin. Psychol.*, **59,** 295–310.

Harrison, J.A., Mullen, P.D. and Green, L.W (1992) A meta-analysis of studies of the health belief model with adults. *Health Education Res.*, **7**, 107–116.

Maes, S., Vingerhoets, A. and van Heck, G. (1987) The study of stress and disease: Some developments and requirements. *Social Sci. Med.*, **6**, 567–78.

Ogden, J. (2000) *Health Psychology: A Textbook*, 2nd edn. Open University Press, Buckingham.

Sheridan, C.L. and Radmacher, S.A. (1992) *Health Psychology: Challenging the Biomedical Model*. John Wiley & Sons, New York.

Stroebe, W and Stroebe, M.S. (2000) *Social Psychology and Health,* 2nd edn. Open University Press, Buckingham.

Trostle, J.A. (1988) Medical compliance as an ideology. *Social Sci. Med.,* **27,** 1299–1308.

INDEX